Thomas E. Gilsdorf
Locally Convex Spaces

Thomas E. Gilsdorf

Locally Convex Spaces

Banach Space Theory, Mathematical Physics, and Distribution Theory Applications

DE GRUYTER

Mathematics Subject Classification 2020
Primary: 46-01, 46A, 46A17; Secondary: 46A19, 46B99

Author
Thomas E. Gilsdorf
Department of Mathematics
Central Michigan University
1200 S. Franklin St.
48859 Mount Pleasant
Michigan
USA

ISBN 978-3-11-139079-6
ISBN 978-3-11-139286-8 (PDF)
ISBN 978-3-11-139415-2 (E-PUB)
DOI https://doi.org/10.1515/9783111392868

Library of Congress Control Number: 2026934791

Bibliographic information published by the Deutsche Nationalbibliothek
The Deutsche Nationalbibliothek lists this publication in the Deutsche Nationalbibliografie;
detailed bibliographic data are available on the Internet at https://dnb.dnb.de.

De Gruyter and Walter de Gruyter GmbH are part of De Gruyter Brill.
www.degruyterbrill.com

Questions about General Product Safety Regulation: productsafety@degruyterbrill.com

Cover illustration: Konstantin Maksimov / iStock / Getty Images Plus

Para mi Famila

Contents

Preface

Welcome to this book! This work provides an updated introductory textbook on the topic of locally convex topological vector spaces. My hope is that readers who enjoy topics such as topology, analysis, and linear algebra will discover appealing relevance in this work. The target audience is anyone who has a typical background of one semester courses in topology, linear algebra, and analysis of normed/Banach spaces. The primary reference I have used for background material on topology is Richmond [102] and that for normed and Banach spaces is Megginson [89]. The first seven chapters are standard. Basic properties of topological vector spaces and local convexity are discussed in Chapters 1 and 2. Constructing locally convex spaces (quotients, products, inductive limits, and more) are in Chapter 3. In Chapter 4, we start wading into the deep and fascinating world of duality theory, where we begin to encounter fundamental results, including the Helly–Hahn–Banach theorem (Section 4.1). Included among the many ramifications of this theorem is the conclusion that the continuous dual of a locally convex space always contains enough elements to allow for nontrivial calculations and results. An independent but important result in Chapter 4 is the Kreĭn–Mil'man Theorem 4.6.4 regarding extreme points of convex compact subsets of locally convex spaces. Chapter 5 includes discussions of several types of locally convex spaces that are basic to the general theory, including those that are (quasi)barreled, bornological, (semi) reflexive, and others. Part of the material of this chapter involves a description of Grothendieck's normed spaces $(E_B, \| \cdot \|_B)$ generated by absolutely convex, bounded subsets. These normed spaces turn out to be quite handy for analyzing various types of locally convex spaces. In Chapter 6, we find that there are useful ways to involve sequences and series in locally convex spaces, including some cases where the spaces are not even metrizable. One example is that of Mackey convergence in which null sequences converge in normed spaces $(E_B, \| \cdot \|)$. Another property of sequences and series is that of $\mathcal{K}$- convergence, which turns out to be connected to metrizable Baire locally convex spaces, among other results. Moreover, in this chapter I give what I hope is an accessible description of spaces with webs (Section 6.3), which exhibit interesting properties of convergent sequences and series. Chapter 7 is about quite general completeness and barreledness properties of locally convex spaces in terms of the normed spaces $(E_B, \| \cdot \|_B)$ alluded to above. Such properties allow for short proofs of deep results such as the Uniform Boundedness Principle and the Banach–Steinhaus theorem. Next, in Chapter 8 the focus is on the Closed Graph Theorem, another significant result in the world of locally convex spaces (and indeed, general topological vector spaces). Webs play a prominent role in the results of this chapter, as they represent the modern context for obtaining closed graph theorems. A few versions of the related and equally significant Open Mapping Theorem are also derived. The Localization Theorem is then presented, followed by a handful of applications of the results of this chapter. Near the end of the chapter is a discussion of a few of the many other versions of the Closed Graph Theorem.

https://doi.org/10.1515/9783111392868-204

Starting in Chapter 9, the flavor of the topics begins to change from the basic concepts of locally convex spaces to a few of the numerous applications of those concepts. Such applications include a brief overview of distributions in Section 9.1, which historically provided an impetus for the systematic development of the theory of locally convex spaces. Indeed, the spaces $\mathcal{D}$ of test functions and $\mathcal{D}'$ of distributions are locally convex spaces that are strictly outside the realm of normed and Banach spaces (not only that, these spaces are intriguingly complicated). The next two sections focus on questions of stepping back from the usual contexts. Traditionally, locally convex spaces involve structures of the scalar field $\mathbb{K}(= \mathbb{C} \text{ or } \mathbb{R})$ and structures of a linear topology. One can ask what happens if we replace either the scalar field or the topology with something different. In the first part of such an exploration, we consider locally convex spaces over scalar fields that do not satisfy the Archimedean property in Section 9.2. Section 9.3 involves replacing the topology of a locally convex space with a more general convergence structure. In each of these sections, some basic definitions and a few results are discussed, the goal being to observe that in such nontraditional contexts, some concepts are basically unchanged while others are drastically distinct. The fun continues in Chapter 10 in which I partake in the impossible task of describing, within the confines of one chapter of a book, some recent and current areas of research that involve locally convex spaces. We get a glimpse at some mathematical physics in the form of distributions having their wavefront sets in a positive cone in Section 10.1. Section 10.2 concerns convenient vector spaces, which are used to define calculus concepts in the context of locally convex spaces. In Section 10.3, we examine a generalization of the concept of duality called abstract duality pairs, and in Section 10.4 a locally convex spaces version of an optimization result known as Ekeland's principle, is studied. The last two chapters are intended to give you the reader, a basic understanding of how the wonderful topic of locally convex spaces can be viewed in some nontraditional and modern contexts. The book ends with a "chapter" on further resources.

The natural question that arises in any endeavor like this book is that of "How is this book not essentially the same as previously published books?" Of course, I could respond with the default reply that every author brings a unique perspective to the topic being written. However, I should be able to say more than this minimum. I have summarized my response below with what I believe sets this book apart from comparable works.

How this book is different...

– *Connections to normed spaces.* The confirmation that every normed space is a locally convex space consists of a 2-minute proof (see Proposition 2.1.1), which motivates a search for locally convex spaces that are not normed. There are plenty of such spaces, however, normed spaces are still involved in important ways. In particular, properties of infinite-dimensional Banach spaces provide examples of spaces from two sides of the topic of locally convex spaces. On the one hand are their interesting properties as normed spaces; on the other hand, infinite-dimensional Banach spaces equipped with their weak topologies turn out to be delightfully complicated, nonmetrizable locally convex spaces. As indicated above, substantial use is made of the normed spaces $(E_B, \| \cdot \|_B)$ (Definition 5.1.3 and Notation 5.1.1). In a variety of contexts, results are obtained assuming that the spaces $(E_B, \| \cdot \|_B)$ are complete; that is, that B is a *Banach disk* (Definition 5.1.4). Moreover, we consider some generalizations of Banach disks to Baire disks and barreled disks (Definition 5.1.4 again).

- *Convergence, convergence, convergence.* First of all, this book contains discussions of traditional convergence in locally convex spaces, those being of filters, sequences, and series. Second, sequential and series convergence appear in several other contexts, including Mackey convergence (Section 6.2), $\mathcal{K}$- convergence (Section 6.1), fast convergence (Definition 6.2.2), and convergence in strands of a web (Section 6.3). Third, the excitement generalizes in Section 9.3 in which convergence is examined in such a way that a topology is not involved.
- *Webs.* The structure of webs was originally defined by De Wilde for proving closed graph theorems, however, webs can used to obtain results connected with other topics such as convergence of sequences and series and barreled spaces. Despite what seems to be a complicated definition, webs are relatively easy to work with and arise intuitively in a variety of contexts. I have included discussions related to the structures of webs in several sections.
- *Nontraditional contexts.* Wandering outside the typical context of locally convex spaces as described above leads to interesting, even surprising, results. I have attempted to give a peek at some such results, specifically in the contexts of non-Archimedean locally convex spaces and convergence vector spaces.
- *Glimpses of some recent research.* Given that the realm of locally convex spaces properly includes all normed spaces, it should not be surprising that the scope of applications of results from this area is vast. I have given a few fleeting glimpses of such applications.

Using this book for courses or research

To use this book as a text for a course, the basic material appears in Chapters 1 through 8. A few sections that are not essential to later development could be skipped here and there if necessary. Regarding use in research, specific items in this book can of course be used as a references. I have generally avoided stating significant results in the exercises. Rather, such items are listed as propositions, corollaries, or theorems for which the relevant proofs are exercises. I have listed all references that I have consulted during the writing of this book. Some of those references were not directly cited in the text but might be useful to readers.

About the exercises

I am not particularly clever when it comes to creating inspiring exercises. Fortunately, there is no shortage of people who are smarter than I am and who can create such exercises. Thus, in addition to my own exercises I have, as traditionally done, adapted numerous exercises from other sources (see Chapter 11). In general, the exercises range from short verifications to creating significant versions of presented results.

How the material for this book got into my mind

This book represents a crockpot of material that has simmered in my brain throughout a variety of experiences. A significant amount of said material comes from courses I have taught at the Centro de Investigación y de Estudios Avanzados (CINVESTAV – Mexico City), the University of North Dakota, and Central Michigan University. The rest of the material has resulted from my own learning, enjoying, and at times publishing on this topic. In particular, I have learned a lot from my collaborators.

Final comments

From time to time, I have tried to step back and take a longer look at some results in order to appreciate, sometimes awe, at their mathematical elegance. It is a kind of "stop and smell the theorems" if you will. In fact, I sometimes pick up an old book that I have not paged through for some time, and smell the pages to relive the experience of that old book fragrance. Independently of an aroma you might perceive from this book (be it on paper or electronic), I hope you find it to be worthwhile.

May 2026   Thomas E. Gilsdorf

Acknowledgments

I appealed to several people for feedback in terms of mathematics, misprints, and errors. The comments, as well as suggested corrections and modifications, significantly elevated the quality of this book. Therefore, I am deeply grateful to the following people who reviewed parts of this work: Angela A. Albanese, Carlos Bosch, Tarun Kumar Chauhan, Juan Carlos Ferrando, Saak Gabriyelyan, David Jornet, Akshay Kumar, Hiruni Kamali Pallage, Thomas Richards, Tom Richmond, Edgar Santos Vega, Rigoberto Vera, and Brooklynn Willett. Any remaining errors are my responsibility.

I would like to give special thanks my professor, mentor, collaborator, and dear friend, Carlos Bosch. As a PhD student in the previous century at Washington State University, I took a course from him, ironically titled "Seminar in applied mathematics," which consisted of an introduction to topological vector spaces. I was quickly fascinated by the topic and that fascination has stayed with me over the years. In fact, writing this book has allowed me to review and relive this topic. Additional gratitude includes the fact that Professor Bosch reviewed effectively the entire manuscript of this book.

https://doi.org/10.1515/9783111392868-205

1 Topological vector spaces: Basic ideas

The first topic of this book focuses on relevant concepts of linear algebra and topology as they relate to what we will call a topological vector space (Definition 1.3.1). We start with vector spaces that are independent of topological structures, and define some geometric concepts, namely those of sets that are balanced, convex, and absorbing (Definitions 1.1.3, 1.1.4, and 1.1.8, respectively). Sets can be enveloped by intersections of all supersets that are balanced and/or convex, called hulls (Definition 1.1.6). Such "shrink wrapping" turns out to be useful as we proceed. We also briefly look at quotient spaces in Definition 1.1.9. For contexts of nonmetrizability, the concept and properties of filters are discussed in Section 1.2. Filters are natural generalizations of sequences by way of Proposition 1.2.1. The fundamental definition of a topological vector space (TVS) appears in Definition 1.3.1. Theorem 1.3.2 lists basic properties of zero neighborhoods that we will make use of in countless (well technically, finitely often) situations. Theorem 1.3.3 states that such properties on a vector space lead to a unique topological vector space topology.

1.1 Vector spaces without a topology

Most of the topics in this book take place in the context of a vector space over a scalar field $\mathbb{F}$, denoted formally as $(E, +, \cdot)$. A vector space is typically written in the more concise form of E. It will seem that most of the emphasis in this book has to do with topological aspects with the vector space concepts always present, seemingly "running in the background." Nevertheless, were we to stop assuming the context of a vector space, many of the results in this book would be quite difficult, if not impossible. Regarding fields, some distinctions will be necessary. For most of what we do, the field $\mathbb{F}$ will be assumed to be either the real numbers $\mathbb{R}$ or the complex numbers $\mathbb{C}$, which we will denote by $\mathbb{K}$. However, when we get to Chapter 10, other types of fields will be considered such as non-Archimedean fields.

1.1.1 Geometric considerations in vector spaces

It may seem unusual to consider geometric aspects of vector spaces, however, it turns out that there are some interesting features to consider. In what follows, we assume the sets are in a vector space over the field $\mathbb{K}$.

If we are able to add elements as well as multiply elements by scalars in a vector space, it seems reasonable that we should be able to define similar operations on sets of elements. This is in fact possible, as seen next.

Definition 1.1.1. Let E be a vector space. Let $A \subset E, B \subset E$. The **set sum** $A + B$ is defined as $A + B = \{x + y : x \in A, y \in B\}$. The **set difference** $A - B$ is defined as $A - B = \{x - y : x \in A, y \in B\}$.

https://doi.org/10.1515/9783111392868-001

Definition 1.1.2. Let E be a vector space. If D is a subset of $\mathbb{K}$ and S is a subset of E, then **the set of scalar multiples by D of S** is $DS = \{ax : a \in D, x \in S\}$. In particular, if $D = \{a\}$, then $aS = \{ax : x \in S\}$.

The set operations defined above often but not always give results that we would expect with numbers. Observe the following simple constructions in the next example.

Example 1.1.1. Sums, scalar multiples, and differences of sets.
(a) In $\mathbb{R}$, let $A = (0,1)$, $B = \{2\}$. Then $A + B = (2,3)$, and $5A = (0,5)$.
(b) In $\mathbb{R}^2$, let $A = \{(1,2),(3,4)\}$. Then $A + A = \{x + y : x, y \in A\} = \{(2,4),(6,8),(4,6)\}$, and $2A = \{2x : x \in A\} = \{(2,4),(6,8)\}$, which tells us that in general, $A + A \neq 2A$.
(c) In $\mathbb{R}$, let $A = \{2,5\}$. Then $A - A = \{x - y : x, y \in A\} = \{0,-3,3\}$, and this tells us that in general, $A - A \neq \{0\}$.

A more satisfactory result regarding part (b) of Example 1.1.1 will be obtained in Proposition 1.1.1 below. Before that result, the next two types of sets will be defined. It will turn out that one or both of the following concepts will be assumed for almost everything we do in this book.

Definition 1.1.3. A set A in a vector space is **balanced** (also **circled**) if for every $x \in A$, and for every $t \in [-1,1]$, we have $tx \in A$. In particular, $tx \in A$ whenever $|t| \leq 1$.

Definition 1.1.4. A set A in a vector space is **convex** (or satisfies the **convexity** property) if it contains all of its line segments; that is, for every $x, y, \in A$ and every $t \in [0,1]$, $tx + (1 - t)y \in A$.

By way of Exercise 1.1.1, it is easy to see that Definition 1.1.3 and Definition 1.1.4 are independent of each other. Meanwhile, if the set A is convex, then the inequality of Example 1.1.1(b) does not arise, as shown in the next proposition.

Proposition 1.1.1. *If the set A is convex then $A + A = 2A$. More generally, if a and b are any positive scalars, then $aA + bA = (a + b)A$.*

Proof. To show that $2A \subset A + A$ is short, and does not require convexity. Conversely, assume that A is convex and let $w \in A + A$. Then $w = x + y$, for some $x, y \in A$. Using $t = 1/2$ from Definition 1.1.4, it follows that $w = x + y = 2(\frac{1}{2}x + \frac{1}{2}y) \in 2A$. Other details of this proof are left as Exercise 1.1.4. $\qquad\square$

We combine Definitions 1.1.3 and 1.1.4 in the following.

Definition 1.1.5. A set in a vector space is **absolutely convex** if it is both convex and balanced.

Proposition 1.1.2. *A set A in a vector space is absolutely convex if and only if $(\forall x, y \in A)(\forall a, b, \in \mathbb{K})$, $ax + by \in A$, whenever $|a| + |b| \leq 1$.*

Proof. ($\Rightarrow$): First, assume A is absolutely convex and let a and b be scalars such that $|a| + |b| \le 1$. Let $x, y \in A$. We need to show that $ax + by \in A$. The assumption that A is balanced implies that if either $a = 0$ or $b = 0$, then $ax + by \in A$. Now assume neither a nor b is zero. Applying the assumption that A is balanced again, $(a/|a|)x$ and $(b/|b|)y$ both belong to A. We apply the assumption that A is convex and observe that for any $t \in [0,1]$, $t(a/|a|)x + (1-t)(b/|b|)y \in A$. Using the clever choice of $t = \frac{|a|}{|a|+|b|}$, we have

$$t\frac{a}{|a|}x + (1-t)\frac{b}{|b|}y = \frac{|a|}{|a|+|b|}\frac{a}{|a|}x + \frac{|b|}{|a|+|b|}\frac{b}{|b|}y = \frac{ax+by}{|a|+|b|} \in A. \tag{1.1}$$

We need one more application of the assumption that A is balanced, in which we have

$$(|a|+|b|)\frac{ax+by}{|a|+|b|} = ax + by \in A, \tag{1.2}$$

because $|a| + |b| \le 1$.

($\Leftarrow$): Conversely, assume the condition that for every $x, y \in A$ and for all $a, b, \in \mathbb{K}$, $ax + by \in A$ whenever $|a| + |b| \le 1$. Letting $b = 0$ shows that A is balanced, and letting $a = t, b = 1 - t$, for $0 \le t \le 1$ shows that A is also convex. $\qquad\square$

The following result reveals facts about how most typical constructions preserve properties of being balanced and/or convex.

Theorem 1.1.1. *In any vector space over $\mathbb{K}$, the following constructions preserve the stated property:*
(a) *Arbitrary intersections of balanced sets are balanced.*
(b) *Arbitrary unions of balanced sets are balanced.*
(c) *Linear combinations of balanced sets are balanced.*
(d) *Arbitrary intersections of convex sets are convex. Hence, arbitrary intersections of absolutely convex sets are absolutely convex.*
(e) *Linear combinations of convex sets are convex. Hence, linear combinations of absolutely convex sets are absolutely convex.*
(f) *Arbitrary products of balanced sets are balanced.*
(g) *Arbitrary products of convex sets are convex. Hence, arbitrary products of absolutely convex sets are absolutely convex.*

Proof. We prove (b), (d), and (g) here, the other items being left as Exercise 1.1.5. For (b), let $B = \bigcup\{B_\alpha : \alpha \in I\}$ be an arbitrary union of balanced sets. Let $t \in \mathbb{K}$ with $|t| \le 1$ and $x \in B$ be given. There exists $\alpha \in I$ such that $x \in B_\alpha$. The set B_α is balanced, hence $tx \in B_\alpha \subset B$; B is therefore balanced.

(d): Let $C = \bigcap\{C_\alpha : \alpha \in I\}$ be an arbitrary collection of convex sets. Let $x, y \in C$ and $t \in [0,1]$ be given. For every $\alpha \in I$, x and y are in the convex set C_α, which implies $tx + (1-t)y \in C_\alpha$. We have shown that C is convex. The second statement is obvious, in light of part (a).

(g): Assume $\{C_\alpha : \alpha \in I\}$ is any collection of convex sets, and let $C = \prod_{\alpha \in I} C_\alpha$ be the corresponding product, per Definition A.1.2. Let $x = (x_\alpha : \alpha \in I), y = (y_\alpha : \alpha \in I) \in C$ and $t \in [0, 1]$ be given. As each C_α is convex, $tx_\alpha + (1 - t)y_\alpha \in C_\alpha$, and the result follows. □

The previous proposition prompts the following definitions.

Definition 1.1.6.
(a) The **convex hull** of a set is the intersection of all convex sets that contain it. The convex hull of A is denoted by conv(A).
(b) The **balanced hull** of a set is the intersection of all balanced sets that contain it. The balanced hull of A is denoted by bal(A).
(c) The **absolutely convex hull** (also **convex, balanced hull**) of a set is the intersection of all convex, balanced sets that contain it. The convex, balanced hull of A is denoted by convbal(A), also denoted by absconv(A).
(d) The **balanced core** of a set is the union of all of its balanced subsets. The balanced core of A is denoted by balcore(A).

We will generally use the notation convbal(A): the convex hull of the balanced hull of a set A. In Exercise 1.1.7, you can verify that the balanced hull of the convex hull is not the same as the convex hull of the balanced hull. If this sounds like double talk, the symbolic statement is that in general, balconv(A) $\neq$ convbal(A). In order to preserve convexity, we will almost always calculate convbal(A) = absconv(A) rather than balconv(A).

Observe the following convenient properties.

Proposition 1.1.3. *For any subset A of a vector space, the following hold:*
(a) *A is balanced if and only if* bal(A) = A.
(b) *A is convex if and only if* conv(A) = A.

Proof. We prove (b) here, the proof of part (a) being similar. First, notice that conv(A) is convex, so if conv(A) = A, then A is convex. On the other hand, if A is convex, then A itself is one of the convex sets used in the formation of conv(A), which implies that conv(A) $\subset A$. By its definition, conv(A) always contains A, so the two sets are equal. □

Definition 1.1.7. For a subset A of E, given $x_1, x_2, \ldots, x_n \in A$ and any combination of scalars $t_1, t_2, \ldots, t_n \in [0, 1]$ such that $\sum_{j=1}^{n} t_j = 1$, the sum $\sum_{j=1}^{n} t_j x_j$ is called a **convex combination** of elements of A.

Proposition 1.1.4. *A subset A of E is convex if and only if A contains all convex combinations of its elements.*

Proof. Exercise 1.1.9. □

Next up is a definition that describes how a set can be expanded to contain a point or another set.

Definition 1.1.8. A set is **absorbing** (also **is absorbent** or **is radial at 0**) if every point in the vector space is contained in all sufficiently large scalar multiples of the set. A set **absorbs** a second set if all sufficiently large scalar multiples of the first set contain the second set.

Thus, a set A is absorbing in a vector space E if and only if $(\forall x \in E)(\exists c \in \mathbb{K})$, $c \neq 0$ such that $x \in aA$ for all $|a| \geq |c|$, and a set A absorbs another set B if and only if $(\exists c \in \mathbb{K})$, $c \neq 0$ such that $B \subset aA$ for all $|a| \geq |c|$.

In the exercises, you will discover that the property of being absorbing is also independent of the properties of being balanced or convex.

Next, we consider linear images of balanced and/or convex sets. Connecting the concept of a linear map with the concepts of balanced and convex sets leads to the following.

Theorem 1.1.2. *A linear map between vector spaces satisfies the following:*
(a) *Linear images and inverse images of balanced sets are balanced.*
(b) *Linear images and inverse images of convex sets are convex.*
(c) *Linear images and inverse images of absolutely convex sets are absolutely convex.*

Proof. The proof of part (a) is shown here. Part (b) is left as an exercise, and part (c) is an obvious consequence of parts (a) and (b). Let E and F be any vector spaces and $T : E \to F$ any linear map. Assume $A \subset E$ is balanced. Given any $y \in T(A)$, there exist $x \in A$ such that $T(x) = y$. Pick any $t \in [-1, 1]$. Then $ty = tT(x) = T(tx)$, by the linearity of T. The conclusion follows from the assumption that A is balanced. For the second part, suppose $B \subset F$ is balanced and put $A = T^{-1}(B)$. For $t \in [-1, 1]$ and any $x \in A$, $T(tx) = tT(x) \in B$, and we conclude that $tx \in T^{-1}(B) = A$. $\qquad\square$

Let us now take a glimpse at quotient vector spaces. Let M be a (linear) subspace of a vector space E. We can define a relation on E by $x \sim y \Leftrightarrow x - y \in M$. It is easy to see (Exercise 1.1.13) that this creates an equivalence relation, and we can express this as $x \sim y$ **modulo** M, or more concisely, $x \sim y \bmod M$. Hence, the relation $\sim$ separates E into equivalence classes $\dot{x}$, whereby for any $x \in E$, $\dot{x} = \{y \in E : x - y \in M\}$.

Definition 1.1.9. For a linear subspace M of E, define the **quotient space** E/M to be the collection of equivalence classes given by

$$E/M = \{\text{equivalence classes } \dot{x}, \text{ where } \dot{x} = \{y \in E : x - y \in M\}\}. \qquad (1.3)$$

Definition 1.1.10. Consider E/M as defined above. Given any equivalence classes $\dot{x}, \dot{y} \in E/M$ and any $a \in \mathbb{K}$, define $\dot{x} + y = \dot{x} + \dot{y}$, and $(\dot{ax}) = a \cdot \dot{x}$.

With the above definition and the observation that the zero of E/M is given by $\dot{0} = \{x \in E : x - 0 \in M\} = M$, it is straightforward to verify that E/M is a vector space, so the word "space" in the definition E/M is justified. Next, to get to and from E/M, we use the following function.

Definition 1.1.11. Consider E/M as defined above. The **canonical quotient map** φ : $E \to E/M$, is defined by $(\forall x \in E)\, \varphi(x) = \overset{\bullet}{x}$. The map φ is also referred to as the **canonical surjection onto** E/M.

Proposition 1.1.5. *The canonical quotient map is linear and surjective (onto).*

Proof. Linearity follows from Definition 1.1.10. The surjective part is left to you as Exercise 1.1.14. $\qquad\square$

In particular, given a linear map $T : E \to F$, we can consider the quotient space of E by its nullspace; that is, by the function $S : E/T^{-1}(\{0\}) \to F$, where $T^{-1}(\{0\}) = \{x \in E : T(x) = 0\}$ is the nullspace (see Definition A.2.3) of T.

One of the most common reasons to use a quotient space is when we have a linear map that is not injective (i. e., not one-to-one, so the nullspace is nontrivial). Specifically, if the linear map $T : E \to F$ is not injective, we can factor through the quotient space, as shown in the diagram below. Note that if T is surjective, then $T(E)$ is replaced by F:

$$
\begin{array}{ccc}
E & \xrightarrow{\;\;T = S \circ \varphi\;\;} & T(E) \\
& \searrow_{\varphi} & \big\uparrow_{S} \\
& & E/T^{-1}(\{0\})
\end{array}
\qquad\qquad (1.4)
$$

Here, $S : E/T^{-1}(\{0\}) \longrightarrow T(E)$ is the bijective linear map such that $T = S \circ \varphi$; that is, such that the diagram commutes. A specific expression for S is given by $S(\varphi(x)) = T(x)$ for each $x \in E$.

The following represents our first encounter with the concept of a dual space.

Definition 1.1.12. A linear map from a vector space to the scalar field is called a **linear functional**. A linear functional is also referred to as a **linear form**.

We will be keenly interested in vector spaces of linear functionals, the first kind of which is defined below.

Definition 1.1.13. The **algebraic dual** of a vector space is the set of all linear functionals on the space. The notation for the algebraic dual is E^{*}.

It is clear that E^{*} forms a vector space with respect to the operations defined by $(\forall f, g \in E^{*})(\forall x \in E)\,(f + g)(x) = f(x) + g(x)$ and $(\forall a \in \mathbb{K})\,(af)(x) = af(x)$. There is an important notational distinction between E^{*} here and E^{*} (this same notation) in the context of normed spaces. For normed spaces, E^{*} typically refers to the set of all *continuous* linear functionals. For consistency, we will use E^{*} to denote the set of *all* linear functionals on any vector space, normed or otherwise.

Exercises

1.1.1. For each of the set properties of being balanced, convex, or absorbing, describe a set that satisfies one of these properties but satisfies neither of the other two.

1.1.2. Again, for each of the set properties of being balanced, convex, or absorbing, describe a set that satisfies two of these properties but not the third.

1.1.3. Prove that if A is a balanced set, then $A - A = A + A$.

1.1.4. Complete the proof of Proposition 1.1.1. In particular, if a and b are not assumed to both be positive, is $aA + bA = (a + b)A$, even if A is convex?

1.1.5. Prove parts (a), (c), (e), and (f), of Theorem 1.1.1.

1.1.6. Prove that if $A \subset B$ and B is convex, then $\operatorname{conv}(A) \subset B$.

1.1.7. Let A be the subset of $\mathbb{R}^2$ defined by $\{(x, y) : x \geq 0, y = -x, y = x, x = 1\}$. Calculate $\operatorname{convbal}(A)$ and $\operatorname{balconv}(A)$. Is $\operatorname{balconv}(A)$ even convex?

1.1.8. Assume A is a nonempty absolutely convex set.
   (a) Prove that $0 \in A$.
   (b) Prove that $aA \subset bA$, whenever $|a| \leq |b|$.

1.1.9. Prove Proposition 1.1.4. Suggestion: For the "$\Leftarrow$" part, proceed by induction.

1.1.10. Describe an example of a set $A \subset \mathbb{R}^2$, such that $A - A \neq \{(0, 0)\}$.

1.1.11. For subsets of a vector space:
   (a) Prove that finite intersections of absorbing sets remain absorbing.
   (b) Construct a collection of absorbing sets such that their intersection is neither empty nor absorbing.
   (c) Construct a nonzero linear map $T : E \to F$ between vector spaces E, F and an absorbing set $A \subset E$ such that $T(A)$ is not absorbing.

1.1.12. Prove part (b) of Theorem 1.1.2.

1.1.13. Verify that the relation on a vector space E given by $x \sim y \Leftrightarrow x - y \in M$, does indeed define an equivalence relation.

1.1.14. Verify that the canonical quotient map is surjective (onto).

1.1.15. Prove that any linear functional (Definition 1.1.12) on a subspace of the algebraic dual (Definition 1.1.13) can be extended to all of the algebraic dual. That is, prove that if f is a linear functional on a subspace F of E, then f can be extended to a linear functional $\tilde{f}$ on E^* such that $\tilde{f}$ agrees with f on F. We will see later on that topologically, continuous linear functionals are not always extendable in this way when convexity is not assumed.

1.2 Filters

It is a well-known fact that if a topological space is not a metric space, then sequences are inadequate for describing some important topological properties. One such property that comes to mind in this context is that the closure of a set cannot be described by way of the convergence of sequences of elements of the set. Later, we will see in Example 3.3.4 a specific example of a set for which the closure cannot be described via sequential convergence. When we find ourselves in such a nonmetrizable topological context, it will therefore be necessary to have some structure of that is more general than sequences. The good news is that there is such a more general structure. In fact, there are two such general structures, namely filters and nets. The structure we will use here is that of filters, as defined next. An accessible discussion of the distinct but comparable concept of a net, along with a comparison to filters, can be found in Wilansky [129].

An intuitive view of filters can be deduced from a review of properties of topological neighborhoods. Given the collection of all neighborhoods of a point x in a topological space $(X, \mathcal{T})$, we observe the following facts:

(i)   The collection $\mathcal{N}_x$ is not empty and every set in the collection is nonempty. Indeed, if all else fails, we can choose X as a member of $\mathcal{N}_x$ and of course, x belongs to every set in $\mathcal{N}_x$.

(ii) Finite intersections of members of the collection $\mathcal{N}_x$ are again members. This follows from the definition of neighborhoods in any topological space.

(iii) Given any set in the collection $\mathcal{N}_x$, all of its supersets are obviously also in the collection.

These three items represent the essence of the concept of a filter, which we now proceed to define formally. The context of a filter is that of any (nonempty) set, although for us the set will typically appear in a vector space. In the definition below, we see that the concept of a filter does not require the presence of a topology, though the example of properties of neighborhoods just reviewed serves as a motivation. A comprehensive, up to date resource on filters is Dolecki [39]. Additional references containing more details about filters include Horváth [67] and Wilansky [129], [130].

Definition 1.2.1. A **filter** is a nonempty collection of nonempty subsets of a universal set that is closed under the formations of finite intersections and supersets. That is, for X denoting a set, a collection $\mathfrak{F}$ of subsets of X is a filter on X if it satisfies the following three properties:

(a)   The collection $\mathfrak{F}$ is nonempty and the empty set is not a member of $\mathfrak{F}$.

(b)   Finite intersections of members of $\mathfrak{F}$ belong to $\mathfrak{F}$.

(c)   Supersets of members of $\mathfrak{F}$ belong to $\mathfrak{F}$; that is, if $A \in \mathfrak{F}$ and $B \supset A$, then $B \in \mathfrak{F}$. In particular, $X \in \mathfrak{F}$.

Example 1.2.1. Examples of filters.

The first item below simply states formally that a collection of neighborhoods of a point x in a topological space represents a filter. The verification of the second two items below follow quickly from the definition of a filter, and you can enjoy writing their proofs in Exercise 1.2.1. Observe that those second two items do not require the presence of a topology.

(a) The collection of all neighborhoods of a point in a topological space is a filter.
(b) If $x \in X$ is a point, then the collection of all subsets of X containing x is a filter. This is called the **discrete filter**.
(c) If A is a nonempty subset of X, then the collection of all subsets of X containing A is a filter.

Two more somewhat similar examples, which are clarified with short proofs, are next. In particular, the first proposition below indicates that sequences can be viewed as filters.

Proposition 1.2.1. *Any infinite sequence can be associated with a filter.*

Proof. Let (x_n) be an infinite sequence in a universal set X. Define $\mathfrak{F}$ to be the collection of all sets that contain all but possibly finitely many elements of the sequence (x_n). Item (a) of Definition 1.2.1 is certainly satisfied. Next, if A and B are members of $\mathfrak{F}$, then each of A and B contains all but finitely many elements of (x_n). Thus, $A \cap B$ contains all but finitely many elements of (x_n), from which we conclude $A \cap B \in \mathfrak{F}$. This proves item (b). Item (c) is obvious. $\qquad\square$

We formalize the concept of a filter that arises from a sequence below.

Definition 1.2.2. The filter associated with a sequence as described above is called a **sequential filter**.

Another term for a sequential filter is that of a **Fréchet filter**. The terminology follows that of [39].

Proposition 1.2.2. *Given any infinite set X, define $\mathfrak{F} = \{A \subset X : A^c \text{ is finite}\}$. Then $\mathfrak{F}$ is a filter.*

Proof. Certainly, the empty set $\emptyset \notin \mathfrak{F}$ by the assumption on X. Moreover, $X^c = \emptyset$ provides proof of membership of X in $\mathfrak{F}$. Now suppose $A, B \in \mathfrak{F}$. By DeMorgan's properties, $(A \cap B)^c = A^c \cup B^c$ is a finite set. Finally, if $A \in \mathfrak{F}$ and $B \supset A$, then $B^c \subset A^c$. Definition 1.2.1 has been satisfied. $\qquad\square$

Two concepts comparable to finer topologies and topological bases, respectively, appear in the following two definitions.

Definition 1.2.3. Consider two filters $\mathfrak{F}_1$ and $\mathfrak{F}_2$ of subsets of a set X. Then $\mathfrak{F}_1$ is **finer than** $\mathfrak{F}_2$ if $\mathfrak{F}_1$ contains all sets of $\mathfrak{F}_2$; that is, $\mathfrak{F}_1 \supset \mathfrak{F}_2$. This can be stated alternatively as $\mathfrak{F}_2$ is **coarser** than $\mathfrak{F}_1$. Filters $\mathfrak{F}_1$ and $\mathfrak{F}_2$ are **equivalent** if each is finer than the other.

Definition 1.2.4. A **filterbase** is a nonempty collection of nonempty subsets of a universal set for which finite intersections of its elements contains another element of the collection. That is, in a set X, a nonempty collection $\mathfrak{B}$ of nonempty subsets of X is a filterbase if given any $A, B \in \mathfrak{B}$, there is a $C \in \mathfrak{B}$ for which $C \subset A \cap B$.

Every filter is itself a filterbase. A given filterbase generates a filter as follows.

Definition 1.2.5. A filterbase $\mathfrak{B}$ **generates a filter** $\mathfrak{F}$ on X by

$$\mathfrak{F} = \{A \subset X : A \supset B, \text{ for some } B \in \mathfrak{B}\}. \tag{1.5}$$

The notation for the filter generated by a filterbase is $[\mathfrak{B}]$.

Example 1.2.2. An example of a filterbase.

Let $\{x_n : n \in \mathbb{N}\}$ be an infinite sequence in a set X, which we know to be associated to a filter from Proposition 1.2.1. Define $\mathfrak{B}$ to be the collection of all tails of the sequence (x_n), that is, $B \in \mathfrak{B} \Leftrightarrow B = \{x_k, x_{k+1}, \ldots\}$, for some $k \in \mathbb{N}$. We check the requirements of Definition 1.2.4. First, observe that any such set B either contains all elements of the sequence or does not contain elements $x_1, x_2, \ldots, x_{k-1}$ for some positive integer $k > 1$, so $\mathfrak{B}$ is a nonempty collection of nonempty sets. For the second requirement, given any sets A and B from $\mathfrak{B}$, if either or both A and B contain the entire sequence, then $C = A \cap B$ satisfies the requirement. Otherwise, assume $x_1, x_2, \ldots, x_k \notin A$ and $x_1, x_2, \ldots, x_m \notin B$. For $l = \max\{k, m\}$, we have that $C = \{x_{l+1}, x_{l+2}, \ldots\} \in \mathfrak{B}$ is the desired subset of $A \cap B$. Hence, $\mathfrak{B}$ is a filterbase, and the filter generated by $\mathfrak{B}$ consists of subsets of the sequence (x_n).

We have already seen the first and nearly obvious connection between filters and topologies, as described in Example 1.2.1(a). A corresponding statement can be made about topological bases and filterbases, as follows.

Proposition 1.2.3. *Any topological base of a topology that does not include the empty set is also a filterbase.*

Proof. Exercise 1.2.5. $\qquad\qquad\qquad\qquad\qquad\qquad\qquad\qquad\qquad\qquad\qquad\qquad\qquad\square$

1.2.1 Convergence of filters

To motivate the definition of convergence of filters, let us review sequential convergence in metric spaces as an example. Such convergence is a process by which the collection of neighborhoods of a point (the limit point) "shrink down" toward that point while containing all but a finite number of terms of the sequence. The collection of all relevant neighborhoods contains balls of arbitrarily small radii that in turn contain the limit point. Convergence of a filter involves the same concept, whereby the sets of a filter (or a filterbase) "shrink down" to the limit point by way of being contained in arbitrary neighborhoods of the limit point. The definitions of convergence of filters and filterbases that follow may seem obscurely abstract, however, we will subsequently verify that such convergence is a natural extension of the concept of convergence of sequences that we have just reviewed.

Definition 1.2.6. In a topological space, a filter **converges** to a point if it is finer than the filter of neighborhoods of that point; that is, in a topological space $(X, \mathcal{T})$, a filter $\mathfrak{F}$ **converges to** a point x if for every $\mathcal{T}$-neighborhood U of x there exists $A \in \mathfrak{F}$ such that $x \in A$ and $A \subset U$. The notation is $\mathfrak{F} \to x$.

We can also define convergence with respect to a filterbase.

Definition 1.2.7. A filterbase **converges** to an element of a topological space if every neighborhood of the element contains a set of the filterbase. In other words, in $(X, \mathcal{T})$, a filterbase $\mathfrak{B}$ converges to x if for every neighborhood U of x, there exists $B \in \mathfrak{B}$ such that $B \subset U$. The notation is $\mathfrak{B} \to x$.

It is straightforward to prove that we can consider convergence of either filters or filterbases whenever and whichever is convenient, as stated below.

Proposition 1.2.4. *A filter generated by a filterbase converges if and only if the filterbase converges to the same point.*

Proof. Exercise 1.2.9. $\qquad\square$

That sequential convergence in any topological space is in fact an example of filter convergence is shown in the next proposition. Thus, filter convergence does indeed generalize sequential convergence.

Proposition 1.2.5. *Sequential convergence in a topological space represents an example of filter convergence. That is, a sequence converges in a topological space if and only if the associated filter of the sequence converges to the same point.*

Proof. Proposition 1.2.4 is convenient here. Assume the sequence (x_n) converges to a point x in a topological space $(X, \mathcal{T})$. For this sequence, define B_k for each $k \in \mathbb{N}$ using the filterbase of tails of sequences given in Example 1.2.2; namely, $B_k = \{x_k, x_{k+1}, \dots\}$, $k \in \mathbb{N}$. Let U be any neighborhood of x. There exists $N \in \mathbb{N}$, depending on U, such that for all $n \geq N$, $x_n \in U$. In other words, for $k \geq N$, $B_k \subset U$. By Proposition 1.2.4, the filter generated by $\mathfrak{B}$ converges to x as well. Conversely, if the associated filter of a sequence converges to a point, then the sequence clearly converges to the same point, too. $\quad\square$

In Hausdorff topologies (see Definition A.3.14), limits of sequences are unique. This property generalizes to filters.

Theorem 1.2.1. *In Hausdorff topological spaces, limits of filters are unique.*

Proof. Exercise 1.2.10. □

Another well-known property of metric spaces is that a point x is in the closure of a set A if and only if a sequence of points from A converges to x. It should come as no surprise to know, as will be seen in the next result, that this property also holds in the context of filters. In the result below, we use filterbases rather than filters because the result does not hold for general filters; see [129, Exercise 3.1.4, p. 28].

Theorem 1.2.2. *In any topological space, closure points of a set are limits of filterbases originating from the set. That is, if $A \subset X$ is a nonempty set in $(X, \mathcal{T})$, then $x \in \overline{A}$ if and only if there is a filterbase $\mathfrak{B}$ of subsets of A such that $\mathfrak{B} \to x$.*

Proof. ($\Rightarrow$): Suppose $A \subset X$ is nonempty and $x \in \overline{A}$. By definition of closure, for every neighborhood U of x, $U \cap A \neq \emptyset$. This motivates us to define $\mathfrak{B}$ by $\mathfrak{B} = \{U \cap A :$ U is a neighborhood of $x\}$. Because $A \neq \emptyset$, $\mathfrak{B}$ is nonempty and each $B \in \mathfrak{B}$ is nonempty. Furthermore, given any $B, C \in \mathfrak{B}$, $B = U \cap A$ for some neighborhood U of x and $C = V \cap A$ for some neighborhood V of x, hence, $B \cap C = W \cap A$, where $W = U \cap V$ is yet another neighborhood of x, verifying that $\mathfrak{B}$ is a filterbase of subsets of A. The next task is to prove that $\mathfrak{B} \to x$. To this end, let U be any neighborhood of x. Then $B_U = U \cap A \in \mathfrak{B}$. In addition, $B_U \subset U$, verifying that $\mathfrak{B}$ is finer than the collection of neighborhoods of x. We conclude that $\mathfrak{B} \to x$.

($\Leftarrow$): Assume there is a filterbase $\mathfrak{B}$ of subsets of A such that $\mathfrak{B} \to x$. In order to show that $x \in \overline{A}$, let U be any neighborhood of x. Then there exists $B \in \mathfrak{B}$ such that $B \subset U$ and $B \subset A$. We now realize that $B \subset U \cap A$. As $B \neq \emptyset$, we conclude that $x \in \overline{A}$. □

In the context of metric spaces, a useful fact is that a function is continuous if and only if it preserves the convergence of sequences. This result generalizes to the context of filters as follows, where we first define the image of a filter.

Definition 1.2.8. Consider a function $f : X \to Y$, and a filter $\mathfrak{F}$ on X. Define $f(\mathfrak{F})$ by

$$f(\mathfrak{F}) = \{B \subset Y : B \supset f(A), \text{ for some } A \in \mathfrak{F}\}. \tag{1.6}$$

Working out Exercise 1.2.11 allows you to verify that the expression for $f(\mathfrak{F})$ above does in fact define a filter.

Theorem 1.2.3. *For topological spaces $(X, \mathcal{T})$ and (Y, σ), a function $f : X \to Y$ is continuous at $x \in X$ if and only if whenever $\mathfrak{F}$ is a filter such that $\mathfrak{F} \to x$ in X, then $f(\mathfrak{F}) \to f(x)$ in Y.*

Proof. ($\Rightarrow$): Assume the function f as described in the statement, is continuous at x. Let $\mathfrak{F}$ be any filter in X that converges to x. We plan to show that $f(\mathfrak{F}) \to f(x)$ in Y. Let V be any open neighborhood of $f(x)$ in Y. The continuity of f implies that $f^{-1}(V)$ is an open neighborhood of x. Denote this open neighborhood of x by U. We assumed that $\mathfrak{F} \to x$, which means there is a set $A \in \mathfrak{F}$ such that $A \subset U$. Properties of images composed with preimages (see Theorem A.1.1), imply

$$f(A) \subset f(U) = f(f^{-1}(V)) \subset V. \tag{1.7}$$

By Definition 1.2.6 of convergence of filters, we have that $f(\mathfrak{F}) \to f(x)$.

$(\Leftarrow)$: We now assume that for every filter $\mathfrak{F}$ that converges to x in X the image filter $f(\mathfrak{F})$ converges to $f(x)$. The most convenient choice of filters to choose is that of the set of all neighborhoods of x, which we will denote here by $\mathfrak{N}$. The description of the convergence in this case goes like this: $\mathfrak{N} \to x$ by the definition of neighborhoods in a topology. Next, by assumption, $f(\mathfrak{N}) \to f(x)$. This implies that given any neighborhood V of $f(x)$, there is a set $U \in \mathfrak{N}$ such that $f(U) \subset V$. Well, this is exactly Definition A.3.22 of continuity at x, and signals the end of the proof. $\qquad\square$

We will return to the ideas of filters from time to time. Meanwhile, one more comment for this section is that, in general, filters can be used to prove versions of many of the results in this book, as well as results that might also arise in the context of general topology. We will consider such a general context in Section 9.3.

Exercises

1.2.1. Verify that the two items (b) and (c) of Example 1.2.1 are indeed filters.

1.2.2. Given any nonempty set X, define $\mathfrak{F}$ to contain only the set X. Prove that $\mathfrak{F}$ is a filter.

1.2.3. Suppose A is a proper subset of universal set X, and let $\mathfrak{F}$ be a filter defined on A. Prove that $\mathfrak{F}$ is a filterbase in X but not a filter on X.

1.2.4. Prove that for any point x in a set X, the discrete filter corresponding to x converges to x. In particular, in topological spaces the corresponding discrete filter converges to a point x regardless of the topology.

1.2.5. Prove Proposition 1.2.3.

1.2.6. Verify that the collection $\mathfrak{F}$ of Definition 1.2.5 is indeed a filter.

1.2.7. Prove that if X is a nonempty set with the trivial topology (see Definition A.3.9), then every filter converges to every point of X.

1.2.8. Suppose $(X, \mathcal{T})$ is a topological space and A is a topological subspace equipped with the relative topology (see Definition A.3.3) from X. Prove that a filterbase on A converges to a point x of A if and only if the filterbase converges to x in X.

1.2.9. From Definition 1.2.7, consider a filter $\mathfrak{F}$ generated by a filterbase $\mathfrak{B}$ in a topological space. Prove that $\mathfrak{F} \to x$ if and only if for every neighborhood U of x there exists $B \in \mathfrak{B}$ such that $B \subset U$. Deduce that a filter generated by a filterbase converges to a point if and only if the filterbase converges to the same point.

1.2.10. Prove Theorem 1.2.1. A proof by contradiction works well here.

1.2.11. Prove that the structure $f(\mathfrak{F})$ of Definition 1.2.8 does indeed define a filter.

1.2.12. Prove that if $\mathfrak{F}_1$ and $\mathfrak{F}_2$ are filters on a set X, then $\mathfrak{F}_1 \cap \mathfrak{F}_2$ is also a filter on X.

1.3 Topological vector spaces

The phrase "topological vector space" rather obviously implies there is some relation between a topological structure and a vector space structure. The first definition of this section, which is also a fundamental concept that will be used in almost everything else we do, defines such a relation.

Definition 1.3.1. A **topological vector space** is a vector space having a topology for which vector addition and scalar multiplication are continuous; that is, a topological vector space is a vector space E over a field $\mathbb{F}$ endowed with a topology $\mathcal{T}$ for which the following functions are continuous:
(a) Vector addition: The function $+ : E \times E \to E$ given by: for every $(x,y) \in E \times E$, $+(x,y) = x + y$.
(b) Scalar multiplication: The function $\cdot : \mathbb{F} \times E \to E$ given by: for every $(a, x) \in \mathbb{F} \times E$, $\cdot(a, x) = a \cdot x$.

Note that we generally assume $\mathbb{F}$ to be $\mathbb{K}$, but in Chapter 10 we will consider other choices of the field $\mathbb{F}$. Also, rather than write "vector addition" every time we add vectors, we simply write "addition", unless the context requires more clarity.

Notation 1.3.1. The pair $(E, \mathcal{T})$ represents a topological vector space and will be denoted by **TVS**. When topological confusion is not anticipated, we simply refer to a TVS as E. The abbreviation TVS also stands for the plural (rather than "TVSs").

The definition above is also described with the following vocabulary.

Definition 1.3.2. A topology on a vector space under which addition and scalar multiplication are continuous is called **compatible (with vector addition and scalar multiplication)**.

Notation 1.3.2. If $(E, \mathcal{T})$ is a TVS and $a \in E$, then $\mathcal{N}_a$ represents a filterbase of neighborhoods of a with respect to $\mathcal{T}$. We will shorten this to "base of neighborhoods" so as to follow common terminology. In particular, $\mathcal{N}_0$ represents a base of neighborhoods of the zero vector 0. Whenever we need to specify which space or topology is involved with zero neighborhoods, we will use the notation of $\mathcal{N}_{0,E}$ or $\mathcal{N}_{0,\mathcal{T}}$, respectively.

Topological vector spaces are abundant. They appear, at least indirectly, in material as early as pre-calculus courses when lively discussions about intervals of real numbers take place. Rigorously, in beginning analysis or advanced calculus courses, one proves that the properties convergence of sequences of real numbers are preserved by sums and scalar multiples. This proves that addition and scalar multiplication of real numbers are continuous functions. In the context of normed spaces, the claim that TVS are abundant is shown in the next proposition.

Proposition 1.3.1. *Every normed space is a topological vector space.*

Proof. Using the fact that a norm generates a metric, we can prove this property using sequences, by way of Theorem A.4.2. To this end, let $(E, \| \cdot \|)$ be any normed space and assume $x_n \to x, y_n \to y$. Proving that $+ : E \times E \to E$ is continuous requires proving that $x_n + y_n \to x + y$. This is accomplished by observing that for any $\varepsilon > 0$, we have $\|x_n + y_n - (x + y)\| \leq \|x_n - x\| + \|y_n - y\|$ by the triangle inequality, and then applying a routine "$\varepsilon/2$" argument leads to $\|x_n + y_n - (x + y)\| \leq \varepsilon/2 + \varepsilon/2 = \varepsilon$, for all n large enough to ensure that $\|x_n - x\| < \varepsilon/2$ and $\|y_n - y\| < \varepsilon/2$. The proof of the continuity of scalar multiplication is similarly short. $\qquad\square$

Throughout the various topics to be covered, we will encounter a wide variety of TVS, including many types of intriguing TVS that are not normed spaces.

Most of the rest of this section is about which kinds of sets can be assumed in a neighborhood base of a TVS. As a start, the continuity of addition and scalar multiplication in a TVS allows us to move conveniently around the space by way of the conclusions of the next theorem. For the concept of a homeomorphism, see Definition A.3.23.

Theorem 1.3.1. *In any TVS addition and nonzero scalar multiplication are homeomorphisms; specifically, in any TVS E, the following hold:*
(a) *For any $y \in E$, the function $f_y : E \to E$ given by $f_y(x) = x + y$ is a homeomorphism.*
(b) *For any $a \in \mathbb{K} \setminus \{0\}$, the function $g_a : E \to E$ given by $g_a(x) = a \cdot x = ax$ is a homeomorphism.*

Proof. For addition, clearly f_y is bijective (one-to-one and onto), with $f_y^{-1}(y) = y - x$. Continuity of the functions f_y and f_y^{-1} follow from the continuity of addition. Similarly, for any nonzero $a \in \mathbb{K}$, g_a is bijective, with $g_a^{-1}(x) = \frac{1}{a}x$, and continuity following from the continuity of scalar multiplication. $\qquad\square$

Below is the convenience that was previously alluded to. Each item follows directly from Theorem 1.3.1. Observe that an expression of the form $x + U$ is a sum of sets, technically, $\{x\} + U$.

Corollary 1.3.1. *If E is a TVS, then*
(a) *For $x \in E$, a collection $\mathcal{N}_x$ is a base of neighborhoods of x if and only if $\mathcal{N}_0$ is a base of neighborhoods of 0. In particular, $U \in \mathcal{N}_0 \Leftrightarrow x + U \in \mathcal{N}_x$, for any $x \in E$.*
(b) *For any $a \in \mathbb{K} \setminus \{0\}$, $U \in \mathcal{N}_0 \Leftrightarrow aU \in \mathcal{N}_0$.*
(c) *If U is an open set, then aU is open for any scalar $a \neq 0$.*
(d) *If U is a closed set, then aU is closed for any scalar $a \neq 0$. In particular, part (a) implies that for any set A in E the closure of A is described as follows:*

$$x \in \overline{A} \Leftrightarrow (\forall U \in \mathcal{N}_0) \, (x + U) \cap A \neq \emptyset. \tag{1.8}$$

The properties of parts (a) and (b) above have special names.

Definition 1.3.3. A topology on a vector space that satisfies part (a) of the previous corollary is referred to as **translation invariant**. A topology on a vector space that satisfies part (b) of the previous corollary is referred to as **dilation invariant**.

Theorem 1.3.2. *Let E be a TVS. Then given any base $\mathcal{N}_0$ of zero neighborhoods and any $U \in \mathcal{N}_0$ the following properties hold:*
(a) *U is absorbing.*
(b) *U contains a balanced zero neighborhood.*
(c) *U contains a zero neighborhood V such that $V + V \subset U$.*
(d) *U contains a closed zero neighborhood.*

Proof. Assume $\mathcal{N}_0$ is any base of zero neighborhoods of a TVS E.

(a) By Theorem 1.3.1(b), given any $x \in E$ and any $U \in \mathcal{N}_0$, there exists $a > 0$ such that $ax \in U$; that is, U is absorbing.

(b) Given any $U \in \mathcal{N}_0$, Corollary 1.3.1(b) implies that aU also belongs to $\mathcal{N}_0$, for any $a > 0$. In particular, this means the balanced core $V = \mathrm{balcor}(U) \subset U$ is a zero neighborhood as well. Of course, $\mathrm{balcor}(U)$ is balanced.

(c) From the definition of continuity of addition in Definition 1.3.1(a), given any $U \in \mathcal{N}_0$, there is a $V \in \mathcal{N}_0$ such that $+(V \times V) \subset U$. In other words, $+(V \times V) = V + V \subset U$.

(d) First, apply parts (b) and (c) to find a balanced $V \in \mathcal{N}_0$ such that $V + V \subset U$. We will show that $\overline{V} \subset U$. For any $x \in \overline{V}$, $(x + V) \cap V \neq \emptyset$, so there is some $w \in V$ for which $x + w \in V$. Because V is balanced $x \in -w + V \subset V + V \subset U$, and we are done. $\qquad\square$

A bigger picture of properties (b), (c), and (d) in Theorem 1.3.2 is that whenever convenient, we may assume any of these properties hold for zero neighborhoods of a TVS. Hence, the remark that follows.

Remark 1.3.1. In any TVS, we may assume that a zero neighborhood U satisfies properties of being balanced, closed, and that U contains another zero neighborhood V for which $V + V \subset U$.

1.3.1 Properties of subsets of a topological vector space

A plethora of properties of subsets of a TVS will now be established. Such properties provide convenient ways of dealing with open and closed sets as well as the closure of a set.

Proposition 1.3.2. *Let E be a TVS and $A, B \subset E$ any subsets. The following properties hold:*
(a) *$A + U$ is open whenever U is open.*
(b) *For any $x \in E$, $\overline{x + A} = x + \overline{A}$. For any nonzero scalar a, $\overline{(aA)} = a\overline{A}$.*
(c) *$\overline{A} + \overline{B} \subset \overline{A + B}$.*
(d) *$A + B$ is compact whenever A and B are compact.*

(e) *If $\mathcal{N}_0$ is a base of zero neighborhoods, then*

$$\overline{A} = \bigcap \{A + U : U \in \mathcal{N}_0\}, \tag{1.9}$$

(f) *The closure of a balanced set is balanced.*
(g) *If $0 \in \mathring{A}$ (where $\mathring{A}$ is the interior of A) and A is balanced, then $\mathring{A}$ is also balanced.*

Proof. Parts (b), (d), and (g) follow quickly from the definitions and are left as Exercise 1.3.3.

(a) Let A be any nonempty subset of E. If U is an open set, then for every $x \in A, x + U$ is open by translation invariance. Thus, $A + U = \bigcup \{x + U : x \in A\}$, which is open as a result of being a union of open sets.

(c) Starting with the topological property that $\overline{A} \times \overline{B} = \overline{A \times B}$ holds in any product of two topological spaces, we apply the equivalence of continuity of the form: For any $A \subset X, f(\overline{A}) \subset \overline{f(A)}$ in Y from Theorem A.3.6(d). Specifically, the continuity of addition, $+ : E \times E \to E$ satisfies the property that

$$\overline{A} + \overline{B} = +(\overline{A} \times \overline{B}) = +(\overline{A \times B}) \subset \overline{+(A \times B)} = \overline{A + B}. \tag{1.10}$$

(e) Given any $x \in E$, the homeomorphism properties of Theorem 1.3.1 allow us to express a neighborhood base $\mathcal{N}_x$ of x by $\{x - U : U \in \mathcal{N}_0\}$. Put $C = \bigcap \{A + U : U \in \mathcal{N}_0\}$ and note that for any $x \in C$, every set of $\mathcal{N}_x$ as just expressed, intersects A. By Equation (1.8), $x \in \overline{A}$. Hence, $C \subset \overline{A}$. Conversely, suppose $x \in \overline{A}$. Then for every $U \in \mathcal{N}_0, x \in A + U$, and we conclude that $x \in C$, that is, $\overline{A} \subset C$, as desired.

(f) Assume A is balanced and consider $x \in \overline{A}$. Given any $U \in \mathcal{N}_0, x + U$ is a neighborhood of x. Moreover, there exists a balanced $W \in \mathcal{N}_0$ with $W \subset U$. By Equation (1.8), there is a $y \in A \cap (x + W)$, and for $a \in \mathbb{K}$ such that $|a| \leq 1$, we have $ax \in aA$ as well as $ax \in ax + aW$. This implies that $ax \in aA \cap (ax + aW) \subset A \cap (ax + W) \subset A \cap (ax + U)$, where we have applied the assumptions of A and W being balanced. The observation that $ax \in \overline{A}$ proves that $\overline{A}$ is indeed balanced. $\quad\square$

For the corollary below, recall from Definition A.3.15 that a function is **open** if it sends open sets to open sets.

Corollary 1.3.2. *The quotient map (see Definition 1.1.11) $\varphi : E \to E/M$ of a TVS onto a subspace M, is open.*

Proof. By Proposition 1.3.2(a), the open sets of E/M are precisely the sets $\varphi(A)$ such that $A + M$ is open in E. Because a set $U + M$ is open in E whenever U is, $\varphi(A)$ is open in E/M for every open subset A of E. $\quad\square$

Hausdorff topological spaces are often preferable, for example, to ensure uniqueness of limits. Because almost all of the TVS we will encounter in this book are assumed to be Hausdorff, this mentality is formulated below.

Assumption 1.3.1. Unless otherwise indicated, a topological vector space will be assumed to be Hausdorff.

In particular, Assumption 1.3.1 applies to any TVS that is not explicitly labeled as Hausdorff. The next result shows how to check if a TVS is Hausdorff. The proof represents an example of the usefulness of the property that in any zero neighborhood U, we can find a zero neighborhood V such that $V + V \subset U$.

Proposition 1.3.3. *A TVS is Hausdorff if and only if the intersection of all sets of a base of zero neighborhoods results in the single element of zero; that is, if $\mathcal{N}_0$ is a base of zero neighborhoods in a TVS E, then E is Hausdorff if and only if*

$$\bigcap\{U : U \in \mathcal{N}_0\} = \{0\}. \tag{1.11}$$

Proof. ($\Rightarrow$): Assume E is Hausdorff and let $x \in E \setminus \{0\}$ be arbitrary. By the Hausdorff assumption, there exist neighborhoods U, V of 0 and x, respectively, such that $U \cap V = \emptyset$. That $0 \in U$, with U being a zero neighborhood implies that $x \notin U$, and even more so, $x \notin \bigcap\{U : U \in \mathcal{N}_0\}$. Hence, $\bigcap\{U : U \in \mathcal{N}_0\} = \{0\}$.

($\Leftarrow$): Now assume $\bigcap\{U : U \in \mathcal{N}_0\} = \{0\}$. Let $x, y \in E$ with $x \neq y$. Of course, this means $x - y \neq 0$, which in turns implies that $x - y \notin \bigcap\{U : U \in \mathcal{N}_0\}$. In particular, there is a $U \in \mathcal{N}_0$ for which $x - y \notin U$. We find a balanced $V \in \mathcal{N}_0$ such that $V + V \subset U$. Certainly, $x + V \in \mathcal{N}_x$ and $y + V \in \mathcal{N}_y$. The proof will be completed if we show that $(x + V) \cap (y + V) = \emptyset$. To this end, suppose there is some $w \in (x + V) \cap (y + V)$. That would mean there is some $z_1 \in V$ and $z_2 \in V$ having the property that $w = x + z_1$ and $w = y + z_2$. As V is balanced, we use Exercise 1.1.3 to obtain $x - y = z_2 - z_1 \in V - V = V + V \subset U$, leading to a contradiction. $\square$

Corollary 1.3.3. *A TVS is Hausdorff if and only if, given any nonzero element, there exists a zero neighborhood that does not contain that element. That is, the TVS E is Hausdorff if and only if, for any $x \neq 0$ in E, there exists $V \in \mathcal{N}_0$ such that $x \notin V$.*

At this juncture, we state an important result, namely that if we happen to find a filterbase of sets in a vector space that satisfy the first three properties (a), (b), and (c) of Theorem 1.3.2, then there exists a compatible topology on the space with those sets as a base of zero neighborhoods. In fact, such a topology is unique. The proof of following result can be found in [67, Theorem 2.3.1, p. 81]. A proof of a corresponding version of this result for when $\mathcal{N}_0$ consists of convex sets will be given in Theorem 2.1.1 of Chapter 2.

Theorem 1.3.3. *Every TVS has a filterbase $\mathcal{B}$ of zero neighborhoods, the sets of which satisfy the properties of being:*
(a) *Absorbing.*
(b) *Balanced.*
(c) *Every set $U \in \mathcal{B}$ contains a subset $V \in \mathcal{B}$ such that $V + V \subset U$.*

Conversely, suppose the existence on a vector space E over $\mathbb{K}$ of a filterbase $\mathfrak{B}$ consisting of sets that satisfy (a)–(c) above. Then there exists a unique compatible topology on E for which $\mathfrak{B}$ forms a base of zero neighborhoods.

The next result represents an example of how concepts from general topology are used in the context of a TVS. Refer to the Appendix for results about compact sets in the context of general topology. A detailed discussion of compactness in TVS will take place in Subsection 2.5.1.

Theorem 1.3.4. *Suppose in a TVS E, K, and C are nonempty sets such that K is compact and C is closed, and that satisfy $K \cap C = \emptyset$. Then there exists $U \in \mathcal{N}_0$ for which $(K + U) \cap (C + U) = \emptyset$.*

Proof. The assumptions that C is closed and $K \cap C = \emptyset$ imply that $K \subset C^c$, the set C^c being open. In particular,

$$(\forall x \in K)(\exists V_x \in \mathcal{N}_0) \quad \text{such that } (x + V_x) \cap C = \emptyset. \tag{1.12}$$

Thus, for each $x \in K$, properties of zero neighborhoods from Theorem 1.3.2 allow us to choose $W_x \in \mathcal{N}_0$ such that W_x is open, balanced, and satisfies $W_x + W_x + W_x \subset V_x$. We then have that $\bigcup \{x + W_x : x \in K\}$ is an open cover of K, which by compactness can be reduced to a finite subcover of the form $\{x_j + W_{x_j} : j \in \underline{n}\}$, for some $n \in \mathbb{N}$. Put

$$U = \bigcap_{j=1}^{n} W_{x_j}, \tag{1.13}$$

which is of course, a zero neighborhood. The proof will be completed once we prove the statement below.

Claim 1.3.1. $(K + U) \cap (C + U) = \emptyset$.

Proof of Claim 1.3.1. Suppose $(K + U) \cap (C + U) \neq \emptyset$. We will show this leads to a contradiction. Let $y \in (K + U) \cap (C + U)$. Then $y = x_K + u_1 = x_C + u_2$, for some $x_K \in K, x_C \in C$, and $u_1, u_2 \in U$. The finite subcover of K indicated above tells us that for some integer j between 1 and n, $x_K \in x_j + W_{x_j}$, which in turn implies that $x_K + u_1 - u_2 \in (x_j + W_{x_j}) + W_{x_j} + W_{x_j}$, by the construction of U and the observation that $-u_2 \in W_{x_j}$ because W_{x_j} is balanced. Next, we have $x_j + W_{x_j} + W_{x_j} + W_{x_j} \subset x_j + V_{x_j}$ from the general assumption of $W_x + W_x + W_x \subset V_x$. Meanwhile, $x_K + u_1 - u_2 = x_C \in C$. Because $W_{x_j} \subset V_{x_j}$, we have that $(x_j + V_{x_j}) \cap C \neq \emptyset$. This is impossible, in light of Equation (1.12). $\square$

Corollary 1.3.4. *For any $K, C \subset E$, if K is compact and C is closed, then $K + C$ is closed.*

Proof. Exercise 1.3.6. $\square$

Exercises

1.3.1. Prove that on any vector space E, the trivial topology (see Definition A.3.9) is compatible.

1.3.2. Prove that on any nonzero vector space over $\mathbb{K}$, the discrete topology (see Definition A.3.9) is not compatible. *Hint:* Consider scalar multiplication, keeping in mind that we assume the field $\mathbb{K}$ is either $\mathbb{R}$ or $\mathbb{C}$, which are assumed to have their standard topologies.

1.3.3. Prove parts (b), (d), and (g) of Proposition 1.3.2.

1.3.4. Prove that for any normed space $(E, \|\cdot\|)$, a base for $\mathcal{N}_0$ is given by $\{n^{-1}B : n \in \mathbb{N}\}$, where B is either the open or closed unit ball of E.

1.3.5. Prove that in any TVS the intersection of all neighborhoods of 0 is the closure of $\{0\}$, that is, $\overline{\{0\}}$.

1.3.6. Prove Corollary 1.3.4.

1.3.7. Prove that every TVS is connected. See Definition A.3.18 for the definition of a connected topological space.

2 Locally convex spaces: The beginning

The seemingly unremarkable definition of convexity is in fact of fundamental importance in the world of TVS. This will become evident as we go through the rest of the chapters. We already have the terminology to state the fundamental definition below. This chapter begins with the foundational Definition 2.1.1 of a locally convex space (LCS). Easy Proposition 2.1.1 shows that all normed spaces are LCS. Some examples of LCS are displayed in Subsection 2.1.2. Theorems 1.3.3 and 2.1.1 describe basic properties of locally convex topologies. In Definition 2.1.4 of a seminorm, we see explicitly how LCS are generalizations of normed spaces. Theorem 2.1.4 shows us that the topology of every LCS can be described by seminorms. Theorem 2.2.1 and Corollary 2.2.1 contain conditions and equivalences for metrizability. Next comes the important Definition 2.2.3 of a bounded subset of a TVS. In Definition 2.3.2(c), the continuous dual appears for the first time. A discussion of normed spaces as viewed from the perch of LCS then ensues, and Kolomogorov's Theorem 2.4.1 implies that if a TVS ever has a bounded zero neighborhood, then it is in fact a normed space. Theorem 2.5.2 of W. Robertson shows a condition for completeness of comparable locally convex topologies, and Theorem 2.5.5 is about a characterization of compactness in LCS. This chapter ends with Example 2.6.1 of a non-locally convex TVS, revealing that the world of TVS is strictly larger than that of LCS. Here we go.

2.1 Locally convex spaces

2.1.1 Definition and basic properties of locally convex spaces

Definition 2.1.1. A **locally convex space** is a topological vector space that has a base of zero neighborhoods consisting of convex sets. We write **LCS** to denote a locally convex space. The abbreviation LCS also stands for the plural (rather than "LCSs"). The phrase **locally convex topology** refers to such a topology.

Obviously, every LCS is a TVS, and later on in this chapter we will see an example of a TVS that is not a LCS. Throughout much of what we do from here onward, results that are valid in general TVS will usually be stated in that context, while those that are valid in LCS but not necessarily in general TVS will be stated in the context of LCS.

2.1.2 Examples of locally convex spaces

Before proceeding, we observe that there are nontrivial examples of LCS, as the following two items reveal.

Proposition 2.1.1. *Every normed space is a LCS.*

https://doi.org/10.1515/9783111392868-002

Proof. From Exercise 1.3.4, we know that a base of zero neighborhoods of any normed space $(E, \|\cdot\|)$ is given by $\{n^{-1}B : n \in \mathbb{N}\}$, where B is the closed unit ball centered at zero. Hence, it suffices to prove that B is convex, and this is easily done. Indeed, given $x, y \in B$ and $0 \leq t \leq 1$, applying the triangle inequality yields $\|tx + (1 - t)y\| \leq t\|x\| + (1 - t)\|y\| \leq t(1) + (1 - t)(1) = 1$. $\qquad\qquad\square$

Example 2.1.1. The locally convex space $C(\Omega)$.

Let Ω be an open subset of the complex plane $\mathbb{C}$ and let $K \subset \Omega$ be a compact set. Let $C(\Omega)$ denote the set of all complex valued continuous functions on Ω. A base of zero neighborhoods is given by: For a given compact set K and any $\varepsilon > 0$, define $U_{K,\varepsilon} = \{f \in C(\Omega) : |f(x)| \leq \varepsilon, \text{ for every } x \in K\}$. As finite unions of compact sets are compact, the collection of all $U_{K,\varepsilon}$ as K and ε vary, and forms a base of zero neighborhoods for a compatible topology on $C(\Omega)$. In Exercise 2.1.2, you can verify that these sets are all convex.

The next example is historically important as it represents part of the structure that was used by L. Schwartz to give a rigorous mathematical description of distributions. Schwartz was one of the early researchers in the area of LCS, and was awarded a Fields Medal in 1950 for his work that involved using the properties of, you guessed it—*locally convex spaces*, to develop said rigorous description of distributions. We will see some of the basic ideas of distributions in Section 9.1. Descriptions of Schwartz and his mathematics can be found in [10], in [120] (which includes some information about S. Sobolev who was also instrumental in the mathematical development of distributions), and in [113]. In order to define the seminorms in the sequel, we first need the notation that appear in the next two definitions.

Definition 2.1.2. The **support of a** complex valued function f defined on a topological space is

$$\operatorname{supp}(f) = \overline{\{x : f(x) \neq 0\}}. \tag{2.1}$$

Definition 2.1.3. A **multiindex** is an n-tuple of positive integers denoted by $p = (p_1, p_2, \ldots, p_n)$. We define the **order of** p to be $|p| = p_1 + p_2 + \cdots + p_n$.

Now suppose f is defined on $\mathbb{R}^n$ and satisfies the property that its partial derivatives of all orders exist and are continuous. To symbolically represent the partials, we represent the orders of partial differentiation using the notation above, namely for a multiindex $p = (p_1, p_2, \ldots, p_n)$, the nth order partial derivative is expressed using Leibniz notation as follows:

$$\partial^p = \partial_1^{p_1} \partial_2^{p_2} \cdots \partial_n^{p_n} = \frac{\partial^{p_1}}{\partial x_1^{p_1}} \frac{\partial^{p_2}}{\partial x_2^{p_2}} \cdots \frac{\partial^{p_n}}{\partial x_n^{p_n}} = \frac{\partial^{|p|}}{\partial x_1^{p_1} \partial x_2^{p_2} \cdots \partial x_n^{p_n}}. \tag{2.2}$$

We are now ready to define the corresponding LCS.

Example 2.1.2. The locally convex space $\mathcal{D}(K)$.

Suppose $\Omega \subset \mathbb{K}^n$ is a fixed, nonempty open set and $K \subset \Omega$ is compact. Define $\mathcal{D}(K)$ to be the set of all functions $f : \Omega \to \mathbb{K}$, which have continuous partial derivatives of all orders and such that $\mathrm{supp}(f) \subset K$. Observe that $\mathcal{D}(K)$ is a vector space. A base of zero neighborhoods is defined for every multiindex p and every $\varepsilon > 0$ as

$$U_{p,\varepsilon} = \{f \in \mathcal{D}(K) : |\partial^p f(x)| \le \varepsilon, \text{for all } x \in K\}. \tag{2.3}$$

As with $C(\Omega)$, you can prove in Exercise 2.1.2 that every $U_{p,\varepsilon}$ is convex; hence, $\mathcal{D}(K)$ is a locally convex space under the topology of these zero neighborhoods.

More examples of locally convex spaces will be given in Subsection 2.1.4. The next goal will be to obtain a description of the types of sets that can be assumed as zero neighborhoods in any LCS, as done in Chapter 1 for general TVS. We saw earlier that in a TVS every zero neighborhood contains a closed zero neighborhood (Theorem 1.3.2(d)). For LCS, we would like to make a similar statement. This can be accomplished after we make use of the continuity of addition and scalar multiplication to prove the next result.

Proposition 2.1.2. *In a TVS, the closure of a convex set remains convex.*

Proof. Assume A to be a convex subset of a TVS E. Let $x, y \in \overline{A}$ be arbitrary and consider $tx + (1 - t)y$, for $0 \le t \le 1$. For any U that is a convex neighborhood of $tx + (1 - t)y$, the continuity of addition and scalar multiplication allow us to find $V \in \mathcal{N}_x$, $W \in \mathcal{N}_y$ for which $tV + (1 - t)W \subset U$. The definition of the closure of a set tells us that $V \cap A \neq \emptyset$ and $W \cap A \neq \emptyset$. Hence, for any $v \in V \cap A$ and $w \in W \cap A$, $tv + (1 - t)w \in U \cap A$, and this means $tx + (1 - t)y \in \overline{A}$. $\qquad\square$

As with the situation of general TVS, we now describe properties of sets in a base of zero neighborhoods in a LCS.

Theorem 2.1.1. *In any LCS, there exist bases of zero neighborhoods having the following properties:*
(a) *The sets in the base are absolutely convex.*
(b) *The sets in the base are absolutely convex and closed.*
(c) *The sets in the base are absolutely convex and open.*

Proof. (a) Consider any zero neighborhood U in a LCS E. Then U must contain a convex neighborhood; call it V. By Theorem 1.3.2(b), V in turn must contain a balanced neighborhood; call this one W. That is, we have $W \subset V \subset U$. By Exercise 1.1.6, $\mathrm{conv}(W) \subset V$. This allows us to conclude that $\mathrm{conv}(W) \subset V \subset U$. The set $\mathrm{conv}(W)$ is the desired absolutely convex zero neighborhood.

(b) Start with the base of zero neighborhoods obtained in part (a), and note that each set in that base contains a closed neighborhood. Now apply Proposition 2.1.2 to the sets in that base.

(c) This part follows by using the interiors of the sets of part (a). ☐

Next, a LCS version of Theorem 1.3.3 can be obtained.

Theorem 2.1.2. *Suppose that on a vector space E over $\mathbb{K}$, $\mathfrak{B}$ is a filterbase consisting of absorbing, absolutely convex sets. Consider the collection $\mathcal{N} = \{aB : a > 0, B \in \mathfrak{B}\}$. Then there exists a unique compatible locally convex topology on E for which $\mathcal{N}$ forms a base of zero neighborhoods.*

Proof. Regarding Theorem 1.3.3, part (a) is already assumed and part (b) is satisfied by the balanced nature of absolutely convex sets. It is only necessary to do a quick calculation to verify item (c) of Theorem 1.3.3. Indeed, given any U of $\mathcal{N}$, the convexity of U allows us to apply Proposition 1.1.1 after observing that $\frac{1}{2}U \subset U$. We conclude that $\frac{1}{2}U + \frac{1}{2}U \subset U$. ☐

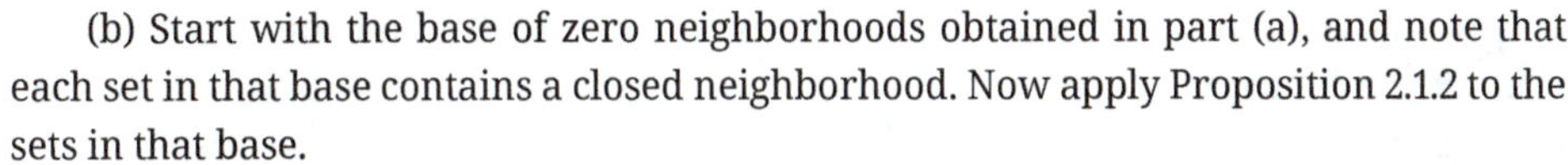

Example 2.1.3. The topology generated by all absolutely convex, absorbing sets.

In any vector space, we can consider the collection of *all* absolutely convex, absorbing sets. If $\mathfrak{B}$ represents the set of all positive multiples of such sets, then $\mathfrak{B}$ is a filterbase, and Theorem 2.1.2 applies: We obtain a locally convex topology, aptly named the **finest locally convex topology** on the given vector space.

For the record, we state one more result involving compact convex sets in a TVS.

Theorem 2.1.3. *In any TVS, the convex hull of a finite union of compact convex sets remains compact.*

Proof. See [91, Theorem 4.4.4, p. 82]. ☐

2.1.3 Seminorms: Generalizations of norms

At the beginning of this section, an easy calculation in Proposition 2.1.1 showed that every normed space is a LCS. Hence, Definition 2.1.1 of a LCS represents a generalization of normed spaces. In Example 2.4.1, we will see an example of a LCS that is not a normed space, verifying that normed spaces are not the only LCS. Meanwhile, to see more precisely how LCS generalize normed spaces, a short review of the definition of a normed space will be illustrative (See also Section A.5). Recall that a norm $\| \cdot \|$ on a vector space E satisfies four properties, those being: For every $x, y \in E$ and every $a \in \mathbb{K}$, the norm satisfies: (i) $\|x\| \geq 0$, (ii) $\|x\| = 0 \Leftrightarrow x = 0$, (iii) $\|ax\| = |a| \cdot \|x\|$, and (iv): $\|x+y\| \leq \|x\| + \|y\|$. The norm, as we saw in Proposition 2.1.1, generates a locally convex topology on the vector space E. In what comes next, we define a natural generalization of a norm.

Definition 2.1.4. Let E be a vector space. A **seminorm** on E is a function p such that for every $x, y \in E$ and every $a \in \mathbb{K}$:
(a) $p(x) \geq 0$.
(b) $p(ax) = |a|p(x)$.
(c) $p(x + y) \leq p(x) + p(y)$.

A vector space endowed with a seminorm is called a **seminormed space**.

Notice that $p(0) = p(0 \cdot 0) = |0|p(0) = 0$. Furthermore, if $p(x) = 0$ implies $x = 0$, then p is a norm, so a seminorm is a generalization of a norm. For a simple example of a seminorm that is not a norm, consider E to be the vector space of all continuous functions defined on the interval $[-1, 1]$ and define p for every $f \in E$ by $p(f) = |f(0)|$.

2.1.4 Examples of seminorms

Some examples will illustrate some of the ways seminorms are defined and used.

Example 2.1.4. Seminorms on the vector space $C(\Omega)$.

Let $C(\Omega)$ denote the set of all complex valued continuous functions on $\Omega \subset \mathbb{K}^n$ of Example 2.1.1. For each compact $K \subset \Omega$, define p_K by $p_K(f) = \sup\{|f(x)| : x \in K\}$.

To check that each p_K is a seminorm, we first note that items (a) and (b) of Definition 2.1.4 of a seminorm are clearly satisfied. For the triangle inequality of item (c), apply the usual triangle inequality in $\mathbb{K}$ and properties of maxima, as follows: ($\forall x \in K$) $|f(x)+g(x)| \leq |f(x)|+|g(x)| \leq \max\{|f(x)| : x \in K\}+\max\{|g(x)| : x \in K\} = p_K(f)+p_K(g)$. Now take the maximum over all $x \in K$, to get $p_K(f + g) \leq p_K(f) + p_K(g)$, as desired.

Example 2.1.5. Seminorms on the vector space $\mathcal{H}(\Omega)$.

For Ω and K as above, let $\mathcal{H}(\Omega)$ denote the collection of holomorphic functions defined on Ω. The vector space $\mathcal{H}(\Omega)$ is a linear subspace of $C(\Omega)$, which implies that the seminorms of Example 2.1.4 are seminorms on $\mathcal{H}(\Omega)$.

Example 2.1.6. Let E be the vector space of all Riemann integrable functions on the interval $[a, b]$, where $a < b$. Define p by: For every $f \in E$ $p(f) = \int_a^b |f(x)|\, dx$. Then p is a seminorm.

You can prove this statement as Exercise 2.1.5. Next, we describe a collection of seminorms on the LCS $\mathcal{D}(K)$.

Example 2.1.7. Seminorms on the vector space $\mathcal{D}(K)$.

Definition 2.1.5. Consider $\mathcal{D}(K)$, the set of all functions $f : \Omega \to \mathbb{K}$, which have continuous partial derivatives of all orders and such that $\operatorname{supp}(f) \subset K$, of Example 2.1.2. We define the following: For every $f \in \mathcal{D}(K)$ and each $p = (p_1, p_2, \ldots, p_n)$,

$$\psi_p(f) = \max\{|\partial^p f(x)| : x \in K\}. \tag{2.4}$$

For each multiindex $p = (p_1, p_2, \ldots, p_n)$, the function ψ_p defines a seminorm. The verification of this statement is left to you as Exercise 2.1.6.

Note that there are countably infinitely many seminorms corresponding to all possible multiindices p. This fact will come up again in Section 2.2 when we look more closely at the connections between seminorms and LCS.

Example 2.1.8. Seminorms on the vector space $\mathcal{E}(\Omega)$.

Let $\Omega \subset \mathbb{K}^n$ be as in Example 2.1.2 and in the previous example. We define the vector space $\mathcal{E}(\Omega)$ to be the set of all functions $f : \Omega \to \mathbb{K}$, which have continuous partial derivatives of all orders. Next, we define seminorms on $\mathcal{E}(\Omega)$ by: For every compact subset K of Ω, every $f \in \mathcal{E}(\Omega)$, and any multiindex p for which $|p| = m$ for some nonnegative integer m,

$$q_{K,p}(f) = \max\{|\partial^p f(x)| x \in K\}. \tag{2.5}$$

Here, $|p| = 0$ refers to the original function. Notice that vector space $\mathcal{E}(\Omega)$ contains the vector space $\mathcal{D}(K)$, and we are using the same definition of seminorm. In Exercise 2.1.11, you can verify that the definition above also defines a seminorm on $\mathcal{E}(\Omega)$. By generalizing the fact that $\mathbb{K}$ can be written as a countable union of compact subsets to the context of $\mathbb{K}^n$, we notice that there is a countably infinite collection of seminorms on $\mathcal{E}(\Omega)$ of the form in Equation (2.5).

Example 2.1.9. Seminorms on the space $\mathcal{S}_k^m$.

Assume m and k are fixed positive integers. Consider the set $\mathcal{S}_k^m$ of all real valued functions f whose derivatives of order $j = 0, 1, \ldots, m$ exist, are continuous, and satisfy: for any $j = 0, 1, \ldots, m$ and $k \in \mathbb{N}$, the following condition holds: $(\forall \varepsilon > 0)(\exists r > 0)$ such that $\sup\{|(1 + |x|^2)^k f^{(j)}(x)| : x \in \mathbb{R}\} < \varepsilon$, whenever $|x| > r$. Observe that this condition implies that each $f \in \mathcal{S}_k^m$ decreases toward zero faster than any polynomial. Hence, $\mathcal{S}_k^m$ is referred to as a space of **rapidly decreasing functions**. Define for each $f \in \mathcal{S}_k^m$ and $j = 0, 1, \ldots, m$,

$$s_{k,j}(f) = \sup\{|(1 + |x|^2)^k f^{(j)}(x)| : x \in \mathbb{R}\}. \tag{2.6}$$

In Exercise 2.1.10, you can gleefully show that the expression above defines a seminorm.

And now for something not quite completely different. We consider what are essentially generalizations of the l_p Banach spaces (see Definition A.5.1). As a prelude, we need to define a certain collection of infinite matrices.

Definition 2.1.6. A **Köthe matrix** is represented as a sequence (a_n) of functions a_n : $\mathbb{N} \to [0, \infty)$ such that

(a) $(\forall j, n \in \mathbb{N})\, a_n(j) \le a_{n+1}(j)$.
(b) $(\forall j \in \mathbb{N})(\exists n \in \mathbb{N})$ such that $a_n(j) > 0$.

A specific example of a Köthe matrix is defined by letting $a_n(j) = n^j$. In this case, the entry in the nth row and jth column is n^j. To create a LCS from a Köthe matrix, one defines a collection of seminorms on a Köthe matrix, as detailed below.

Example 2.1.10. Köthe echelon spaces.

Definition 2.1.7. Given $p \in [1, \infty)$ and a Köthe matrix $K = (a_n)$, the **Köthe echelon space** over $\mathbb{K}$ of order p is defined as

$$\lambda_p(K) = \left\{ x = (x_j) : (\forall j \in \mathbb{N})\, x_j \in \mathbb{K},\ \text{and}\ q_n(x) = \left[\sum_{j=1}^{\infty} a_n(j)|x_j|^p \right]^{\frac{1}{p}} < \infty,\ \forall n \in \mathbb{N} \right\}. \quad (2.7)$$

It is straightforward to check that each $\lambda_p(K)$ is a vector space. Exercise 2.1.7 gives you the opportunity to prove that each q_n is a seminorm.

It is worth pointing out that Köthe echelon spaces can also be defined for $p = \infty$. Details of the $p = \infty$ case as well as further details about Köthe echelon spaces can be found [20, Section 2.4.2, pp. 62–67].

Horváth [67, pp. 439–442] provides excellent discussions of most of the examples described above, as well as many more examples of vector spaces and corresponding definitions of seminorms. Moreover, those descriptions include explanations of relations that exist between most of the spaces.

2.1.5 How seminorms generate a locally convex topology

We now examine how seminorms can generate a topology. Connecting seminorms with the topology of a LCS will be obtained using properties of the Minkowski functional, which is defined below.

Definition 2.1.8. Given an absolutely convex set A in a vector space E, the **Minkowski functional** (also called the **gauge**) of A is μ_A, given by

$$\mu_A(x) = \inf\{a > 0 : x \in aA\}, \tag{2.8}$$

for each $x \in E$.

If for each $a > 0$ $x \notin aA$, then we define $\mu_A(x) = +\infty$. If $0 \in A$, then $\mu_A(0) = 0$.

Now, observe the following.

Proposition 2.1.3. *The Minkowski functional of an absolutely convex, absorbing set is in fact a seminorm.*

Proof. We verify the three parts of the definition of a seminorm: Let A be an absolutely convex, absorbing subset of E.

(a) The definition of $\mu_A(x)$ implies that for every $x \in E$, $\mu_A(x) \geq 0$, and the assumption that A is absorbing implies $\mu_A(x) < \infty$ for every $x \in E$.

(b) We wish to show that for every scalar c, $\mu_A(cx) = |c|\mu_A(x)$. First of all, if $c = 0$, then $\mu_A(cx) = |c|\mu_A(x)$ holds trivially. So, assume $c \neq 0$. By Theorem 1.1.1, aA is balanced for every scalar $a \neq 0$. Let us calculate $\mu_A(cx) = \inf\{a > 0 : cx \in aA\}$ using the balanced assumption of A:

$$cx \in aA \Leftrightarrow -cx \in -aA \Leftrightarrow -cx \in aA \Leftrightarrow |c|x \in aA, \tag{2.9}$$

from which we deduce that

$$\mu_A(cx) = \inf\{a > 0 : |c|x \in aA\} = \inf\{a|c| > 0 : x \in aA\} \tag{2.10}$$
$$= |c| \inf\{a > 0 : x \in aA\}$$
$$= |c|\mu_A(x),$$

by properties of infima.

(c) To show: For every $x, y \in E$, $\mu_A(x + y) \leq \mu_A(x) + \mu_A(y)$. Making use of Proposition 1.1.1, the convexity of A allows us to write (for $a, b > 0$), $aA + bA = (a + b)A$. Given $x \in aA$, and $y \in bA$, we have $x + y \in (a + b)A$, which means $\mu_A(x + y) = \inf\{t > 0 : x + y \in tA\} \leq a + b$. Taking the infima over all such a and b results in $\mu_A(x+y) \leq \mu_A(x) + \mu_A(y)$. $\square$

Corollary 2.1.1. *If $\mathcal{N}_0$ consists only of convex sets that are also balanced, then for each $U \in \mathcal{N}_0, \mu_U$ is a seminorm.*

Proof. In addition to the assumption that the set U is absolutely convex, we only need to state that every zero neighborhood is absorbing. $\square$

The following indicates a concise description of the relation between a set and its Minkowski functional. For convenience, we assume the set is absolutely convex.

Proposition 2.1.4. *For any absolutely convex set A of a LCS, one has $\{x : \mu_A(x) < 1\} \subset A \subset \{x : \mu_A(x) \leq 1\}$.*

Proof. Exercise 2.1.13. $\square$

We also have the following properties of the Minkowski functional.

Proposition 2.1.5. *If A and B are absolutely convex and absorbing in E, then:*
(a) *If $a \neq 0$, then $\mu_{aA} = \frac{1}{|a|}\mu_A$.*
(b) *If $A \subset B$, then $\mu_A \geq \mu_B$.*

Proof. Exercise 2.1.14. $\qquad\square$

In general, μ_A is not a norm, as will be seen in Chapter 3 when we discuss products of TVS.

Observe that a seminorm p is continuous at x if and only if for every neighborhood V of $p(x)$ in $\mathbb{R}$ there is a neighborhood U of x such that $p(U) \subset V$. That is, $(\forall \varepsilon > 0)(\exists U \in \mathcal{N}_0)$ such that $p(x) < \varepsilon$ whenever $x \in U$. With this information, we prove some equivalences to the continuity of seminorms. As part of the process, we will make use of a seminorm version of the "backward triangle inequality," which is of some interest of its own. The details are next.

Proposition 2.1.6. *If p is a seminorm on a TVS E, then for every $x, y \in E$ $|p(x) - p(y)| \leq p(x - y)$.*

Proof. Exercise 2.1.3. $\qquad\square$

Proposition 2.1.7. *If $p : E \to [0, \infty)$ is a seminorm on a TVS E, then the following are equivalent:*
(a) *p is continuous on E.*
(b) *$\{x : p(x) < 1\} = p^{-1}([0, 1)) \in \mathcal{N}_0$.*
(c) *$\{x : p(x) \leq 1\} = p^{-1}([0, 1]) \in \mathcal{N}_0$.*
(d) *p is continuous at 0.*

Proof. (a) $\Rightarrow$ (b) $\Rightarrow$ (c) follow from general topological equivalences of continuity (see Theorem A.3.6), and an obvious set inclusion.

(c) $\Rightarrow$ (d): For efficiency of notation, let $U = \{x : p(x) \leq 1\}$. By assumption (c), $U \in \mathcal{N}_0$. This implies that for any $a > 0$, $aU \in \mathcal{N}_0$. Given any $x \in aU$, $p(x) \leq a$, hence, $p(x) \in [0, a]$, and we conclude that p is in fact, continuous at 0.

(d) $\Rightarrow$ (a): Consider any $x \in E$ and let $\varepsilon > 0$ be arbitrary. Our goal is to prove that p is continuous at x. As p is continuous at 0, $(\exists U \in \mathcal{N}_0)(\forall z \in U)\, p(z) < \varepsilon$. The translation invariance of TVS topologies from Definition 1.3.3 tells us that $V = x + U$ is a neighborhood of x. Let $y \in V$ be such that $y = x + z$, for $z \in U$. The only task left to do is show that $|p(x) - p(y)| < \varepsilon$, which is done using z in the calculation below:

$$|p(x) - p(y)| \leq p(x - y) = p(x - (x + z)) = p(-z) = p(z) < \varepsilon. \tag{2.11}$$

As $\varepsilon > 0$ was arbitrary, we conclude that p is indeed continuous at x, and the proof is done. $\qquad\square$

We now apply the above result to the Minkowski functional.

Proposition 2.1.8. *In a TVS, the Minkowski functional of an absolutely convex absorbing set is continuous if and only if the set is a zero neighborhood.*

Proof. ($\Rightarrow$): Let V denote an absolutely convex absorbing set and assume μ_V is continuous. By Proposition 2.1.7, $\mu_V^{-1}([0,1))$ is open in E. Meanwhile, $\mu_V(x) < 1$ implies that $x \in V$, including that $0 \in V$. Thus, $V \in \mathcal{N}_0$.

($\Leftarrow$): Exercise 2.1.15. $\qquad\square$

The proof of the following useful observation is left as an exercise.

Proposition 2.1.9. *For any seminorm p, the sets $\{x : p(x) < 1\}$, and $\{x : p(x) \le 1\}$ are absolutely convex and absorbing.*

Proof. Exercise 2.1.16. $\qquad\square$

We need another property of continuous seminorms that will be useful in an upcoming result (Theorem 2.1.4 below).

Proposition 2.1.10. *If p is a continuous seminorm on the vector space E, then*

$$\overline{\{x \in E : p(x) < 1\}} = \{x \in E : p(x) \le 1\}. \tag{2.12}$$

Proof. Put $V_p = \{x \in E : p(x) < 1\}$ and $A_p = \{x \in E : p(x) \le 1\}$. We wish to prove that $\overline{V_p} = A_p$. On one hand, because p is continuous, A_p is closed, being the inverse image of a closed set. As $V_p \subset A_p$, we have that $\overline{V_p} \subset A_p$.

To prove the other inclusion, assume $x \notin \overline{V_p}$. Then there exists a neighborhood U of x such that $U \cap V_p = \emptyset$. In particular, there is a scalar $0 < a < 1$ (sufficiently close to 1) for which $ax \notin V_p$. Thus, $p(ax) > 1$, which means $p(x) > 1/a > 1$, and we conclude that $x \notin A_p$. $\qquad\square$

At this point, we take a first look at initial topologies in order to prove an important result. More details about initial topologies will come up again in Chapter 3.

Definition 2.1.9. Given a collection $\{f : X \to Y\}$ of functions from a set X to a topological space $(Y, \mathcal{T}_Y)$, the **initial topology** on X is the coarsest topology for which every f is continuous. In this case, the initial topology is generated using $\mathcal{S} = \{f^{-1}(V) : V \in \mathcal{T}_Y\}$ as a subbase (see Definition A.3.6) of $\mathcal{T}_Y$.

The next theorem is fundamental. It tells us how to construct a LCS from a collection of seminorms, and it also tells us that the topology of a LCS can *always* be generated by a collection of seminorms, generalizing the fact that a normed space topology is generated by a norm.

Theorem 2.1.4. *A given collection of seminorms on a vector space can be used to generate a locally convex topology. Conversely, the topology of every locally convex space can be generated by a collection of seminorms.*

Proof. For the first statement, let $\mathcal{P}$ denote a collection of seminorms. We use the initial topology of Definition 2.1.9 to obtain the coarsest topology such that all of the seminorms of $\mathcal{P}$ are continuous. This topology is generated by the subbase consisting of positive multiples of sets of the form $U_p = \{x \in E : p(x) \leq 1\}$, for each seminorm $p \in \mathcal{P}$. That is, a base of the generated topology consists of sets of the form

$$V = \bigcap_{j=1}^{n} aU_{p_j}, \tag{2.13}$$

for $p_j \in \mathcal{P}$ and $a > 0$. By the way, each set of the form U_p is convex, so the base of the topology consists of convex sets. What is left to do is to verify that the topology generated by a subbase of sets of the form shown in Equation (2.13) is in fact a compatible topology. To verify this, we would like to apply Theorem 1.3.3, and we can do so if we verify that the collection of sets of the form V above is a filterbase that satisfies the three conditions of Theorem 1.3.3. Let $\mathfrak{B}$ denote the collection of all sets of form V from Equation (2.13) above. First, we check the three properties (a), (b), and (c) of Theorem 1.3.3.

Proposition 2.1.9 tells us that each such set V is both (a): absorbing and (b): balanced. As for property (c), the convexity of each set V allows us to state that $\frac{1}{2}V + \frac{1}{2}V \subset V$.

We now show that $\mathfrak{B}$ is a filterbase. The verification does not take long. Indeed, $\mathfrak{B} \neq \emptyset$, because we can choose $U = E$ as a default choice if necessary. Second, each $V \in \mathfrak{B}$ is nonempty by virtue of being absorbing. Third, consider any sets of the form $A = aU_p$, $B = bU_q$, for some $a, b > 0$ and corresponding to some $p, q \in \mathcal{P}$, respectively. By letting $c = \min\{a, b\}$, it is clear that $cU_p \subset aU_p$ and $cU_q \subset bU_q$, from which we deduce that for any two sets $V_1, V_2 \in \mathfrak{B}$, we can find, via appropriate minima, a third set $V_3 \in \mathfrak{B}$ for which $V_3 \subset V_1 \cap V_2$. Thus, by Theorem 1.3.3, we conclude that the topology generated by the subbase is indeed a compatible locally convex topology.

The proof of the second statement amounts to answering the question of where a person could find the seminorms that would describe a locally convex topology. Theorem 2.1.1(b) asserts that a base of zero neighborhoods in any LCS can be formed using absolutely convex, closed sets. Proposition 2.1.10 provides what is needed to complete the proof by way of the observation that each such zero neighborhood represents a continuous seminorm. In other words, the seminorms that we seek are precisely those that are continuous with respect to the locally convex topology. $\square$

Equation (2.13) can be restated in the following form.

Corollary 2.1.2. *Given a collection $\mathcal{P}$ of seminorms on a vector space and the locally convex topology generated by those seminorms, any zero neighborhood contains a set of the form*

$$\{x : p_j(x) \leq \varepsilon, j \in \underline{n}, \varepsilon > 0\}. \tag{2.14}$$

Example 2.1.11. The seminorm topology on $C(\Omega)$.

We saw in Example 2.1.1 that for each compact $K \subset \Omega$, $p_K(f) = \sup\{|f(x)| : x \in K\}$ defines a seminorm. The locally convex topology generated by these seminorms is called the **topology of uniform convergence on compact sets**. This topology has a few aliases. In particular, it also goes by the name **compact-open topology**, as well as the name **topology of compact convergence**.

Next, it will be useful to express continuity of linear maps in terms of seminorms, which is the gist of the next item.

Proposition 2.1.11. *Let E and F be LCS, with topologies generated by collections of continuous seminorms $\mathcal{P} = \{p_\alpha : \alpha \in I\}$ and $\mathcal{Q} = \{q_\beta : \beta \in J\}$, respectively. A linear map $T : E \to F$ is continuous if and only if for every $q_\beta \in \mathcal{Q}$, there exist a seminnorm $p_\alpha \in \mathcal{P}$ and a positive number M such that $q_\beta(T(x)) \le M p_\alpha(x)$, for all $x \in E$.*

Proof. ($\Rightarrow$): Assume T is continuous. As each q_β is continuous, the composition $q_\beta \circ T$ is continuous, which implies that for each $\beta \in \mathcal{Q}$, there exist an $\alpha \in \mathcal{P}$ and a constant $C = C_{\alpha,\beta} > 0$ such that

$$q_\beta(T(x)) \le 1 \quad \text{whenever } p_\alpha(x) \le C. \tag{2.15}$$

We intend to show that $M = \frac{1}{C}$ will suffice for proving the conclusion. For this, start with an arbitrary $x \in E$. There are two cases:

<u>Case 1</u>: $p_\alpha(x) = 0$. In this case, for every $a > 0$, $p_\alpha(ax) = a p_\alpha(x) = a \cdot 0 = 0 \le C$. Equation (2.15) then indicates that $a q_\beta(T(x)) = q_\beta(T(ax)) \le 1$, for every $a > 0$, which in turn implies that $q_\beta(T(x)) = 0 \le \frac{1}{C} p_\alpha(x)$.

<u>Case 2</u>: $p_\alpha(x) \ne 0$. Properties of seminorms tell us that $p_\alpha\left(\frac{Cx}{p_\alpha(x)}\right) = C$. From this, we invoke Equation (2.15) again and calculate

$$q_\beta\left(T\left(\frac{Cx}{p_\alpha(x)}\right)\right) = \frac{C}{p_\alpha(x)} q_\beta(T(x)) \le 1 \Rightarrow q_\beta(T(x)) \le \frac{1}{C} p_\alpha(x). \tag{2.16}$$

Thus, for all $x \in E$, $q_\beta(T(x)) \le M p_\alpha(x)$, where $M = \frac{1}{C}$.

($\Leftarrow$): Now assume that for every $q_\beta \in \mathcal{Q}$, there is a seminnorm $p \in \mathcal{P}$ and a positive number M for which $q_\beta(T(x)) \le M p_\alpha(x)$, for all $x \in E$. We will use this to prove that T is continuous, which of course, we only need to establish for $x = 0$. First, let $W \in \mathcal{N}_{0,F}$ be arbitrary. Corollary 2.1.2 reveals that W contains a set of the form $\{y : q_{\beta_j}(y) \le \varepsilon\}$, where $q_{\beta_j} \in \mathcal{Q}, j \in \underline{n}$ and $\varepsilon > 0$. By the assumption that all seminorms $q_\beta \in \mathcal{Q}$ are continuous, we observe that $q = \max\{q_{\beta_j} : j \in \underline{n}\}$ is continuous. Using $T(x)$ in place of y, we have $q(T(x)) \le M p_\alpha(x)$. Put $U = \{x : q(T(x)) \le M\}$. Then $U \in \mathcal{N}_{0,E}$, and $U \subset T^{-1}(W)$. Therefore, T is continuous. $\qquad\square$

Corollary 2.1.3. *On a LCS whose topology is generated by a collection $\mathcal{P} = \{p_\alpha : \alpha \in I\}$ of continuous seminorms, a linear functional T is continuous if and only if, there exists $\alpha \in I$ and a constant $M > 0$ such that for every $x \in E$, $|T(x)| \le Mp_\alpha(x)$.*

The following notion will appear a few times.

Definition 2.1.10. A collection of seminorms is **saturated**, if for any finite collection p_j, $j \in \underline{n}$ of the collection, the seminorm $p = \max\{p_j : j \in \underline{n}\}$ also belongs to the collection.

Having a saturated collection of seminorms offers some conveniences. For example, using a saturated collection of seminorms to generate a locally convex topology means we need only use positive multiples of zero neighborhoods rather than also calculating finite intersections. This is because if $\mathcal{P}$ is a saturated collection of seminorms on a vector space E, then for any finite collection of seminorms $p_j, j \in \underline{n}$ from $\mathcal{P}$, $p = \max\{p_j : j \in \underline{n}\}$ also belonging to $\mathcal{P}$ means the zero neighborhood U_p corresponding to p satisfies $U_p = \bigcap\{U_{p_j} : j \in \underline{n}\}$. Of course, any locally convex topology can always be generated by a saturated collection of seminorms by way of using the collection of all seminorms that are continuous with respect to the topology. A consequence of this short discussion is the following.

Assumption 2.1.1. We can assume without loss of generality that a collection of seminorms that generates a locally convex topology is of seminorms that are either saturated or continuous.

See [67] or [91] for more details. Hence, by Assumption 2.1.1 we may just as well assume a collection of seminorms to be saturated rather than continuous.

Now is a good time to compare how results get established in LCS to how results get established in normed spaces. Basically, we can proceed either by geometric or analytic perspectives. In normed spaces, the geometric perspective amounts to using geometric aspects of sets (open or closed balls, etc.). Meanwhile using the analytic perspective of norms amounts to using properties of the given norm $\|\cdot\|$ to deduce results. In the context of LCS, one can also choose to use either the geometric perspective of neighborhoods or the analytic perspective of seminorms, directly generalizing the context of normed spaces.

As you learn about LCS, you will find that some people prefer to work analytically with seminorms while others prefer working geometrically with neighborhoods. You will notice that this author's preference is to work mostly from the geometric perspective.

As simple examples of these two perspectives, first consider describing how a sequence (x_n) converges to zero in a normed space. One can either explain geometrically that any ball centered at 0 contains all but finitely many terms of the sequence, or explain analytically that for any $\varepsilon > 0$, $\|x_n\| < \varepsilon$ for all sufficiently large n. An example from the context of LCS is determining when a LCS is Hausdorff. One could view this geometrically as in Corollary 1.3.3, by checking to see if for every $x \ne 0$ there is a zero neighborhood U that does not contain x, or one could check if a LCS is Hausdorff using seminorms, which is the essence of the next result.

Proposition 2.1.12. *Suppose $\mathcal{P} = \{p_\alpha : \alpha \in I\}$ is a collection of seminorms that generates the topology of the LCS E. Then E is Hausdorff if and only if, for any $x \neq 0$, there is a $p_\alpha \in \mathcal{P}$ such that $p_\alpha(x) \neq 0$.*

Proof. ($\Rightarrow$): Assume E is Hausdorff and let x be any nonzero element of E. By Corollary 1.3.3, there exists a zero neighborhood U such that $x \notin U$. Corollary 2.1.2 implies that U contains a finite intersection of sets of the form $\{y \in E : p_{\alpha_j}(y) < a, j \in \underline{k}\}$, for some $a > 0$ and $p_{\alpha_j} \in \mathcal{P}$. As $x \notin U$, at least one of the $p_{\alpha_j}(x)$ must be nonzero.

($\Leftarrow$): Conversely, suppose for some $p_\alpha \in \mathcal{P}$ and some $x \neq 0$, it turns out that $p_\alpha(x) = a$, where $a > 0$. Then $\{y \in E : p_\alpha(y) \leq a/2\}$ is a zero neighborhood from which x is excluded. Corollary 1.3.3 now applies again. $\qquad\square$

Corollary 2.1.4. *A seminorm that generates the topology of a Hausdorff TVS is a norm.*

Proof. Observe that if a single seminorm generates the topology of a TVS such that the topology is Hausdorff, then it must be a norm. $\qquad\square$

Here is another example of the connection between seminorms and locally convex topologies.

Proposition 2.1.13. *The finest locally convex topology of Example 2.1.3 is generated by the collection of all seminorms.*

Proof. Exercise 2.1.17. $\qquad\square$

Going forward, it behooves us to make the definition and observation below regarding the convergence of sequences in a TVS. Keep in mind Definition A.3.10 of convergence of sequences in a topological space. For the record, let us formally state the definition of a null sequence in a TVS.

Definition 2.1.11. A sequence (x_n) in a TVS is a **null sequence** if it converges to zero.

Proposition 2.1.14. *In a TVS a sequence (x_n) **converges** to a point x if and only if $(x_n - x)$ is a null sequence in the TVS.*

Proof. Exercise 2.1.18. $\qquad\square$

Next is one more item before we move on to the next exciting topic. We wish to connect seminorms with the concept of convergence of filters and sequences, and this is done in a straightforward way.

Proposition 2.1.15. *Suppose E is a LCS with $\mathcal{P}$ as a family of continuous seminorms that generates the topology of E. Then a filter $\mathfrak{F}$ converges to a point x if and only if for each $\varepsilon > 0$ and each $p \in \mathcal{P}$, there exists $A \in \mathfrak{F}$ such that for every $y \in A$, $p(x - y) < \varepsilon$. In particular, a sequence (x_n) converges to 0 if and only if, $(\forall p \in \mathcal{P})\, p(x_n) \to 0$.*

Proof. Exercise 2.1.19. $\qquad\square$

Exercises

2.1.1.   We saw in Proposition 2.1.2 that the closure of a convex set is again convex. On the other hand, the convex hull of a closed set need not be closed. Describe an example of such a set in $\mathbb{R}^2$.

2.1.2.   Verifying some convexity:

(a)   Prove that each set $U_{K,\varepsilon}$ of Example 2.1.1 is convex.

(b)   Prove that each set $U_{p,\varepsilon}$ of Example 2.1.2 is convex.

2.1.3.   Prove Proposition 2.1.6. The proof is similar to that of the comparable result in $\mathbb{R}$: $||x| - |y|| \leq |x - y|$, for all $x, y \in \mathbb{R}$.

2.1.4.   Suppose $\{p_1, p_2, \ldots, p_n\}$ is a finite collection of seminorms on a vector space E. Prove that the following are also seminorms:

(a) $p(x) = \sup\{p_j(x) : j \in \underline{n}\}$.

(b)

$$p(x) = \sum_{j=1}^{n} a_j p_j(x), \tag{2.17}$$

where $a_j \geq 0$ in $\mathbb{K}$.

2.1.5.   Prove that the description in Example 2.1.6 constitutes a seminorm. While you are at it, verify that this seminorm is not a norm.

2.1.6.   Prove that the definition of the function ψ_p in Example 2.1.7 does in fact define a seminorm. Also, verify that for each multiindex p, the seminorm ψ_p is not a norm.

2.1.7.   Prove that each q_n of Example 2.1.10 is indeed a seminorm.

2.1.8.   Consider $\mathbb{R}$ as the countable union of compact sets:

$$\mathbb{R} = \bigcup_{m=1}^{\infty} \{\text{compact } K_m : K_m \subset \overset{\circ}{K}_{m+1}, (\forall m \in \mathbb{N})\}; \tag{2.18}$$

(e. g., one could have $K_m = [-m, m]$). Let $\mathcal{D}(K_m)$ denote the set of all infinitely differentiable functions from $\mathbb{R}$ to $\mathbb{R}$ whose support is contained in K_m.

(a)   Prove that p_j, defined by $(\forall f \in \mathcal{D}(K_m)) \, p_j(f) = \sup\{|f^{(j)}(x)| : x \in \mathbb{R}\}$ defines a semi-norm on $\mathcal{D}(K_m)$, for each $j \in \mathbb{N} \cup \{0\}$. Here, $f^{(j)}$ denotes the jth derivative of f.

(b)   Determine which, if any, of these seminorms is a norm.

2.1.9.   Consider $\Omega \subset \mathbb{C}$ an open set, and

$$C(\Omega) = \{f : \Omega \to \mathbb{C}, \text{ such that } f \text{ is continuous}\}, \tag{2.19}$$

and the collection $\mathcal{U}$ of sets $U_{K,\varepsilon}$ defined by

$$(\forall K \subset \Omega \text{ compact})(\forall \varepsilon > 0) \quad U_{K,\varepsilon} = \left\{f \in C(\Omega) : \sup\{|f(x)| : x \in K\} \leq \varepsilon\right\}. \tag{2.20}$$

(a)   Prove that each $U_{K,\varepsilon}$ is absorbing.

(b)   Prove that $(\forall U_{K,\varepsilon} \in \mathcal{U})(\exists V \in \mathcal{U})$ such that $V + V \subset U$.

2.1.10.   From Example 2.1.9, on the linear space S_k^m, prove that

$$s_{k,j}(f) = \sup\left\{\left|\left(1 + |x|^2\right)^k f^{(j)}(x)\right| : x \in \mathbb{R}\right\} \tag{2.21}$$

defines a seminorm, where $f^{(j)}$ represents the jth derivative of f, $j = 0, 1, \ldots, m$.

2.1.11.   Prove that the construction in Equation (2.5) does in fact define a seminorm on $\mathcal{E}(\Omega)$.

2.1.12. Suppose p and q are seminorms on a TVS E. Prove that if $q \leq p$ and p is continuous, then q is continuous.

2.1.13. Prove Proposition 2.1.4.

2.1.14. Prove Proposition 2.1.5.

2.1.15. Prove the ($\Leftarrow$) part of Proposition 2.1.8.

2.1.16. Prove Proposition 2.1.9.

2.1.17. Prove Proposition 2.1.13.

2.1.18. Prove Proposition 2.1.14.

2.1.19. Prove Proposition 2.1.15.

2.1.20. Suppose p and q are seminorms on E. Prove that if $\{x : p(x) \leq 1\} = \{x : q(x) \leq 1\}$, then $p = q$.

2.2 Metrizable spaces and bounded sets

This section begins with the will to determine a condition that indicates when the topology of a TVS can be generated by a metric. The first item is to formalize what is meant by a TVS whose topology is generated by a metric, per Definition A.4.2.

Definition 2.2.1. A TVS is **metrizable** if its topology can be generated by a metric.

Being in a metric space allows us to do what we want to do topologically by using sequences when we more generally would need to use filters. We seek a condition on TVS such that its topology is indeed generated by a metric. We could consider the more general pseudometrizable TVS as well, however, because we will mainly focus on Hausdorff spaces, we will stay focused on metrizable TVS. References in which more general pseudometrics on TVS are discussed include [91] and [130]. In the main result below, we assume the TVS to be locally convex. Deeper discussions related to the metrizability of general TVS can be found in [75, pp. 124–125, 186–187] or [67, pp. 110–116]. Because all compatible topologies are translation invariant, a metric that induces such topologies must satisfy the same property, which we formally denote in the following definition.

Definition 2.2.2. A metric d on a vector space is **translation invariant** if for every x, y, z in the vector space, $d(x + z, y + z) = d(x, y)$.

Here comes the main result of this section.

Theorem 2.2.1 (Metrization of a LCS). *A LCS is metrizable if and only if it is Hausdorff and has a countable base of zero neighborhoods. In the case where the space is metrizable, the metric is translation invariant.*

Proof. ($\Rightarrow$): Assume the LCS is metrizable. Then it is Hausdorff because every metric space is Hausdorff. Moreover, every metric space has a countable base at each of its points ([102, p. 240]).

($\Leftarrow$): Assume the LCS $(E, \mathcal{T})$ is Hausdorff and has a countable base $\{U_n : n \in \mathbb{N}\}$ of zero neighborhoods. We may safely assume that each U_n is absolutely convex and open. Of course, each U_n is also absorbing. Thus, we can appeal to the Minkowski functional of each U_n, and will it denote by μ_n. Note that by Proposition 2.1.3, each μ_n is a seminorm.

We must do two things: Construct a metric on E, and prove that the constructed metric generates a topology equivalent to $\mathcal{T}$.

Define a function $f : E \to \mathbb{R}$ at each $x \in E$ by

$$f(x) = \sum_{n=1}^{\infty} \frac{1}{2^n} \inf\{\mu_n(x), 1\}. \tag{2.22}$$

Claim 2.2.1. *The function $d : E \times E \to [0, \infty)$ given by: for each $x, y \in E$,*

$$d(x,y) = f(x - y) = \sum_{n=1}^{\infty} \frac{1}{2^n} \inf\{\mu_n(x - y), 1\}, \tag{2.23}$$

defines a translation invariant metric on E.

Proof of Claim 2.2.1. Observe that $f(x)$ exists as a finite value for each $x \in E$, and in fact, $0 \le f(x) \le 1$. We intend to show that d is the desired translation invariant metric. Notice that if we check the properties of a metric for each μ_n, then routine calculations lead to the same conclusion for d. Let $x, y, z \in E$ be arbitrary. First, it is clear that $d(x,y) \ge 0$. Second, if $x = y$, then certainly $d(x,y) = 0$. If $d(x,y) = 0$, then we have $0 = \sum_{n=1}^{\infty} 2^{-n} \inf\{\mu_n(x - y), 1\}$; hence, $\mu_n(x - y) = 0$ for every $n \in \mathbb{N}$. We assumed $(E, \mathcal{T})$ is Hausdorff and that $\{U_n : n \in \mathbb{N}\}$ is a base of zero neighborhoods, so applying Proposition 2.1.12 tells us that $x - y = 0$, which in turn allows us to conclude that $x = y$. Third, by properties of seminorms, $\mu_n(y - x) = \mu_n(x - y)$; hence, $d(y, x) = d(x, y)$. Fourth, the triangle inequality of d follows from the calculation $\mu_n(x - y) = \mu_n((x - z) + (z - y)) \le \mu_n(x - z) + \mu_n(z - y)$. Last but not least, $\mu_n((x + z) - (y + z)) = \mu_n(x - y)$, which leads to the translation invariance of d. This proves the claim.

Claim 2.2.2. *The metric d generates a topology equivalent to $\mathcal{T}$ on E.*

Proof of Claim 2.2.2. To begin with, observe that a base of neighborhoods generated by the metric d is of the form $\{V_n : n \in \mathbb{N}\}$, where each V_n is the open ball of radius 2^{-n}, centered at 0. Here, we have made use of the translation invariance of d. Meanwhile, for $\mathcal{T}$ we have $U_n = \{x : \mu_n(x) < 1\}$, for each $n \in \mathbb{N}$. First, we show that for each $n \in \mathbb{N}$, $V_n \subset U_n$. To see this, consider any $n \in \mathbb{N}$ and suppose x is such that $x \notin U_n$. Then $\mu_n(x) \ge 1$, which implies $d(x, 0) = f(x) = \sum_{j=1}^{\infty} \frac{1}{2^j} \inf\{\mu_j(x), 1\}$ must be greater than or equal to 2^{-n}. Hence, $x \notin V_n$, and we see that the topology generated by d is finer than $\mathcal{T}$.

On the other hand, for each $n \in \mathbb{N}$, $V_n = 2^{-n}\{x : d(x, 0) < 1\} = 2^{-n}\{x : f(x) < 1\}$. By the definition of U_n and because μ_n is continuous for each $n \in \mathbb{N}$, $\{x : f(x) < 1\}$ is an open $\mathcal{T}$-neighborhood of 0. Thus, the topology $\mathcal{T}$ is finer than the topology generated by d. This proves the claim and completes the proof of the theorem. $\qquad\square$

The corresponding metrizability result for general TVS in which the neighborhoods U_n are assumed to be balanced can be found in [107, Theorem 1.24]. As a byproduct of the previous theorem, we obtain another example of how one can view LCS either in terms of seminorms or in terms of neighborhoods in the following.

Corollary 2.2.1. *For a Hausdorff LCS, the following are equivalent:*
(a) *The LCS is metrizable.*
(b) *The LCS has a countable base of zero neighborhoods.*
(c) *The topology of the LCS is generated by a countable collection of seminorms.*

When a TVS is metrizable, we can arrange the sequence of zero neighborhoods in a variety of convenient ways, as see in the proposition below. These particular formations will be useful from time to time.

Proposition 2.2.1. *Suppose E is a metrizable TVS. Then*

(a) *There exists a base $\{U_n : n \in \mathbb{N}\}$ of zero neighborhoods such that the following hold:*
   (i) *Each U_n is balanced and closed.*
   (ii) *For each $n \in \mathbb{N}$, $U_{n+1} + U_{n+1} \subset U_n$. In particular, the base is decreasing: $U_1 \supset U_2 \supset \cdots$.*
   (iii) *For each $n \in \mathbb{N}$, U_{n+1} is absorbing in U_n.*

(b) *If E is locally convex, then the base in part (a) can be simplified to:*
   (i) *Each U_n is absolutely convex and closed.*
   (ii) *For each $n \in \mathbb{N}$, $2U_{n+1} \subset U_n$. In particular, the base is decreasing: $U_1 \supset 2U_2 \supset \cdots$.*
   (iii) *The topology of the LCS is generated by a countable collection $\{p_n : n \in \mathbb{N}\}$ of seminorms such that for each $n \in \mathbb{N}$, $p_n \le 2p_{n+1}$.*
   (iv) *For each $n \in \mathbb{N}$, U_{n+1} is absorbing in U_n.*

Proof. Exercise 2.2.2. □

2.2.1 Bounded sets

The concept of a bounded set should express the idea of such sets being "small" in some sense. In a TVS, this can be done in a natural way. The definition is next.

Definition 2.2.3. A subset of a TVS is **bounded** if it is absorbed by sufficiently large scalar multiples of any zero neighborhood. That is, a subset B of a TVS E is bounded if for any zero neighborhood U, there exists $r > 0$ such that for every $|a| \ge r$, $B \subset aU$.

Note that an equivalent statement is that a set B is bounded if sufficiently small multiples of B are absorbed by any zero neighborhood. Figure 2.1 shows an example of a bounded set B (that looks like a calculator) being absorbed by an arbitrary zero neighborhood U (that looks like an ink blot), in the LCS $\mathbb{R}^2$.

We can use part (b) of Theorem 1.3.2 to assume for any set of a base of zero neighborhoods of a TVS, we can find a balanced subset. In particular, the proposition below gives us a slightly shorter route to proving that a subset of a TVS is bounded.

Proposition 2.2.2. *A subset B of a TVS is bounded if and only if for every $U \in \mathcal{N}_0$, there exists $a > 0$ such that $B \subset aU$.*

Proof. Exercise 2.2.5. □

As with other concepts, it is useful to list the main properties, in this case, of bounded sets.

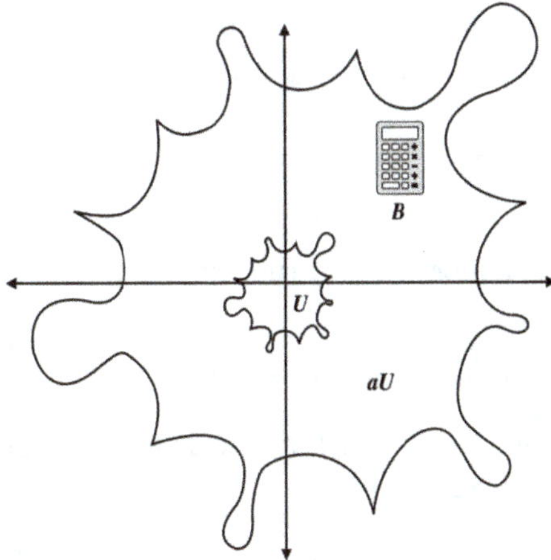

Figure 2.1: A bounded set in a TVS.

Proposition 2.2.3. *The following properties hold for bounded sets in a TVS:*
(a) *Subsets of bounded sets are bounded.*
(b) *Arbitrary intersections and finite unions of bounded sets are bounded.*
(c) *Scalar multiples of bounded sets remain bounded.*
(d) *Finite sums of bounded sets are bounded.*
(e) *The balanced hull of a bounded set is bounded.*
(f) *The closure of a bounded set is bounded.*

Proof. Part (a) is obvious.

For part (b), the statement about intersections follows from part (a). As for finite unions, suppose $B_1, B_2, \ldots, B_k$ are bounded sets. Let $U \in \mathcal{N}_0$ be arbitrary. By Proposition 2.2.2, there exist $a_1, a_2, \ldots, a_k > 0$ such that $B_j \subset a_j U$ for each $j \in \underline{k}$. Choosing $a = \max\{a_1, a_2, \ldots, a_k\}$ implies $\bigcup_{j=1}^{k} B_j \subset aU$.

(c) This follows directly from the definition of boundedness.

(d) As in part (b), suppose $B_1, B_2, \ldots, B_k$ are bounded sets and let $U \in \mathcal{N}_0$ be arbitrary. Applying the continuity of addition, we find a balanced $V \in \mathcal{N}_0$ such that $\underbrace{V + V + \cdots + V}_{k \text{ times}} \subset U$. Just like in part (a), there exist $a_1, a_2, \ldots, a_k > 0$ such that $B_j \subset a_j V$ for each $j \in \underline{k}$. Choose again $a = \max\{a_1, a_2, \ldots, a_k\}$, and we observe that because V is balanced,

$$B_1 + B_2 + \cdots + B_k \subset \underbrace{aV + aV + \cdots + aV}_{k \text{ times}} \tag{2.24}$$

$$= a(\underbrace{V + V + \cdots + V}_{k \text{ times}})$$

$$\subset aU.$$

(e) If B is bounded and $U \in \mathcal{N}_0$ is arbitrary, we find a balanced $V \in \mathcal{N}_0$ such that $V \subset U$. Certainly, $B \subset aV$ for some positive a. Note that aV is balanced if V is, and we obtain $\mathrm{bal}(B) \subset aV \subset aU$.

(f) This part is similar to part (e), in which we can assume $U \in \mathcal{N}_0$ to be closed and observe that for and $a > 0$, aU is also closed. $\qquad\square$

The case of a LCS includes an additional and important construction that preserves boundedness.

Proposition 2.2.4. *In a LCS, the convex hull of a bounded set is bounded.*

Proof. Assume B is bounded and let $U \in \mathcal{N}_0$, where we assume U to be convex. Then for an $a > 0$ for which $B \subset aU$, we have $\mathrm{conv}(B) \subset \mathrm{conv}(aU) = aU$. $\qquad\square$

The properties above allow us to form bounded sets that have convenient properties. In particular, given any bounded subset of a LCS, the closed, absolutely convex hull of that set is also bounded. This inspires the following definition, which is a variant of that of [70, 8.1, p. 151].

Definition 2.2.4. An absolutely convex (respectively, closed absolutely convex) bounded set in a LCS is called a **disk** (resp., **closed disk**). Additionally, if B is a bounded subset of a LCS, then $D = \overline{\mathrm{conv}\,\mathrm{bal}(B)}$ is the closed disk formed by B.

Some authors define a disk to be any absolutely convex set. Next, consider a collection that can be useful in describing all bounded sets. The formal definition follows.

Definition 2.2.5. A **system of bounded sets** in a TVS is a collection $\mathcal{B}$ of bounded sets such that any bounded subset of the TVS is contained in some set from $\mathcal{B}$. If $\mathcal{B}$ is countable, then we call $\mathcal{B}$ a **fundamental sequence of bounded sets**. The abbreviation for a fundamental sequence of bounded sets is f. s. b.

Example 2.2.1. An example of a f. s. b.

In any normed space, $\mathcal{B} = \{nB : n \in \mathbb{N}\}$, where B is the unit ball of the space, is a f. s. b.

Given any f. s. b. $\{A_n : n \in \mathbb{N}\}$ in a LCS, we can form another f. s. b. as follows: Let $C_n = \overline{\mathrm{absconv}(A_n)}$, then create the new f. s. b. by

$$B_n = \overline{\mathrm{absconv}(A_1 \cup A_2 \cup \cdots \cup A_n)}. \tag{2.25}$$

In this way, we obtain an ordered f. s. b. of closed disks $B_1 \subset B_2 \subset \cdots$, and we can even adjust the f. s. b. such that $B_n \subset \frac{1}{2}B_{n+1}$, for each $n \in \mathbb{N}$, for example.

Next up, we have an alternate method of determining if a set is bounded in a TVS, using sequences.

Proposition 2.2.5. *A subset B of a TVS is bounded if and only if whenever (x_n) is a sequence of elements of B and (a_n) is a sequence in $\mathbb{K}$, if $a_n \to 0$ in $\mathbb{K}$, then $a_n x_n \to 0$ in E.*

Proof. ($\Rightarrow$): Assume B to be bounded. Let (x_n) be any sequence of elements of B and let (a_n) be any null sequence (see Definition A.3.10) in $\mathbb{K}$. Let $U \in \mathcal{N}_0$ be arbitrary. Of course, we may assume U is balanced. The assumption of boundedness of B implies that there exists some $a > 0$ such that $B \subset aU$. Thus, for each $n \in \mathbb{N}$, $a_n x_n \in a_n U$. We also assumed $a_n \to 0$ in $\mathbb{K}$. This tells us that there exists $N \in \mathbb{N}$ such that for all $n \geq N$, $|a_n| < 1$. Hence, for all $n \geq N$, $a_n x_n \in a_n U \subset U$, so $a_n x_n \to 0$ in E.

($\Leftarrow$): Assume that for any sequence (x_n) of elements of B and any null sequence (a_n) in $\mathbb{K}$, the sequence $(a_n x_n)$ is a null sequence in E. Suppose B is not bounded. Then there must be a $U \in \mathcal{N}_0$ that does not absorb B. In particular, for every $n \in \mathbb{N}$, there exists $x_n \in B$ such that $x_n \notin nU$. If $a_n = \frac{1}{n}$, then (a_n) is a null sequence in $\mathbb{K}$, however, for every $n \in \mathbb{N}$, $a_n x_n \notin U$, contradicting the assumption that $a_n x_n \to 0$ in E. We conclude that B must be bounded after all. $\qquad\square$

As we have seen several times, there is a seminorm version of the concept we are currently considering. In this case, the concept is of bounded subsets of a LCS. The precise statement follows.

Proposition 2.2.6. *Suppose the topology of a LCS E is generated by the collection $\mathcal{P}$ of seminorms. A subset B of E is bounded if and only if $p(B)$ is bounded for every $p \in \mathcal{P}$.*

Proof. Exercise 2.2.8. $\qquad\square$

Recall that in the context of normed spaces a linear map is continuous if and only if it maps bounded sets to bounded sets. An example of a linear map that maps bounded sets to bounded sets but is not continuous appears in Exercise 2.3.9. In the general LCS context, we will see more instances of discontinuous linear functions that map bounded sets to bounded sets. One instance will be shown in Example 5.1.3. We will also observe how in LCS that are not normable, it is possible to create normed subspaces using bounded sets. In fact, it is possible to define a structure on a vector space that has the properties of bounded sets, even without the presence of a topology. Such a construction is called a **bornology** (from the French word *borné*). For reference, the general definition is given below.

Definition 2.2.6. A **bornology** on a set is a collection of subsets that covers the set and is closed under the formation of subsets and finite unions. That is, a bornology on the set S is a collection $\mathcal{B}$ of subsets of S such that the union of all sets in $\mathcal{B}$ is S, and for which subsets and finite unions of elements of $\mathcal{B}$ are also in $\mathcal{B}$. In this case, we refer to the elements of $\mathcal{B}$ as the **bounded sets** of $\mathcal{B}$.

On a vector space, we can define a bornology that is comparable to a TVS topology.

Definition 2.2.7. A bornology $\mathcal{B}$ on a vector space is a **vector bornology** if the functions of vector addition and scalar multiplication map elements of $\mathcal{B}$ to elements of $\mathcal{B}$; that is, these functions map bounded sets to bounded sets. In this case, we refer to the vector space and bornology as a **bornological vector space.**

A more detailed discussion of bornologies and bornological vector spaces can be seen in [91, Section 6.7]. Bornologies and bornological vector spaces can be used to obtain results that do not come directly or easily from topological methods. Examples of such uses are found in [13] and Chapter 1 of [82]. Earlier developments regarding bornologies can be seen in [66] and [128], among others.

The following property was originally defined by George Mackey (see [85, p. 182]), who was one of the pioneers of the study of LCS. The property below appears from time to time in interesting contexts such as in that of Smooth Extension Theorems; see [82, 22.17, p. 236].

Definition 2.2.8. A LCS is **Mackey first countable** (also: satisfies **Mackey's countability condition**) if given any sequence (B_n) of bounded subsets, there exists a sequence (a_n) of scalars, infinitely many of which are nonzero, such that $\bigcup_{n=1}^{\infty} a_n B_n$ remains bounded.

It is easy to see that every normed space is Mackey first countable. Generalizing this statement to metrizable LCS is not so evident.

Theorem 2.2.2. *Every metrizable LCS is Mackey first countable.*

Proof. Let (B_n) be any sequence of bounded subsets of the metrizable LCS E. Because E is metrizable, let $\{U_n : n \in \mathbb{N}\}$ be a base of $\mathcal{N}_0$ in E. Proposition 2.2.1(b)(ii) is useful: We may assume $U_1 \supset U_2 \supset \cdots$. As each B_k is bounded, for every $k \in \mathbb{N}$ there exists $a_k \in (0,1)$ for which $a_k B_k \subset U_k$. Let $B = \bigcup_{k=1}^{\infty} a_k B_k$. We will show that B is bounded. Let $V \in \mathcal{N}_0$ be given. Without loss of generality, $V = U_n$, for some $n \in \mathbb{N}$. Thus, $a_n B_n \subset U_n$. Now, for all $k \geq n$, $a_k B_k \subset U_n$ because we assumed the base $\mathcal{N}_0$ of E is decreasing. For the lower values, $k = 1, 2, \ldots, n-1$, there exist $b_k > 0$ for which $a_k B_k \subset b_k U_n$. Put $a = \max\{b_1, b_2, \ldots, b_{n-1}, 1\}$. We conclude that for every $k \in \mathbb{N}$, $a_k B_k \subset a U_n$, which implies $\bigcup_{k=1}^{\infty} a_k B_k \subset aV$, and we have achieved the desired result. $\qquad\square$

Kąkol and Saxon [71] and [72] wrote the definition below and studied some of its properties.

Definition 2.2.9. A LCS is **docile** if every infinite-dimensional subspace contains an infinite-dimensional bounded set.

In Exercise 2.2.14, you can write a direct proof that every infinite-dimensional normed space is docile. A connection between docile LCS and Mackey first countable spaces is given next.

Proposition 2.2.7. *Every Mackey first countable Hausdorff LCS is docile.*

Proof. Assume L is an infinite-dimensional linear subspace of a Mackey first countable LCS. From a Hamel basis of L, we find an infinite set $\{x_n : n \in \mathbb{N}\}$ of linearly independent vectors. By assumption, there exist infinitely many nonzero scalars a_n such that $B =$

$\{a_n x_n : n \in \mathbb{N}\}$ is a bounded set. As L is a vector space, $B \subset L$, and we have found an infinite-dimensional bounded subset of L. $\qquad\square$

Corollary 2.2.2. *Every metrizable LCS is docile.*

Thus, for example, the LCS $\mathcal{D}(K)$ (see Definition 2.1.5) is Mackey first countable, hence also docile.

A natural question that arises is whether the converse is true, that is, if every docile LCS is Mackey first countable. It turns out that there exists a docile LCS that is not Mackey first countable. Details can be seen in [26] and [24].

Exercises

2.2.1. In the proof of Theorem 2.2.1, prove that $f(-x) = f(x)$.

2.2.2. Prove Proposition 2.2.1.

2.2.3. Use Corollary 2.2.1 determine which of the spaces of Examples 2.1.7, 2.1.8, and 2.1.9 are metrizable.

2.2.4. Suppose E is a metrizable TVS. Prove that if U is a subset of E that absorbs every null sequence, then U is a zero neighborhood.

2.2.5. Prove Proposition 2.2.2. *Hint:* It is the balanced property of the sets that are needed.

2.2.6. Prove that every finite set in a TVS is bounded.

2.2.7. Prove that every convergent sequence in a Hausdorff TVS is bounded.

2.2.8. Prove Proposition 2.2.6.

2.2.9. Prove that a subset B of a LCS is bounded if it satisfies the following condition: Given a sequence (a_n) of nonzero scalars, for every sequence (x_n) of elements of B, the sequence $(a_n x_n)$ is bounded.

2.2.10. Prove that a set is bounded if all of its countable subsets are bounded.

2.2.11. Prove that a subset of a linear subspace of a TVS is bounded if and only if the set is bounded in the entire space.

2.2.12. Suppose a seminorm on a LCS is bounded on an open set. Prove that the seminorm is continuous.

2.2.13. Prove that any subspace of a Mackey first countable LCS is Mackey first countable. Prove that any infinite-dimensional subspace of a docile LCS is docile.

2.2.14. Give a direct proof that every infinite-dimensional normed space is docile.

2.2.15. Prove that a continuous, linear image of a docile LCS is docile. Deduce that a continuous linear image of a Mackey first countable LCS is docile.

2.2.16. Suppose E has a fundamental sequence of bounded sets (see Definition 2.2.5) such that no B_n is absorbing in E. Prove that E is not docile.

2.2.17. Write a proof of Theorem 2.2.2 using seminorms in place of neighborhoods.

2.2.18. The proof of Theorem 2.2.1 can be done via a different approach. Prove the theorem using the following construction: In the metrizable LCS E, suppose $\{p_n : n \in \mathbb{N}\}$ is a collection of continuous seminorms such that for each $n \in \mathbb{N}$ and each $x \in E$, $p_{n+1}(x) \le p_n(x)$. Then for any sequence (a_n) of positive scalars such that $\sum_{n=1}^{\infty} a_n < \infty$, the function $d : E \times E \to [0, \infty)$, defined by

$$d(x, y) = \sum_{n=1}^{\infty} \frac{a_n \, p_n(x - y)}{1 + p_n(x - y)}, \tag{2.26}$$

is a translation invariant metric that generates the topology of E.

2.3 Continuous linear maps

It is probably no surprise to read that linear maps will play an important roll in much of what we do in the world of TVS. We have already seen a few examples of linear images, for example, of balanced and convex sets in Theorem 1.1.2 of Chapter 1. Meanwhile, a typical topological tendency is that, given a function between topological spaces, we are curious about at which points, if any, the function is *continuous*. For us, most of the functions we encounter will be linear. The first result of this section reveals that the linear structures involved allow us to reduce the determination of continuity on an entire TVS to checking if the linear map is continuous at the origin. We will make use of Definition A.3.22 in the Appendix of continuity of a function at a point. Of course, if we happen to verify any of the general equivalences of continuity such as in Theorem A.3.6, we can conclude continuity as well. Let us get to the first result.

Theorem 2.3.1. *Let E, F be TVS and $T : E \to F$, a linear map. The following are equivalent.*
(a) *T is continuous at every point of E.*
(b) *T is continuous at 0.*
(c) *For every zero neighborhood V in F, there exists a zero neighborhood U in E such that $T(U) \subset V$.*

Proof. As the implications (a) $\Rightarrow$ (b) $\Rightarrow$ (c) are obvious, we only need to prove (c) $\Rightarrow$ (a). So, let $x \in E$ be arbitrary. Let W be any neighborhood of $T(x)$. The translation invariance of the topologies allows us to state that for some neighborhood V of zero in F, $W = V + T(x)$. Now apply (c): There is a zero neighborhood U in E for which $T(U) \subset V$. This implies that $W \supset T(x) + T(U)$. The linearity of T reduces the previous observation to: $W \supset T(x + U)$, where $x + U$ is a neighborhood of x. Definition A.3.22 applies to W and $x + U$, specifically, $T(x + U) \subset W$. T is continuous at x, and we are done. $\qquad\square$

We can immediately apply Theorem 1.2.3 from Chapter 1.

Corollary 2.3.1. *For TVS E, F, a linear map $T : E \to F$ is continuous if and only if whenever the filter $\mathcal{F} \to 0$ in E, the filter $T(\mathcal{F}) \to 0$ in F. In particular, when E and F are metrizable TVS, a linear map is continuous if and only if it maps null sequences in E to null sequences in F.*

Fond memories of determining whether a real-valued function (not necessarily linear) is or is not uniformly continuous prompts us to consider uniform continuity in the context of linear maps on TVS. The corollary below clarifies the situation in short order, but first we need to state a linear map definition of uniform continuity.

Definition 2.3.1. A linear map T from a subset A of a TVS E to another TVS F is **uniformly continuous on** A if for every zero neighborhood V in F there exists a zero neighborhood U in E such that whenever $x, y \in A$ and $x - y \in U$, $T(x) - T(y) \in V$. If T is uniformly continuous on E, we simply refer to T as being **uniformly continuous**.

Corollary 2.3.2. *Every continuous linear map is uniformly continuous.*

Proof. Suppose the linear map T is continuous. By Theorem 2.3.1(b), T is continuous at 0. Thus, for any $V \in \mathcal{N}_{0,F}$, there exists $U \in \mathcal{N}_{0,E}$ such that whenever $x - y \in U$, the linearity of T implies that $T(x) - T(y) = T(x - y) \in V$. $\qquad\square$

In the context of a seminorm, uniform continuity is expressed below.

Proposition 2.3.1. *If a seminorm is continuous at 0, then it is uniformly continuous.*

Proof. Exercise 2.3.10. $\qquad\square$

The next result is deceptively important in that linear combinations of continuous linear maps is a concept we will encounter often.

Proposition 2.3.2. *Linear combinations of continuous linear maps are continuous.*

Proof. We prove that the sum of continuous linear maps is continuous, leaving the scalar multiple part as Exercise 2.3.1. Let S and T be continuous linear maps from a TVS E to another TVS F. Let V be an arbitrary zero neighborhood in F. Now is a good time to apply Theorem 1.3.2(c): We find another zero neighborhood W in F such that $W + W \subset V$. By continuity of S and T, there exist zero neighborhoods U_1 and U_2 in E such that $S(U_1) \subset W$ as well as $T(U_2) \subset W$. Put $U = U_1 \cap U_2$ and note that U is a zero neighborhood in E. Hence,

$$(S + T)(U) = S(U) + T(U) \subset S(U_1) + T(U_2) \subset W + W \subset V, \tag{2.27}$$

and we conclude that $S + T$ is continuous by Theorem 2.3.1(c). $\qquad\square$

Therefore, the collection of all continuous linear maps between TVS forms a vector space, and this observation encourages a look at some general examples and notations regarding collections of linear maps.

Definition 2.3.2. Let E and F be TVS. Then
(a) The set of all linear maps from E to F is denoted by $L(E, F)$.
(b) The set of all continuous linear maps from E to F is denoted by $\mathcal{L}(E, F)$.
(c) The set of all sequentially continuous (see Definition A.3.22) linear maps from E to F is denoted by $\mathcal{L}_S(E, F)$.
(d) The set of all continuous linear maps from E to the scalar field $\mathbb{K}$ is called the **(topological) dual** of E, and is denoted by E'.

Certainly, $\mathcal{L}(E, F)$ is a linear subspace of $L(E, F)$. Likewise, the space E' is a linear subspace of E^* (the vector space of all linear functionals from Definition 1.1.13 of Chapter 1). In the realm of normed spaces, E' refers to the set of *all* linear functionals on E, continuous or not. A corresponding notational clarification is that we will use the notation E' to denote the set of all *continuous* linear functionals on a vector space, even for normed spaces. Both the algebraic and continuous dual spaces are important, but the continuous dual is in fact, a fundamental part of the theory of LCS. Starting in Chapter 4, there will be many deep results that have some connection to E'.

Another typical topological tendency is that, given a property of a set or topological space the ensuing question is whether said property is preserved by continuous functions that are defined on such a set or topological space. We follow that practice here, namely whenever we have new property or type of TVS (usually a LCS), we ask the question: "Is this property preserved by continuous *linear* maps?" We can already view a couple of such instances, the first of which leads to a distinction between normed spaces and general LCS.

Proposition 2.3.3. *Continuous linear maps on TVS send bounded sets to bounded sets.*

Proof. Assume E and F are TVS and $T : E \to F$ is a continuous linear map. Let $B \subset E$ be any bounded set and consider $T(B)$ in F. Let V be any zero neighborhood in F. Theorem 2.3.1 informs us that, due to the continuity of T there exists a zero neighborhood U in E such that $T(U) \subset V$. Because B is bounded in E, there exists $a > 0$ such that $B \subset aU$. Therefore, $B \subset aU \Rightarrow T(B) \subset T(aU) = aT(U) \subset aV$, and the proof is completed. $\square$

Surely, you recall (or see Theorem A.5.1) that for linear maps between normed spaces, boundedness, and continuity are equivalent. This is not true in the world of LCS. Exercise 2.3.9 includes one example of a discontinuous linear map that sends bounded sets to bounded sets, and from time to time we will encounter other examples. Such examples set apart the collection of general LCS from that of normed spaces.

We need a few fancy words to describe when two TVS are "essentially the same" in a linear sense.

Definition 2.3.3.
(a) A **continuous isomorphism** between two TVS is a linear bijection such that the function and its inverse are both continuous. When such a function exists, the two TVS are **(continuously) isomorphic.** Usually, we suppress the word "continuous," unless we need to distinguish between such an isomorphism and a linear isomorphism (see A.2.3).
(b) An isomorphism of a TVS onto itself is called an **automorphism.**
(c) A linear map from one TVS into another TVS is a **strict morphism** if it is an isomorphism onto its image.

Observe that an isomorphism could be alternatively described as a linear homeomorphism.

Proposition 2.3.4. *A continuous linear image of a docile LCS is docile.*

Proof. Exercise 2.3.11. $\square$

Exercises

2.3.1. Prove the second part of Proposition 2.3.2, namely that a scalar multiple of a continuous linear map between TVS is continuous.

2.3.2. Write a direct proof that a bounded linear map between two normed spaces is uniformly continuous.

2.3.3. Suppose the LCS E carries the finest locally convex topology (Example 2.1.3). Let F be any other LCS and prove that every linear map $T : E \to F$ is continuous.

2.3.4. Use Corollary 2.1.2 to prove that for a linear map $T : E \to F$ with collections of continuous seminorms as in Proposition 2.1.11, T is continuous if and only if for every $q_\beta \in \mathcal{Q}$, there exist n continuous seminorms $p_{a_j} \in \mathcal{P}, j \in \underline{n}$, and a positive number M such that for all $x \in E$, $q_\beta(T(x)) \le M \max\{p_{a_j}(x) : j \in \underline{n}\}$.

2.3.5. Use the previous Exercise 2.3.4 to prove that if a vector space E has two locally convex topologies, $\mathcal{T}_1$ and $\mathcal{T}_2$, generated by collections $\mathcal{P}$ and $\mathcal{Q}$, respectively, then $\mathcal{T}_1$ is finer than $\mathcal{T}_2$ if and only if for every $q_\beta \in \mathcal{Q}$, there exist n continuous seminorms $p_{a_j} \in \mathcal{P}, j \in \underline{n}$, and a nonnegative number M such that for all $x \in E$, $q_\beta(x) \le M \max\{p_{a_j}(x) : j \in \underline{n}\}$.

2.3.6. Prove that if E and F are Hausdorff TVS and $T : E \to F$ is sequentially continuous (see Definition A.3.22), then

   (a) T maps bounded sets to bounded sets.

   (b) T maps Cauchy sequences to Cauchy sequences.

2.3.7. Prove that if E and F are TVS and $T : E \to F$ is a linear map that sends a zero neighborhood in E to a bounded subset of F, then T is continuous.

2.3.8. Prove that if a linear map between TVS is bounded on some zero neighborhood, then the map is continuous.

2.3.9. Let E be the vector space of continuous, complex valued functions on $[0,1]$. Consider the locally convex space topology $\mathcal{T}_1$ generated by the collection of seminorms: $(\forall f \in E)(\forall x \in [0,1])\, p_x(f) = |f(x)|$.

   Now consider the topology $\mathcal{T}_2$ induced by the metric below. For every $f, g \in E$,

$$d(f,g) = \int_0^1 \frac{|f(x) - g(x)|}{1 + |f(x) - g(x)|} \, dx. \tag{2.28}$$

   (a) Prove that the identity map $\mathrm{id} : (E, \mathcal{T}_1) \to (E, \mathcal{T}_2)$ maps bounded sets to bounded sets by proving that every $\mathcal{T}_1$– bounded set is $\mathcal{T}_2$– bounded.

   (b) Prove that the identity map $\mathrm{id} : (E, \mathcal{T}_1) \to (E, \mathcal{T}_2)$ is sequentially continuous (see Definition A.3.22).

   (c) Prove that the identity map $\mathrm{id} : (E, \mathcal{T}_1) \to (E, \mathcal{T}_2)$ is not continuous, thus verifying the existence of a discontinuous linear map that sends bounded sets to bounded sets.

2.3.10. Prove Proposition 2.3.1.

2.3.11. Prove Proposition 2.3.4.

2.3.12. Let E be an infinite-dimensional vector space and equip E with the finest locally convex topology of Example 2.1.3. Prove that for any LCS F, every linear map $T : E \to F$ is continuous.

2.4 Normed spaces as locally convex spaces

In this section, we take our first look at normed spaces from the perspective of LCS. Let us start with a proposition that summarizes a few observations about normed spaces as seen from this point of view.

2.4.1 Some locally convex properties of normed spaces

Proposition 2.4.1. *Let $(E, \|\cdot\|)$ be any normed space. Let B denote either the open or closed unit ball of E. Then the following hold:*
(a) *The ball B is a bounded subset of E.*
(b) *A base of zero neighborhoods of E is given by $\{\frac{1}{n}B : n \in \mathbb{N}\}$.*
(c) *The Minkowski functional of B generates the topology of the norm of E; hence, the LCS topology of E is generated by one norm.*
(d) *The norm on E generates a translation invariant metric given by $d(x,y) = \|x - y\|$, for every $x, y \in E$; in particular, every normed space is Hausdorff.*
(e) *For every nonzero x in E, there exists $n \in \mathbb{N}$ for which $x \notin \frac{1}{n}B$.*

Proof. Exercise 2.4.1. $\qquad\qquad\square$

If we chance upon a LCS and suspect that it might be a normed space, then the terminology we will use in such contexts is given in the next definition.

Definition 2.4.1. A LCS is **normable** if its topology can be generated by a norm. In this case, we simply refer to the LCS as a **normed space**.

It may seem like a big deal is being made in Proposition 2.4.1 about rather simple facts regarding the boundedness of the unit ball of a normed space. The importance is revealed in the next result in which we conclude that the only LCS that are normable are those that have at least one bounded zero neighborhood.

Theorem 2.4.1 (Kolomogorov's theorem). *A Hausdorff LCS is normable if and only if it contains a bounded zero neighborhood.*

Proof. ($\Rightarrow$): If a LCS is already a normed space, then the unit ball of the space is a bounded zero neighborhood.

($\Leftarrow$): Assume the Hausdorff LCS E has a bounded zero neighborhood U. We will conveniently assume that U is absolutely convex. Let $\mathcal{N}_0$ be any base of zero neighborhoods of E. By Theorem 2.1.1, we assume each set in $\mathcal{N}_0$ is absolutely convex. Let $V \in \mathcal{N}_0$ be any neighborhood other than U. There exists $a > 0$ such that $U \subset aV$, or, $\frac{1}{a}U \subset V$. If $N \in \mathbb{N}$ is such that $\frac{1}{N} < \frac{1}{a}$, then $\frac{1}{N}U \subset V$.

On the other hand, consider any set $\frac{1}{k}U$, $k \in \mathbb{N}$. By assumption, $U \in \mathcal{N}_0$, so $\frac{1}{k}U \in \mathcal{N}_0$, too. We have established that the topology on E that is generated by $\mathcal{N}_0$ is equivalent to the topology on E generated by the collection $\{\frac{1}{n}U : n \in \mathbb{N}\}$. Thus, the topology of E is generated by one seminorm, namely the Minkowski functional μ_U of U. As E is Hausdorff, Corollary 2.1.4 implies that μ_U is in fact, a norm. $\qquad\square$

The next goal is to indicate an example of a nonnormable LCS. As a prelude, imagine $\mathbb{R}^2$ with the standard (Euclidean) topology. It seems intuitively obvious that every nontrivial linear subspace of $\mathbb{R}^2$ is unbounded. This fact is true in any Hausdorff TVS, which is a consequence of the next proposition.

Proposition 2.4.2. *A subspace of a TVS is bounded if and only if it is contained in the closure of the zero vector.*

Proof. ($\Rightarrow$): Assume the subspace L of a TVS E is bounded. If x is an element of L that does not belong to $\overline{\{0\}}$ (the closure of the zero vector), then neither does nx, for every $n \in \mathbb{N}$, because L is a subspace. This implies that $\lim_{n\to\infty}(\frac{1}{n})(nx) = x \neq 0$, because $x \notin \overline{\{0\}}$. By Proposition 2.2.5, L is not bounded, a contradiction. Thus, $L \subset \overline{\{0\}}$.

($\Leftarrow$): Recalling from Exercise 1.3.5 that $\overline{\{0\}}$ is the intersection of all zero neighborhoods of E, if $L \subset \overline{\{0\}}$, then L is bounded. $\qquad\square$

Corollary 2.4.1. *The only subspace of a Hausdorff TVS that is bounded is the trivial subspace that consists of the zero vector.*

Proof. In this case, $\overline{\{0\}} = \{0\}$, because singletons are closed in Hausdorff topological spaces. $\qquad\square$

The utility of the previous result is that it allows us to find an example of a LCS that is not normable.

Example 2.4.1. A metrizable LCS that is nonnormable.

As a simplified version of Example 2.1.1, let $C(\mathbb{R})$ denote the set of all real valued continuous functions $f : \mathbb{R} \to \mathbb{R}$, with seminorms $p_K(f) = \max\{|f(x)| : x \in K\}$, for each compact $K \subset \mathbb{R}$. The locally convex topology generated by these seminorms is the topology of uniform convergence on compact sets (see Example 2.1.11). Because $\mathbb{R}$ can be written as a countable union of compact sets (using, e. g., $[-m, m]$, for each $m \in \mathbb{N}$), we deduce from Corollary 2.2.1 that $(C(\mathbb{R}), \mathcal{T})$ is a metrizable LCS. On the other hand, $(C(\mathbb{R}), \mathcal{T})$ is not normable, as the following calculation reveals: A typical zero neighborhood of functions $f \in C(\mathbb{R})$ are those f such that for any given $m \in \mathbb{N}$, $|f(x)| \leq 1$ on $[-m, m]$. Such neighborhoods contain the subspace of functions that are zero on $[-m, m]$, which we then realize is in fact a nontrivial subspace of $C(\mathbb{R})$. Corollary 2.4.2 tells us that all such subspaces are unbounded. Finally, by tacit Assumption 2.1.1 that collections of seminorms generating a topology are saturated, we conclude that *every* zero neighborhood is unbounded. Kolomogorov's theorem 2.4.1 convinces us that $C(\mathbb{R})$, with this collection of seminorms, $C(\mathbb{R})$ is not normable.

What is striking about this example is that when we imagine a metric space (X, d) and open balls centered at some point x_0 of the space that are of the form $\{x : d(x_0, x) < r\}$, we conveniently draw those sets as appearing to be bounded. In the context of this example, we would need to draw pictures of all sets $\{x : d(x_0, x) < r\}$ as *unbounded*.

In Chapter 3 when we discuss products of LCS, there will be more examples of non-normable LCS. The importance of such examples is that they show us a world of LCS that is significantly larger than that of normed spaces, providing us with lots of new kinds of spaces to explore.

Exercises

2.4.1.   Prove Proposition 2.4.1.

2.4.2.   Prove that a subset of a subspace of a TVS is bounded if and only if the set is bounded in the relative topology on that subspace.

2.4.3.   Prove that a convergent sequence in a TVS is bounded.

2.4.4.   Prove that if a locally convex topology can be generated by a finite number of continuous seminorms then it can be generated by only one continuous seminorm.

2.4.5.   Prove that a sequentially continuous linear map between TVS maps bounded sets to bounded sets.

2.4.6.   Prove that on every vector space a norm can be defined. *Hint:* Consider a Hamel basis.

2.5 Completeness and compactness

If a friend shows us a mysterious sequence from a normed or metric space and asks if it converges, then we need a tool for determining sequential convergence *without knowing a value of the limit in advance*. The main concept in this context is, of course, that of completeness, by way of Cauchy sequences (see Definition A.4.3). For nonmetrizable spaces, two obstacles appear. The first is that sequences generally will not suffice for determining convergence. The second is that of how to determine if "differences of elements are getting arbitrarily small," as is the case for Cauchy sequences. Within the realm of TVS, these obstacles can be dealt with. Filters can be used in place of sequences as we already know, and we have a concept of differences of elements and sets in TVS. Thus, the following definition is valid.

Definition 2.5.1. A sequence (x_n) in a TVS is a **Cauchy sequence** if for every zero neighborhood U there exists $N \in \mathbb{N}$ such that for all $m, n \geq N$, $x_n - x_m \in U$. A filter $\mathfrak{F}$ (resp., filterbase $\mathfrak{B}$) on a subset A of a TVS is a **Cauchy filter** (resp., **Cauchy filterbase**) if for every zero neighborhood U there exists $S \in \mathfrak{F}$ (resp., $B \in \mathfrak{B}$) such that $S - S \subset U$ (resp., $B - B \subset U$).

Notice how the definition implies that the differences $x_n - x_m$, $S - S$, $B - B$ get arbitrarily small by way of being contained in arbitrary zero neighborhoods. Clearly, any filter that is finer than a Cauchy filter is also a Cauchy filter. Similar statements hold for filterbases. Also, it is easy to deduce that continuous linear maps send Cauchy filters to Cauchy filters because all continuous linear maps are uniformly continuous (see Corollary 2.3.2). For LCS, we have the following seminorm version of Cauchy filters and seminorms.

Proposition 2.5.1. *Suppose E is a LCS with $\mathcal{P}$ as a family of continuous seminorms that generates the topology of E. A filter $\mathfrak{F}$ is Cauchy if and only if for each $\varepsilon > 0$ and each $p \in \mathcal{P}$, there exists $A \in \mathfrak{F}$ such that for every $x, y \in A$, $p(x - y) < \varepsilon$. In particular, a sequence (x_n) is Cauchy if and only if, for each $p \in \mathcal{P}$ and each $\varepsilon > 0$ there exists $N \in \mathbb{N}$ such that $p(x_n - x_m) < \varepsilon$, whenever $n, m \geq N$.*

Proof. This follows directly from Proposition 2.1.15. $\qquad\square$

The next result shows that a simpler condition can be used to determine if a filterbase is Cauchy in a TVS.

Proposition 2.5.2. *A filterbase $\mathfrak{B}$ in a TVS is Cauchy if and only if for every $U \in \mathcal{N}_0$, there exist a set $B \in \mathfrak{B}$ and an element $x \in E$ such that $B \subset x + U$. The corresponding result for filters also holds.*

Proof. ($\Rightarrow$): If $\mathfrak{B}$ is Cauchy, then the result follows immediately from the expression $B - B \subset U$.

($\Leftarrow$): Suppose the condition of the statement holds. Given $U \in \mathcal{N}_0$, choose $V \in \mathcal{N}_0$ such that $V - V \subset U$. Then for the $B \in \mathcal{B}$ and $x \in E$ as assumed, we have $B - B \subset (x + V) - (x + V) = V - V \subset U$, and the proof is completed. $\qquad\square$

Connecting Cauchy sequences to Cauchy filters is as follows.

Proposition 2.5.3. *Every Cauchy sequence in a TVS is a Cauchy filter.*

Proof. From Definition 1.2.2, we have that the set of all tails $B_k = \{x_k, x_{k+1}, \ldots\}$, $k \in \mathbb{N}$, of a sequence (x_n) forms a filterbase. Proposition 1.2.5 indicates that (x_n) is a Cauchy sequence if for every zero neighborhood U. There exists $K \in \mathbb{N}$ such that for all $k \geq K$, $B_k - B_k \subset U$. $\qquad\square$

The next proposition generalizes the idea from metric spaces that convergent sequences are Cauchy.

Proposition 2.5.4. *Every convergent filter in a TVS is a Cauchy filter.*

Proof. First, observe that the filter of all neighborhoods of a point in a TVS is a Cauchy filter. For any zero neighborhood U, there is a balanced zero neighborhood V such that $V - V \subset U$. Thus, if a filter converges to a point of a TVS, then the filter is finer than the filter of all neighborhoods of the point, and is therefore a Cauchy filter. $\qquad\square$

We are ready for a definition of the important concept of completeness in the context of TVS. In fact, throughout the various topics to be covered, generalizations and versions of completeness will often be a part of the discussion. Three such versions are in the next item.

Definition 2.5.2. A subset of a TVS is **complete** if every Cauchy filter on the set converges to an element of the set.

A TVS is: **complete** if every Cauchy filter converges to an element of the space; **quasicomplete** if every closed, bounded subset of the space is complete; **sequentially complete** if every Cauchy sequence converges to an element of the space.

There are a couple of unsurprising results relating completeness and closedness that are next, followed by an indication of the relations between the three types of TVS in the definition above.

Proposition 2.5.5.
(a) *Scalar multiples of complete subsets of a Hausdorff TVS are complete.*
(b) *Complete subsets of a Hausdorff TVS are closed.*
(c) *Closed subsets of a complete set are complete in any TVS.*

Proof. (a) This follows directly from the definition.

(b) If A is a complete subset and $x \in \overline{A}$, then $\{U \cap A : U \in \mathcal{N}_x\}$ is a Cauchy filterbase on A and converges to x. The assumption that the TVS is Hausdorff implies that no other point can be a limit of this convergent filterbase, which tells us that $x \in A$.

(c) Suppose the set A is complete and that $B \subset A$ is closed. Let $\mathfrak{F}$ be any Cauchy filter defined on B. Then $\mathfrak{F}$ is a filterbase on A and must converge to a point of A. As B is a closed subset of A, Theorem 1.2.2 requires that the limit of $\mathfrak{F}$ be an element of B. $\qquad\square$

Proposition 2.5.6. *Every complete TVS is quasicomplete and every quasi-complete TVS is sequentially complete.*

Proof. If A is a closed, bounded subset of a complete TVS then the conclusion of the first statement follows from Proposition 2.5.5(b). The proof of the second statement will be done by you as Exercise 2.5.2. $\qquad\square$

The reverse implications are generally false. Counterexamples will be described later when we discuss, for example, semireflexive spaces. Nevertheless, for metrizable TVS, the three types of completeness coincide. Speaking of metrizable TVS, we have the following labels for completeness.

Definition 2.5.3. A complete, metrizable TVS is called an **F-space**; a complete metrizable LCS is called a **Fréchet space**.

Clearly, every Banach space is a Fréchet space. There are Fréchet spaces that are nonnormable, hence not Banach spaces, an example of which is displayed next.

Example 2.5.1. $C(\Omega)$, for $\Omega \subset \mathbb{R}$ open, where $\Omega = \cup\{K_n : K_n$ is compact, $n \in \mathbb{N}\}$. Equip $C(\Omega)$ with the topology $\mathcal{T}$ generated by the collection $\{p_n : n \in \mathbb{N}\}$ of seminorms, where $p_n(f) = \max\{|f(x)| : f \in C(\Omega), x \in K_n\}$, for each $n \in \mathbb{N}$.

Proposition 2.5.7. *The LCS $C(\Omega)$ with the topology $\mathcal{T}$ generated by the seminorms $\{p_n, n \in \mathbb{N}\}$, is a nonnormable Fréchet space.*

Proof. The details of Example 2.4.1 indicate that under $\mathcal{T}$, $C(\Omega)$ is metrizable and nonnormable. Proving completeness will verify that $(C(\Omega), \mathcal{T})$ is a Fréchet space. The metrizability of $(C(\Omega), \mathcal{T})$ allows us to work with sequences. Let (f_k) be a Cauchy sequence in $C(\Omega)$. For any $x \in \Omega$, there is $n \in \mathbb{N}$ such that $x \in K_n$. This implies that $(\forall \varepsilon > 0)(\exists N \in \mathbb{N})(\forall k, m \geq N)$:

$$\left| f_k(x) - f_m(x) \right| \leq p_n(f_k - f_m) < \varepsilon. \tag{2.29}$$

The sequence $(f_k(x))$ is therefore Cauchy in $\mathbb{K}$. As $\mathbb{K}$ is complete, $f_k(x)$ converges to some $f(x)$. Equation (2.29) indicates that convergence is, in fact, uniform. This means the function f is continuous. Because this argument applies to every compact set K_n, we conclude that f is continuous on all of Ω. Hence, $(C(\Omega), \mathcal{T})$ is complete. $\qquad\square$

Every normed space can be considered (up to isometry) as a dense subspace of a Banach space (see [14, 40.15, p. 170]), said Banach space being denoted as the completion of the normed space. Below is the corresponding definition for TVS.

Definition 2.5.4. A **completion** of a TVS E is a complete TVS $\widehat{E}$ that contains a dense, linearly isomorphic copy of E. For a subset A of a TVS, we denote its completion by $\widehat{A}$, as the closure of A in $\widehat{E}$.

Theorem 2.5.1. *Every TVS has a completion. In particular, every Hausdorff TVS has a Hausdorff completion.*

A proof of this statement can be seen in [18, 1.1.17, p. 51].

The next result is significant in that it indicates a criterion for which completeness can be deduced in a finer Hausdorff compatible topology.

Theorem 2.5.2 (W. Robertson, [104]). *Assume $\mathcal{T}_1$ and $\mathcal{T}_2$ are Hausdorff compatible topologies on a vector space E such that $\mathcal{T}_1$ is finer than $\mathcal{T}_2$. If $\mathcal{T}_1$ has a base of zero neighborhoods that are $\mathcal{T}_2$-complete, then E is $\mathcal{T}_1$-complete.*

Proof. Let $\mathcal{N}_{0,\mathcal{T}_1}$ denote a base of zero neighborhoods of E that are $\mathcal{T}_2$-complete. Given any $\mathcal{T}_1$ - Cauchy filter $\mathfrak{F}$, the goal is to prove that the filter is $\mathcal{T}_1$-convergent. To achieve this, let $U \in \mathcal{N}_{0,\mathcal{T}_1}$ be arbitrary. Then there exists $A_1 \in \mathfrak{F}$ such that $A_1 - A_1 \subset U$. Fix $y \in A_1$. We then have that $\mathfrak{B} = \{y - A : A \in \mathfrak{F}\}$ is a $\mathcal{T}_2$-Cauchy filterbase due to $\mathcal{T}_1$ being finer than $\mathcal{T}_2$; that is, the identity $\mathrm{id} : (E, \mathcal{T}_1) \to (E, \mathcal{T}_2)$ is linear and uniformly continuous by Corollary 2.3.2; hence, id sends Cauchy filters to Cauchy filters. Because $y - A_1 \subset U$ and U is $\mathcal{T}_2$-complete, $\mathfrak{B} \to y - x$ with respect to $\mathcal{T}_2$, for some $x \in E$. The Hausdorff assumption implies that x is unique. Hence, $\mathfrak{F} \to x$ with respect to $\mathcal{T}_2$. Now, U is $\mathcal{T}_2$-closed by Proposition 2.5.5(b), so $A_1 - x \subset U$, and this implies that $A_1 \subset x + U$. As U was arbitrary, we observe that $\mathfrak{F}$ is finer than $\mathcal{N}_{0,\mathcal{T}_1}$; that is, by the definition of convergence of filters, $\mathfrak{F} \to x$ with respect to $\mathcal{T}_1$. $\square$

2.5.1 Compactness in topological vector spaces

Refer to Appendix B for results about compact sets in the context of general topology. For arbitrary TVS, there is no assurance that the space is metrizable, so in such cases we appeal to our filter friends. Specifically, we need a few more filter concepts.

Definition 2.5.5. A **cluster point** of a filter (resp., filterbase) is a point that belongs to the closure of every set of the filter (resp., filterbase).

The following item can be compared to the situation in $\mathbb{R}$ wherein if a Cauchy sequence has a convergent subsequence then the original sequence converges (to the same limit as that of the subsequence, of course).

Proposition 2.5.8. *If a Cauchy filter in a Hausdorff LCS has a cluster point, then the filter converges to that point.*

Proof. Let $\mathfrak{F}$ be a Cauchy filter in a TVS E such that x is a cluster point of $\mathfrak{F}$. Let $U \in \mathcal{N}_0$ be given. Then let $V \in \mathcal{N}_0$ such that $V + V \subset U$. There exists $S \in \mathfrak{F}$ such that $S - S \subset V$.

Notice that $S \cap (x + V) \neq \emptyset$ by way of x as a cluster point of $\mathfrak{F}$. Now let $y \in S \cap (x + V)$. If $w \in S$, then $w - y \in V$ and $w \in y + V \subset x + V + V \subset x + U$, and we see that $S \subset x + U$, which means $\mathfrak{F} \to x$. $\qquad\square$

Definition 2.5.6. A filter $\mathfrak{U}$ on a set S is an **ultrafilter** if every finer filter on S is equivalent to $\mathfrak{U}$ (see Definition 1.2.3).

The following is a significant result because its proof includes an application of Zorn's Lemma A.1.1.

Theorem 2.5.3. *Every filterbase on a set is contained in at least one ultrafilter.*

Proof. The proof is somewhat technical, and it can be observed in [129, Theorem 7.3.1, pp. 130–131]. $\qquad\square$

The next result is valid in the context of general topology, but we will make use of it in the context of TVS. The proof can be found in [67, p. 143].

Theorem 2.5.4. *In a Hausdorff topological space the following are equivalent for a subset K:*
(a) *K is compact.*
(b) *Every filter on K has a cluster point.*
(c) *Every ultrafilter on K converges to a point of K.*

The previous two theorems facilitate a proof of the completeness of compact subsets in Hausdorff TVS, as follows.

Proposition 2.5.9. *Compact subsets of Hausdorff TVS are complete.*

Proof. Assume K is a compact subset of a Hausdorff TVS. Let $\mathfrak{F}$ be any Cauchy filter on K. Theorem 2.5.3 asserts that there is an ultrafilter $\mathfrak{U}$ that is finer than $\mathfrak{F}$, and Theorem 2.5.4(c) then implies that $\mathfrak{U}$ converges to some point x. As $\mathfrak{F}$ is a coarser filter than $\mathfrak{U}$, we have that $\mathfrak{F} \to x$, as well. A compact subset of Hausdorff space is closed, which tells us that $x \in K$; that is, K is complete. $\qquad\square$

The next item represents a version of being "small" like a compact set, but not having the completeness property.

Definition 2.5.7. A subset A of a TVS E is **totally bounded** (also: **precompact**) if for every $U \in \mathcal{N}_0$, there exist finitely many elements $x_1, x_2, \ldots, x_n \in E$ such that $A \subset \cup\{x_j + U : j \in \underline{n}\}$.

If you read other books on the topic of TVS, you will find that some authors assume the x_j in the definition above, belong to the set A. This difference of definition does not affect any of the results we will encounter. The concept of total boundedness is part of the string of relations obtained in the next result.

Proposition 2.5.10. *Every compact subset of a TVS is totally bounded and every totally bounded subset is bounded.*

Proof. First, assume K is compact. Let $U \in \mathcal{N}_0$ be arbitrary, and let $V \subset U$ be open. Then $\{x + V : x \in K\}$ is an open cover of K, which reduces to a finite subcover $\{x_j + V : j \in \underline{n}, x_j \in K\}$, for some $n \in \mathbb{N}$. Thus, $K \subset \cup\{x_j + V : j \in \underline{n}\} \subset \cup\{x_j + U : j \in \underline{n}\}$, so K is totally bounded. Second, assume A is a totally bounded set and let $U \in \mathcal{N}_0$ be arbitrary. Find a balanced $V \in \mathcal{N}_0$ for which $V + V \subset U$. There are elements $x_j, j \in \underline{n}$, of E such that $A \subset \cup\{x_j + V : j \in \underline{n}\}$. We know that V is absorbing, which implies that there exists $a \geq 1$ such that $x_j \in aV$, for all $j \in \underline{n}$. Therefore,

$$A \subset aV + V \subset a(V + V) \subset aU, \tag{2.30}$$

and A is bounded. $\qquad\square$

Corollary 2.5.1. *Every compact subset of a TVS is bounded.*

The reverse implications of the proposition above are generally false. Indeed in Exercise 2.5.8 you can explain the details of totally bounded set that is not compact in the space c_{00} (see Example A.5.1). In Chapter 3, Exercise 3.2.6 allows you to describe a set that is bounded but not totally bounded.

Some properties of totally bounded sets are found in the next proposition.

Proposition 2.5.11.
(a) *The closure of a totally bounded set is totally bounded.*
(b) *Every subset of a totally bounded set is totally bounded.*
(c) *Finite unions of totally bounded sets are totally bounded.*
(d) *Finite sums and scalar multiples of totally bounded sets are totally bounded.*

Proof. For part (a), assume A is totally bounded and U is any zero neighborhood. First, find a closed neighborhood $V \subset U$. Second, exist points $x_1, x_2, \ldots, x_n$ such that $A \subset \cup\{x_j + V : j \in \underline{n}\}$, and this leads to the following:

$$\overline{A} \subset \overline{\bigcup_{j\in\underline{n}}(x_j + V)} = \bigcup_{j\in\underline{n}}(x_j + V) \subset \bigcup_{j\in\underline{n}}(x_j + U), \tag{2.31}$$

where we have taken a finite union of closed sets in the equality above. Thus, $\overline{A}$ is totally bounded. Parts (b), (c), and (d) are left as Exercise 2.5.6. $\qquad\square$

We now come to a characterization of compact sets in Hausdorff TVS.

Theorem 2.5.5. *In a Hausdorff TVS, a set is compact if and only if it is both totally bounded and complete.*

Proof. ($\Rightarrow$): Suppose the set K is a compact subset of a Hausdorff TVS E. We have already verified that K is totally bounded (Proposition 2.5.10) and complete (Proposition 2.5.9).

($\Leftarrow$): Assume K is totally bounded and complete. We will use Theorem 2.5.4(c). Let $\mathfrak{U}$ be any ultrafilter on K. The intention is to prove that $\mathfrak{U}$ is Cauchy by way of Proposition 2.5.2, after which we will apply the completeness assumption. Let $V \in \mathcal{N}_0$ be given. As K is totally bounded, there exist finitely many points $x_j, j \in \underline{n}$, for some $n \in \mathbb{N}$, for which $K \subset \cup\{x_j + V : j \in \underline{n}\}$.

Claim 2.5.1. *There is a $j_0 \in \underline{n}$ such that $x_{j_0} + V \in \mathfrak{U}$.*

Proof of Claim 2.5.1. First, we show that for some $j_0 \in \underline{n}$, every $S \in \mathfrak{U}$ intersects $x_{j_0} + V$. Suppose this is not the case. Then for every $j \in \underline{n}$, there is $S_j \in \mathfrak{U}$ such that $S_j \cap (x_j + V) = \emptyset$. Because $S \subset K \subset \cup\{x_j + V : j \in \underline{n}\}$ for every $S \in \mathfrak{U}$, it follows that $\bigcap_{j \in \underline{n}} S_j = \emptyset$. This empty intersection is contradictory because, being a filter, finite intersections of sets in $\mathfrak{U}$ must be nonempty. Hence, there really is a $j_0 \in \underline{n}$ such that $S \cap (x_{j_0} + V) \neq \emptyset$, for every $S \in \mathfrak{U}$. This means $\mathfrak{W} = \{S \cap (x_{j_0} + V) : S \in \mathfrak{U}\}$ constitutes a filterbase of subsets of K. $\mathfrak{W}$ is finer than $\mathfrak{U}$. Meanwhile, $\mathfrak{U}$ being an ultrafilter implies that $\mathfrak{W}$ is equivalent (see Definition 1.2.3) to $\mathfrak{U}$. Hence, $x_{j_0} + V \in \mathfrak{U}$; call it S_0, and we have proved the claim.

Now, there exists $S \in \mathfrak{U}$, namely S_0 of Claim 2.5.1 that satisfies $S_0 \subset x_{j_0} + V$. The neighborhood V was arbitrary, so Proposition 2.5.2 applies, and $\mathfrak{U}$ is Cauchy. By the completeness assumption of K, $\mathfrak{U} \to x$ for some $x \in K$. We have established the compactness of K. $\qquad\square$

The previous theorem can be used to justify why some authors define what we call totally bounded using the word *precompact*.

Theorem 2.5.6. *A subset of a Hausdorff TVS is totally bounded if and only if its completion is compact.*

Proof. ($\Rightarrow$): If A is totally bounded, then its completion $\widehat{A}$ is totally bounded as a closure, by Proposition 2.5.11(a). Theorem 2.5.5 tells us that $\widehat{A}$ is compact in the Hausdorff TVS $\widehat{E}$ by way of being totally bounded and complete.

($\Leftarrow$): Because E is linearly isomorphic to a subspace of $\widehat{E}$, the set A can be considered as a subset of $\widehat{A}$. Proposition 2.5.10 ensures that if $\widehat{A}$ is compact, then $\widehat{A}$ is totally bounded, and A is therefore totally bounded by Proposition 2.5.11(b). $\qquad\square$

The next result addresses hulls of compact and totally bounded sets.

Proposition 2.5.12.
(a) *In a LCS E, the convex hull of a totally bounded set remains totally bounded.*
(b) *The balanced hull of a compact set remains compact; the balanced hull of a totally bounded set remains totally bounded.*

Proof. (a) Assume A is totally bounded. Let $U \in \mathcal{N}_0$, and find $V \in \mathcal{N}_0$ that is open, convex, and $V \subset U$. The total boundedness of A implies that there exist points $x_1, x_2, \ldots, x_n \in E$ such that $A \subset \bigcup\{x_j + V : j \in \underline{n}\}$. We need to look at the convex hull C of $\{x_j : j \in \underline{n}\}$. Observe that C is the image under addition of the convex hull of finitely many points in

$\mathbb{K}^n$. By Carathéodory's theorem for convex sets in finite-dimensional spaces (see [83]), it can be deduced that the closed, convex hull of a finite set in $\mathbb{K}^n$ is compact. Hence, C is the continuous image of a compact set in $\mathbb{K}^n$ and is therefore compact. As a compact set, C is totally bounded, so there exist $y_1, y_2, \ldots, y_k \in E$ such that $C \subset \bigcup\{y_i + V : i \in \underline{k}\}$. Notice that $A \subset C + V$ and $C + V$ is convex. This implies that

$$\operatorname{conv}(A) \subset C + V \subset \bigcup_{i \in \underline{k}}(y_i + V) \subset \bigcup_{i \in \underline{k}}(y_i + U), \tag{2.32}$$

and we conclude that $\operatorname{conv}(A)$ is totally bounded.

Hopefully, you will find it enlightening to prove part (b) as Exercise 2.5.10. $\qquad\square$

By now, you have probably noticed that no mention has been made of the convex hull of a compact set being compact. An example of the convex hull of a compact set (in a normed space, no less) that is not compact can be seen in [91, Example 4.8.8, p. 94]. On the other hand, it is not difficult to prove the following.

Proposition 2.5.13. *In a LCS, the convex hull of a compact set remains compact if and only if the convex hull of the set is complete.*

Proof. Exercise 2.5.11. $\qquad\square$

In Chapter 4, we will make use of the following facts regarding totally bounded subsets of a LCS.

Definition 2.5.8. Let E be a TVS. For a zero neighborhood U of E, a set $A \subset E$ is **small of order** U if $x - y \in U$ for all $x, y \in A$.

The following provides a condition that is equivalent to total boundedness in a LCS.

Proposition 2.5.14. *A subset of a LCS is totally bounded if and only if for every absolutely convex zero neighborhood, the set can be covered by a finite collection of sets that are small of order for that neighborhood. That is, A is totally bounded if and only if for every absolutely convex $U \in \mathcal{N}_0$, there exist sets $A_1, A_2, \ldots, A_n$ that are small of order U and such that*

$$A \subset \bigcup_{j=1}^{n}(a_j + U), \tag{2.33}$$

where $a_j \in A_j$ for each $j \in \underline{n}$.

Proof. Exercise 2.5.12. $\qquad\square$

It is the next proposition that we will need in Chapter 4 in order to prove a result called the Alaoğlu–Bourbaki Theorem 4.4.1.

Proposition 2.5.15. *Suppose $\mathcal{V}$ is a collection of absolutely convex neighborhoods in a Hausdorff LCS E such that finite intersections of $\mathcal{V}$ form a base of zero neighborhoods.*

For a subset A of E, if for each $V \in \mathcal{V}$, A is covered by a finite union of sets that are small of order V, then A is totally bounded.

Proof. Given any $V, W \in \mathcal{V}$, if $U = V \cap W$, then $U \in \mathcal{N}_0$ by assumption. Next, there exist $V_1, V_2, \ldots, V_n \in \mathcal{V}$ that are small of order V, and there exist $W_1, W_2, \ldots, W_m \in \mathcal{V}$ that are small of order W, for which $A \subset \bigcup_{i=1}^{n} V_i$ and $A \subset \bigcup_{j=1}^{m} W_j$. The collection $\{V_i \cap W_j : i \in \underline{n}, j \in \underline{m}\}$ satisfies the assumptions of Proposition 2.5.14, and we conclude that A is totally bounded. $\qquad\square$

Exercises

2.5.1. Prove that every Cauchy sequence in a Hausdorff TVS is bounded. Deduce that every Cauchy sequence in a Hausdorff TVS is contained in a closed, bounded set.

2.5.2. Prove that every quasicomplete TVS is sequentially complete (cf: Proposition 2.5.6).

2.5.3. Prove that continuous linear maps send totally bounded sets to totally bounded sets.

2.5.4. Prove that the space $\mathcal{D}(K)$ of Example 2.1.7 is a metrizable, nonnormable LCS.

2.5.5. Prove that for $p = 1$, the Köthe echelon space $\lambda_1(K)$ of Example 2.1.10 is a nonnormable Fréchet space. *Hint:* For the completeness part, consider a Cauchy sequence $(x^{(k)})$ in $\lambda_1(K)$. For every $n \in \mathbb{N}$ and every $\varepsilon > 0$, there exists $M \in \mathbb{N}$ such that for all $k, l \geq M$, one has

$$\sum_{j=1}^{\infty} a_n(j)\left|x_j^{(k)} - x_j^{(l)}\right| = q_n\left(x^{(k)} - x^{(l)}\right) \leq \varepsilon. \tag{2.34}$$

By the definition of a Köthe matrix from Example 2.1.10, for each $j \in \mathbb{K}$, the limit as $k \to \infty$ of $x_j^{(k)}$ exists. Denote that limit by the sequence $x = (x_j)$. Using the fact that Cauchy sequences are bounded, show that $x = (x_j)$ belongs to $\lambda_1(K)$. Note that by similar steps, the completeness of $\lambda_p(K)$ for $p \in (1, \infty]$ can be established.

2.5.6. Prove parts (b), (c), and (d) of Proposition 2.5.11.

2.5.7. Suppose E is a TVS that is not Hausdorff. Let $x \in \overline{\{0\}}$ be a nonzero element. Prove that the set $A = \overline{\{0\}} \setminus \{x\}$ is compact but not closed. This situation also shows an example of a complete subset of a TVS that is not closed.

2.5.8. Consider any incomplete normed space (e. g., c_{00}) and a Cauchy sequence that does not converge. Prove that the elements of the sequence comprise a totally bounded set that is not compact.

2.5.9. Write another proof of Theorem 1.3.4 using the results of this section.

2.5.10. Prove Proposition 2.5.12(b). *Hint:* For compactness, note that the balanced hull of a set is given by $D \cdot A$, where $D = \{a \in \mathbb{K} : |a| \leq 1\}$ is compact in $\mathbb{K}$. The set $D \cdot A$ is the continuous linear image under scalar multiplication of $D \times \widehat{A}$, and is compact. For the totally bounded part, apply this argument to $\widehat{A}$.

2.5.11. Prove Proposition 2.5.13.

2.5.12. Prove Proposition 2.5.14.

2.6 A topological vector space that is not locally convex

By definition, every LCS is a TVS, and by now you probably suspect that there must be a TVS that is not a LCS. That is the case, and displaying an example of such a TVS is the goal of this section. First, we must acknowledge that we cannot look for a normed space as such an example because they are all LCS. Thus, we will need to "think outside the norm." On the other hand, the example below is a metrizable TVS, and the metric even resembles a norm that you have likely seen before.

Example 2.6.1. A topological vector space that is not locally convex.

Let p be any number such that $0 < p < 1$, and consider the set of all sequences (x_n) of scalars such that $\sum_{n=1}^{\infty} |x_n|^p < \infty$. It is easy to check that this defines a nontrivial vector space. We will use the common notation of this space: l_p, where $0 < p < 1$. Next, define $d : l_p \times l_p \to [0, \infty)$ by: For every $x = (x_n), y = (y_n) \in l_p$,

$$d(x,y) = \sum_{n=1}^{\infty} |x_n - y_n|^p. \tag{2.35}$$

In Exercise 2.6.1, you can verify that the definition above defines a translation invariant metric, and that this metric generates a compatible topology on l_p. Thus, we conclude that (l_p, d) is a TVS. We will now show the following.

Proposition 2.6.1. *The TVS (l_p, d), for $0 < p < 1$, is not locally convex.*

Proof. To establish that (l_p, d) is not locally convex, assume (l_p, d) *is* locally convex, and we set out to reveal a contradiction. If (l_p, d) is locally convex, then there is a base of convex zero neighborhoods. In particular, there exists a convex zero neighborhood U such that $U \subset \{x : d(x, 0) \le 1\}$. Moreover, there is some $a > 0$ such that $\{x : d(x, 0) \le a\} \subset U$. Now let $e^{(j)}$ denote the element of l_p that has a 1 in the jth entry and zeros elsewhere; that is, $e^{(j)} = (e_n^{(j)}) = (0, 0, \ldots, 0, 1, 0, 0, \ldots)$, where $e_j^{(j)} = 1$ and $e_n^{(j)} = 0$ whenever $n \ne j$. We calculate

$$d(a^{1/p} e^{(j)}, 0) = \sum_{n=1}^{\infty} |\, a^{1/p} e_n^{(j)} \,|^p = a, \tag{2.36}$$

which implies that $a^{1/p} e^{(j)} \in U$, for every $j \in \mathbb{N}$. We now define for each $m \in \mathbb{N}$ an element $y^{(m)} = (y_n^{(m)}) \in l_p$ as shown below:

$$y^{(m)} = \left(\underbrace{\frac{a^{1/p}}{m}, \frac{a^{1/p}}{m}, \ldots, \frac{a^{1/p}}{m}}_{m \text{ times}}, 0, 0, \ldots \right) \tag{2.37}$$

$$= \frac{1}{m} \sum_{j=1}^{m} a^{1/p} e^{(j)}.$$

Observe that $0 \le \frac{1}{m} \le 1$ and $\sum_{j=1}^{m} \frac{1}{m} = 1$; in other words, $y^{(m)}$ is a convex combination (Definition 1.1.7) of elements $e^{(j)}$ of U. By our convexity assumption of U, $y^{(m)}$ is one of its elements. Now let us calculate $d(y^{(m)}, 0)$:

$$d(y^{(m)}, 0) = \sum_{n=1}^{\infty} |y_n^{(m)}|^p = \sum_{n=1}^{m} \left| \frac{a^{1/p}}{m} \right|^p = m \left| \frac{a^{1/p}}{m} \right|^p = a \cdot m^{1-p}. \tag{2.38}$$

As $1 - p > 0$, for sufficiently large m, $d(y^{(m)}, 0) > 1$, implying that for large enough m, $y^{(m)} \notin U$, and we have arrived at a contradiction. We can only conclude that (l_p, d) is not, in fact, locally convex. $\qquad\square$

With relevant adjustments, it can also be shown that following space is a non-locally convex TVS: $L_p(\Omega, \mathcal{A}, \mu)$, $0 < p < 1$, where Ω is a subset of $\mathbb{R}$, Σ is a σ-algebra of subsets of Ω, and μ is a positive measure on Σ. Work through Exercise 2.6.3 or see [89, 2.2.6, pp. 164–165] for details. Another type of example, which involves continuous functions, can be seen in [67, Example 2, pp. 86–87].

Remark 2.6.1 (Nonlocally convex topological vector spaces). The focus throughout this book is on spaces that are locally convex. Two classic books on the topic of metrizable nonlocally convex spaces are those of Rolewicz [106] and of Kalton, Peck, and Roberts [73]. In the general context, Sánchez Ruiz [109] focused on TVS that are not locally convex. The motivation for restricting our attention to LCS will be made profoundly clear in Chapter 4.

Exercises

2.6.1. Prove that the definition of d given in Equation (2.35) is indeed a metric and that it is furthermore translation invariant. Deduce that d generates a TVS topology on l_p, $0 < p < 1$. *Hint:* For $0 < p < 1$, the inequality $(a + b)^p \le a^p + b^p$ holds for positive scalars a and b.

2.6.2. For $p = \frac{1}{2}$, sketch the unit ball in $\mathbb{R}^2$ of the metric d of Equation (2.35), and prove that it is not convex. Does this really represent an example of a Hausdorff TVS of finite dimension that is not locally convex? Explain.

2.6.3. Consider the vector space $L_p(\mathbb{R})$, $0 < p < 1$, of Lebesgue measurable (equivalence classes via almost everywhere equality of) functions for which $\int_{\mathbb{R}} |f|^p \, d\lambda < \infty$, where λ is Lebesgue measure. Prove that $(L_p(\mathbb{R}), d)$ is a nonlocally convex TVS with the metric d given by: For every $f, g \in L_p(\mathbb{R})$,

$$d(f, g) = \int_{\mathbb{R}} |f - g|^p \, d\lambda. \tag{2.39}$$

3 Creating locally convex spaces

Suppose in a park we find a box of seminorms along with a vector space on which the seminorms are defined. In that case, we know how to use those seminorms to construct a LCS: Theorem 2.1.4 directs us to apply the initial topology to obtain a locally convex topology generated by the given seminorms. In this chapter, we consider something of the opposite situation: If we have one or more known LCS laying around, what are some ways that we can use some parts (like subspaces) or put collections of LCS together to create new LCS? Possibilities we will consider here include subspaces (including hyperplanes, Definition 3.1.4), quotients (Definition 1.1.9), products, (Definition 3.3.1), projective limits (Definition 3.4.3), and inductive limits (Definition 3.5.3). Along the way, we study finite-dimensional TVS and discover by way of Riesz's Theorem 3.2.2 that such spaces are equivalent to $\mathbb{K}^n$ for some $n \in \mathbb{N}$, making them mostly uninteresting from the perspective of TVS. The first item to consider is that of subspaces, quotients, and hyperplanes.

3.1 Subspaces, quotients, and hyperplanes

Whenever there is no other indication, a subspace of a TVS will be equipped with the relative topology (Definition A.3.3). Some easy observations about subspaces include that the relative topology on a subspace preserves the compatibility of the topology; that is, a subspace of a TVS is a TVS. In addition, a subspace is itself a convex set, so the intersection of a subspace with convex neighborhoods of a LCS are again convex, indicating that a subspace of a LCS is also a LCS. In particular, if the topology of the LCS is generated by a collection of seminorms, then the relative topology of a subspace is given by the restrictions of those seminorms to the subspace. Subspaces of metrizable spaces are metrizable; subspaces of normed spaces are normed. The next item pertains to the closure of a subspace.

Proposition 3.1.1. *In a TVS, the closure of a subspace is again a subspace.*

Proof. We apply parts (b) and (c) of Proposition 1.3.2: If F is a vector subspace of E and a, b are any two scalars, then applying Theorem A.3.6(d) leads to

$$a\overline{F} + b\overline{F} = \overline{(aF)} + \overline{(bF)} \subset \overline{(aF + bF)} \subset \overline{F}, \tag{3.1}$$

which verifies that $\overline{F}$ is closed under linear operations of addition and scalar multiplication. $\qquad\square$

https://doi.org/10.1515/9783111392868-003

3.1.1 Quotients

A context in which subspaces play an important role is that of quotient spaces, which is the next topic.

Recall Definition 1.1.9 of a quotient of a vector space, including the notation $\overset{\bullet}{x}$ of equivalence classes. In typical behavior, whenever we come across a vector space we wish to equip it with a compatible topology, and this is accomplished as follows. Let M be a subspace of a TVS E and consider the canonical quotient map $\varphi : E \to E/M$. The next definition uses this information about φ to define a topology on E/M.

Definition 3.1.1. The **quotient topology** is the finest topology for which the canonical quotient map is continuous.

The definition above implies that a set U belongs to the quotient topology on E/M if and only if $\varphi^{-1}(U)$ belongs to the topology of E. The next proposition clarifies that φ is even an open mapping (see Definition A.3.15) and that E/M is a TVS under the quotient topology.

Proposition 3.1.2. *Consider the quotient space E/M for a subspace M of a TVS E. Then:*
(a) *The canonical quotient map $\varphi : E \to E/M$ is open.*
(b) *The vector space E/M is a TVS under the quotient topology defined by φ.*
(c) *If the TVS is locally convex, then the quotient space is also a LCS.*

Proof. To prove part (a), let U be an open subset of E. By definition of φ, $\varphi^{-1}(\varphi(U)) = U + M$. The assumption that U is open and the translation invariance property of the topology of E imply that for every $x \in M$, $x + U$ is open. Proposition 1.3.2(a) alerts us to the fact that $\bigcup \{x + U : x \in M\} = M + U$ is open. From the comment following Definition 3.1.1, we conclude that $\varphi(U)$ is open.

(b) On the vector space E/M, consider first addition: For any two $\overset{\bullet}{x}, \overset{\bullet}{y} \in E/M$, the surjectivity of φ indicates that there exist $x, y \in E$ such that $\overset{\bullet}{x} = \varphi(x), \overset{\bullet}{y} = \varphi(y)$. Therefore, if U is a neighborhood of $\overset{\bullet}{x} + \overset{\bullet}{y}$ in E/M, then $\varphi^{-1}(U)$ is a neighborhood of $x + y$ in E, and there exist neighborhoods V of x and W of y such that $V + W \subset \varphi^{-1}(U)$. Then $\varphi(V)$ and $\varphi(W)$ are neighborhoods of $\overset{\bullet}{x}$ and $\overset{\bullet}{y}$, respectively, for which $\varphi(V) + \varphi(W) \subset U$, proving the continuity of addition in E/M. Now, on to scalar multiplication. If U is a neighborhood of $a \cdot \overset{\bullet}{x}$ in E/M for a scalar a, then there is a neighborhood D of a in $\mathbb{K}$ as well as a neighborhood W of x in E such that $DW \subset \varphi^{-1}(U)$, where DW is as defined in Definition 1.1.2. The proof of this part is completed by noticing that $\varphi(W)$ is a neighborhood of $\overset{\bullet}{x}$ for which $D\varphi(W) \subset U$.

(c) Consider any zero neighborhood U in E/M. Then $\varphi^{-1}(U)$, being a zero neighborhood in E, contains a convex zero neighborhood V. Hence, $\varphi(V) \subset U$ and, being the linear image of a convex set, $\varphi(V)$ is a convex zero neighborhood in E/M. $\qquad\square$

Next, we find a way to determine whether a quotient space is Hausdorff.

Proposition 3.1.3. *A quotient of a TVS by a subspace is Hausdorff if and only if the subspace is closed. In particular, a TVS is Hausdorff if an only if its zero vector is closed.*

Proof. ($\Rightarrow$): If the quotient E/M is Hausdorff, then singletons are closed. In particular, $\{\overset{\bullet}{0}\}$ is closed. The continuity of the canonical quotient map φ implies that $\varphi^{-1}(\{\overset{\bullet}{0}\})$ is closed in E. Conveniently, $\varphi^{-1}(\{\overset{\bullet}{0}\}) = M$.

($\Leftarrow$): Now assume M is a closed subspace of E. Consider any element $\overset{\bullet}{x} \neq \overset{\bullet}{0}$. This means $\overset{\bullet}{x} = \varphi(x)$ for some nonzero $x \in E$, and $x \notin M$. The closedness of M allows us to find a neighborhood U of x such that $U \cap M = \emptyset$. Translate back to the origin: $\overset{\bullet}{x} - \varphi(U)$ is a $\overset{\bullet}{0}$-neighborhood that does not contain $\overset{\bullet}{x}$. An application of Corollary 1.3.3 provides the finishing touch to conclude that E/M is Hausdorff. The second statement follows from the recollection that $\{0\}$ is a subspace of E. $\qquad\square$

Observe that if a TVS E is not Hausdorff, then $E/\overline{\{0\}}$ is Hausdorff, considering that $\overline{\{0\}}$ is also a subspace of E by way of Proposition 3.1.1. The following definition thus makes sense.

Definition 3.1.2. The **Hausdorff space associated with** a TVS is the quotient space by the closure of the zero vector. That is, the Hausdorff space associated with E is $E/\overline{\{0\}}$.

Predictably, the next topic is about how to represent a quotient LCS in terms of seminorms.

Definition 3.1.3. Given a seminorm p on a LCS E and a subspace M, define $\overset{\bullet}{p}$ on E/M by: $(\forall \overset{\bullet}{x} \in E/M)\overset{\bullet}{p}(\overset{\bullet}{x}) = \inf\{p(x) : x \in \overset{\bullet}{x}\}$.

Proposition 3.1.4. *The function $\overset{\bullet}{p}$ of Definition 3.1.3 above is a seminorm on E/M. Moreover, if $\{p_\alpha : \alpha \in I\}$ is a collection of seminorms that generates the topology on E, then $\{\overset{\bullet}{p}_\alpha : \alpha \in I\}$ is a collection of seminorms that generates the topology on E/M.*

Proof. First, it is clear from the definition that $\overset{\bullet}{p}(\overset{\bullet}{x}) \geq 0$ for any $\overset{\bullet}{x} \in E/M$. Next, for any $a \in \mathbb{K}$, $a\overset{\bullet}{x} \in E/M$, hence

$$\overset{\bullet}{p}(a\overset{\bullet}{x}) = \inf\{p(ax) : x \in \overset{\bullet}{x}\} = |a|\inf\{p(x) : x \in \overset{\bullet}{x}\} = |a|\overset{\bullet}{p}(\overset{\bullet}{x}). \tag{3.2}$$

For the triangle inequality, consider any $\overset{\bullet}{p}(\overset{\bullet}{x}), \overset{\bullet}{p}(\overset{\bullet}{y}) \in E/M$. Properties of infima indicate that for any $\varepsilon > 0$, there exists $x \in \overset{\bullet}{x}$ such that $p(x) \leq \overset{\bullet}{p}(\overset{\bullet}{x}) + \varepsilon/2$. Similarly, there exists $y \in \overset{\bullet}{y}$ such that $p(y) \leq \overset{\bullet}{p}(\overset{\bullet}{y}) + \varepsilon/2$. As p is a seminorm, the following calculation ensues:

$$p(x + y) \leq p(x) + p(y) \leq \overset{\bullet}{p}(\overset{\bullet}{x}) + \overset{\bullet}{p}(\overset{\bullet}{y}) + \varepsilon. \tag{3.3}$$

The arbitrariness of ε leads to

$$\overset{\bullet}{p}(\overset{\bullet}{x} + \overset{\bullet}{y}) = \inf\{p(x + y) : x + y \in \overset{\bullet}{x} + \overset{\bullet}{y}\} \leq \overset{\bullet}{p}(\overset{\bullet}{x}) + \overset{\bullet}{p}(\overset{\bullet}{y}), \tag{3.4}$$

and we have verified that $\overset{\bullet}{p}$ is a seminorm on E/M. To complete the proof, it suffices to prove that if p is a continuous seminorm on E, then $\overset{\bullet}{p}$ is continuous on E/M. For any such seminorm $\overset{\bullet}{p}$ and any $\varepsilon > 0$, if $U = \{x : p(x) < \varepsilon\}$ in E, then $\varphi(U) = \{\overset{\bullet}{x} : \overset{\bullet}{p}(x) < \varepsilon\}$, which in view of Proposition 2.1.7, implies that $\overset{\bullet}{p}$ is continuous on E/M. $\qquad\square$

A couple of observations about quotients of metrizable spaces is in order. First, by the definition of the quotient topology (Definition 3.1.1) a quotient of a metrizable space by a subspace is again metrizable. The second observation, that quotients by closed subspaces of complete, metrizable spaces are complete requires more than just a comment, so it appears as the next result.

Proposition 3.1.5. *A quotient of a complete metrizable TVS by a closed subspace is a complete metrizable TVS.*

Proof. Let E be a complete, metrizable TVS and M a closed subspace of E. Certainly, E/M is metrizable by the comments above. In particular, let $\{U_n : n \in \mathbb{N}\}$ be a zero neighborhood base for E. By Proposition 2.2.1(a)(ii), we may assume that for every $n \in \mathbb{N}$, $U_{n+1} + U_{n+1} \subset U_n$. For each $n \in \mathbb{N}$, let $\overset{\bullet}{U}_n = \varphi(U_n)$ be a corresponding zero neighborhood in E/M; that is, $\{\overset{\bullet}{U}_n : n \in \mathbb{N}\}$ is a zero neighborhood base in E/M. Suppose $(\overset{\bullet}{x}_n)$ is a Cauchy sequence in E/M. We can find a subsequence $(\overset{\bullet}{x}_{n_k})$ of $(\overset{\bullet}{x}_n)$ such that $\overset{\bullet}{x}_{n_k} - \overset{\bullet}{x}_{n_m} \in \overset{\bullet}{U}_n$, whenever $k, m \geq n$. Let $y, w \in E$ such that $y \in \overset{\bullet}{x}_{n_k}$ and $w \in \overset{\bullet}{x}_{n_m}$. Then $y - w \in U_n + M$, by the definition of a quotient vector space. Let us calculate as follows: First, $y \in \overset{\bullet}{x}_{n_k} \Rightarrow y \in w + U_n + M$, so $y = w + u + z$, where $u \in U_n$ and $z \in M$. Next, $y - z = w + u \in w + U_n$. Moreover, $y - z \in \overset{\bullet}{x}_{n_m}$, because $y - z \in \overset{\bullet}{y}$. This shows that for any $w \in \overset{\bullet}{x}_{n_m}$,

$$\overset{\bullet}{x}_{n_m} \cap (w + U_n) \neq \emptyset, \tag{3.5}$$

whenever $k, m \geq n$. To apply the completeness assumption, we construct a sequence (x_n) in E inductively: Let $x_0 \in \overset{\bullet}{x}_0$ be arbitrary. Assume we have chosen $x_1, x_2, \ldots, x_l$ such that $x_j \in \overset{\bullet}{x}_{n_j}$ and $x_j \in x_{j-1} + U_{j-1}, j \in \underline{l}$. Now choose $x_{l+1} \in \overset{\bullet}{x}_{n_{l+1}}$ using Equation (3.5):

$$x_{n_{l+1}} \in \overset{\bullet}{x}_{n_{l+1}} \cap (x_l + U_l) \subset x_l + U_l. \tag{3.6}$$

Thus, for each $p \in \mathbb{N}$,

$$
\begin{aligned}
x_{l+p} &\in x_{l+p-1} + U_{l+p-1} \subset x_{l+p-2} + U_{l+p-2} + U_{l+p-1} \\
&\cdots \subset x_l + U_l + U_{l+1} + \cdots + U_{l+p-1} \\
&\subset x_l + U_{l-1},
\end{aligned}
\tag{3.7}
$$

by our assumption that $U_{n+1} + U_{n+1} \subset U_n$ for each $n \in \mathbb{N}$. Therefore, (x_n) is Cauchy in E and converges to some $x \in E$, by the completeness of E. The continuity of φ implies that $\overset{\bullet}{x}_{n_k} \to \overset{\bullet}{x}$, where $\overset{\bullet}{x} = \varphi(x)$. To finish, observe that $\overset{\bullet}{x}$ is a cluster point of the filter associated with the sequence $(\overset{\bullet}{x}_n)$. An application of Proposition 2.5.8 tell us that $\overset{\bullet}{x}_n \to \overset{\bullet}{x}$.

Note that the proof for the case of Fréchet spaces only requires the citation of the fact that the linear image of a convex set is convex. □

3.1.2 Hyperplanes

Get ready for an important concept. It has to do with proper subspaces that are as close as possible to being the entire vector space.

Definition 3.1.4. A **hyperplane** is a maximal proper subspace of a vector space. That is, H is a hyperplane in a vector space E if and only if, H is a subspace such that $\dim(E/H) = 1$.

An example of a hyperplane that is easy to visualize is that of a plane through the origin in $\mathbb{R}^3$. The quotient statement $\dim(E/M) = 1$ expresses the concept that a hyperplane is but one dimension away from being the whole vector space. In fact, there is a name for the dimension of a quotient vector space, which is now stated.

Definition 3.1.5. The **codimension** of a subspace M of a vector space E is the dimension of the quotient vector space E/M. The notation is $\mathrm{codim}(M) = \dim(E/M)$.

Thus, a subspace H is a hyperplane in E if and only if $\mathrm{codim}(H) = 1$. The next proposition connects the geometric concept of a hyperplane with the analytic concept of a linear functional.

Proposition 3.1.6. *A subspace of a vector space is a hyperplane if and only if it is the nullspace of a nonzero linear functional.*

Proof. ($\Rightarrow$): Assume H is a hyperplane in E. Then $\mathrm{codim}(H) = 1$ implies $H \neq E$. Let y be an element of $E \setminus H$, and let $L = \mathrm{span}(\{y\})$ be the linear span of y. It turns out that E is the linear direct sum (see Definition A.2.1): $E = H \oplus L$, as we now verify. On one hand, $H \oplus L \subset E$ is obvious. On the other hand, for any $x \in E$, either $x \in H$ or $x \notin H$. In the latter case, x must be in L, proving that $E = H \oplus L$. In particular, every $x \in E$ is of the form $x = h + ay$ for $y \in E \setminus H$, some $h \in H$, and some $a \in \mathbb{K}$. Define a functional f on $E = H \oplus L$ by $(\forall x = h + ay \in E)\, T(h + ay) = a$. If $x = h + ay$, $w = k + by \in E$ are arbitrary, then for any scalar c, $T(cx + w) = T((ch + k) + (ca + b)y) = ca + b = cT(x) + T(w)$. Hence, T is in fact a linear functional. Furthermore, $T(y) = 1$, and that means T is nonzero. One more item of note is that $T^{-1}(\{0\}) = H$, and we have found the linear functional whose nullspace is H.

($\Leftarrow$): Suppose $T : E \to \mathbb{K}$ is a nonzero linear functional for which $T^{-1}(\{0\}) = H$. We set out to show that H is a hyperplane. For starters, as a nullspace H is a subspace of E. To complete the proof, it is necessary to show that $\mathrm{codim}(H) = 1$. Define a function $g : E/H \to \mathbb{K}$ by, $g(\dot{x}) = g(x + H) = T(x)$, for $x \in \dot{x}$. The function g is injective because if $\dot{x} \neq \dot{y}$, then $x - y \notin T^{-1}(\{0\}) \Leftrightarrow T(x - y) \neq 0 \Leftrightarrow T(x) \neq T(y)$, by linearity of T. Meanwhile, because T is nonzero, there is at least one $w \in E$ such that $T(w) \neq 0$, which implies that g is surjective. We conclude that $E/T^{-1}(\{0\})$ is linearly isomorphic to $\mathbb{K}$. This last statement leads to the observation that $\operatorname{codim}(H) = \dim(E/T^{-1}(\{0\})) = \dim(\mathbb{K}) = 1$, and H is indeed a hyperplane. $\qquad\square$

The next result shows that hyperplanes can come in only two flavors: closed or dense.

Proposition 3.1.7. *A hyperplane in a TVS is either closed or dense.*

Proof. Let $\overline{H}$ be the closure of a hyperplane H in a TVS E. The superset $\overline{H}$ of H either equals H or it does not. If $\overline{H} = H$, then of course, H is closed. Otherwise, Proposition 3.1.1 reveals that $\overline{H}$ is a subspace of E. Therefore, $\overline{H}$ cannot be a proper subset of E because that would contradict the assumption that H is already a maximal proper subspace of E. Hence, $\overline{H} = E$. $\qquad\square$

A related result is that continuity distinguishes whether a hyperplane is closed or dense. To prove that result, we need the lemma below.

Lemma 3.1.1. *Suppose E is a Hausdorff TVS and $\dim(E) = 1$. If $w \in E$ is a nonzero element, then the function $f : \mathbb{K} \to E$ given by $f(c) = cw$, $c \in \mathbb{K}$ is an isomorphism.*

Proof. Clearly, f is linear, continuous, and bijective. What needs to be proved is that f^{-1} is also continuous. Given $\varepsilon > 0$, there is $U \in \mathcal{N}_{0,E}$ such that $\varepsilon w \notin U$, by the assumption that E is Hausdorff. We can assume U to be balanced. We observe that for any scalar a, $aw \in U \Rightarrow |a| < \varepsilon$, which you can verify in Exercise 3.1.6. This proves that f^{-1} is continuous. $\qquad\square$

Proposition 3.1.8. *A hyperplane is closed if and only if it is the nullspace of a continuous linear functional.*

Proof. ($\Rightarrow$): Assume $H = T^{-1}(\{0\})$ is a closed hyperplane. The intention is to prove that the linear functional T is continuous. By Proposition 3.1.3, E/H is a Hausdorff quotient. Let $\varphi : E \to E/H$ be the canonical quotient map. By Diagram 1.4, $S : E/T^{-1}(\{0\}) \to \mathbb{K}$ is injective, where $T = S \circ \varphi$. Moreover, as $T \neq 0$, we have that S is also surjective. Thus, S is a bijection. As $\dim(E/H) = 1$, S can be described as in Lemma 3.1.1, and we conclude that S is in fact an isomorphism. Taking this one step further, $T = S \circ \varphi$ is a composition of continuous functions and is therefore continuous.

($\Leftarrow$): If $H = T^{-1}(\{0\})$ for a continuous linear functional T, then H is closed as the inverse image of $\{0\}$, which is closed in $\mathbb{K}$. $\qquad\square$

At this moment, it is worthwhile to see examples of the two types of hyperplanes. For the first one, we will need to glance ahead to the next section for a necessary result.

Example 3.1.1. An example of a closed hyperplane.

In the next section, Proposition 3.2.1 will reveal that every linear map from a finite-dimensional LCS to $\mathbb{K}$ is continuous. Hence, every hyperplane in such spaces is closed.

Example 3.1.2. An example of a dense hyperplane.

Let E be an infinite-dimensional normed space. On the unit sphere $S = \{x \in E : \|x\| = 1\}$, choose points $x_n \in S, n \in \mathbb{N}$ that are linearly independent, and let $\mathcal{B}$ denote a base of the vector space E that includes every x_n. Let $a_0 \neq 0$ be a fixed element of $\mathbb{K}$ and define f on $\mathcal{B}$ such that for every $n \in \mathbb{N}, f(x_n) = a_0 n$ and $f(x) = 0$ for all other $x \in \mathcal{B}$. From properties of bases in vector spaces, f can be extended to a linear map on all of E. Thus, f is a linear functional. It is discontinuous because it maps a bounded set to an unbounded set (see Theorem A.5.1). Thus, the nullspace of f is a dense hyperplane.

Exercises

3.1.1. Consider a normed space $(E, \| \cdot \|)$ and a closed subspace M of E. Define on E/M the following: $(\forall \overset{\bullet}{x} \in E/M) \| |\overset{\bullet}{x}| \| = \inf\{\|x\| : x \in \overset{\bullet}{x}\}$. Give a direct proof that this defines a norm on E/M.

3.1.2. Prove that if E is a normed space and M is a closed subspace of E such that M and E/M are Banach spaces, then E is a Banach space.

3.1.3. Let E be an infinite-dimensional vector space and equip E with the finest locally convex topology of Example 2.1.3.
   (a) Prove that every subspace of E is closed.
   (b) Prove that every quotient of E carries its finest locally convex topology.

3.1.4. Suppose p is a seminorm on a vector space E. Prove that for the subspace $M = p^{-1}(\{0\})$, $\overset{\bullet}{p}$ of Definition 3.1.3 is in fact a norm on E/M.

3.1.5. Suppose U is an absolutely convex, absorbing subset of a vector space E, and that M is a subspace of E. Prove that the quotient seminorm $\overset{\bullet}{p}_U$ in E/M is in fact the Minkowski functional of $\overset{\bullet}{U} = \varphi(U)$. *Hint:* Let $\overline{g}$ denote the Minkowski functional of $\overset{\bullet}{U}$ in E/M. Let $\overset{\bullet}{x} \in \overset{\bullet}{U}$ and $\varepsilon > 0$ be arbitrary. Show that if $\overset{\bullet}{p}_U(\overset{\bullet}{x}) = a$, then $\overline{g}(\overset{\bullet}{x}) \le a + 2\varepsilon$, and also that if $\overline{g}(\overset{\bullet}{x}) = a$, then $\overset{\bullet}{p}_U(\overset{\bullet}{x}) < a + \varepsilon$.

3.1.6. Prove the statement $aw \in U \Rightarrow |a| < \varepsilon$, from the proof of Lemma 3.1.1.

3.1.7. Prove that Lemma 3.1.1 is false if the assumption that the TVS is Hausdorff is omitted.

3.1.8. Assume f and g are two nontrivial linear functionals such that the hyperplane H is defined by both $f^{-1}(\{0\})$ and $g^{-1}(\{0\})$. Prove that there exists a scalar $a \ne 0$ such that $g = af$.

3.1.9. Prove that the value $a \in \mathbb{K}$ of the proof of Proposition 3.1.6 is unique.

3.1.10. Consider $E = C([0,1])$, the space of all continuous functions on $[0,1]$. Let $H = \{f \in E : f(0) = 0\}$. Prove that H is a hyperplane in E and determine a linear functional g such that $H = g^{-1}(\{0\})$.

3.2 Finite-dimensional topological vector spaces

The next goal is to prove that finite-dimensional Hausdorff TVS are equivalent in a topological vector space sense, to the normed space $\mathbb{K}^n$ (with any norm), for some $n \in \mathbb{N}$. This is accomplished by proving that there is essentially only one Hausdorff compatible topology on a finite-dimensional vector space. The result takes some work, but along the way there is plenty of enjoyable topology that gets applied. The details are next. For the results of this section, recall Definition 1.3.2 of a compatible topology.

Theorem 3.2.1. *On a finite-dimensional vector space there exists, up to topological equivalence, a unique compatible Hausdorff topology.*

Proof. Start with any vector space E over the field $\mathbb{K}$ that is of dimension n, for some $n \in \mathbb{N}$. Let $\{e_j : j \in \underline{n}\}$ be a Hamel basis of E. Define $\varphi : \mathbb{K}^n \to E$ by: For each $x = (x_1, x_2, \ldots, x_n) \in \mathbb{K}^n$, $\varphi(x) = \sum_{j=1}^n x_j e_j$. Clearly, φ is linear and bijective, so it is at least a *linear* isomorphism (see Definition A.2.3). On $\mathbb{K}^n$, let us put the topology generated by the max norm: $\|x\| = \max\{|x_j| : j \in \underline{n}\}$, for x as above. With this topology, $\mathbb{K}^n$ is, of course, a Hausdorff normed TVS. Define $\mathcal{T}$ to be the topology generated by the images under φ of the sets of the topology on $\mathbb{K}^n$. Seeing as how φ is a linear isomorphism, $(E, \mathcal{T})$ is a Hausdorff TVS. We intend to prove that, up to topological equivalence (see Definition A.3.13), $\mathcal{T}$ is unique. Thus, let σ be any other Hausdorff TVS topology on E. The objective is to show that σ is equivalent to $\mathcal{T}$.

Claim 3.2.1. *The topology $\mathcal{T}$ is finer than σ.*

Proof of Claim 3.2.1. Let $U \in \mathcal{N}_{0,\sigma}$ be given. Exploiting the continuity of addition, there exists a $V \in \mathcal{N}_{0,\sigma}$ such that $\underbrace{V + V + \cdots + V}_{n \text{ times}} \subset U$. As a neighborhood, V is absorbing, which means that for each $j \in \underline{n}$ there is $a_j > 0$ such that for each $x_j \in \mathbb{K}$, $x_j e_j \in V$, whenever $|x_j| < a_j$. A consequence of this is that

$$\sum_{j=1}^n x_j e_j \in \underbrace{V + V + \cdots + V}_{n \text{ times}} \subset U. \tag{3.8}$$

Now put $a = \min\{a_j : j \in \underline{n}\}$ and let $W = \{\sum_{j=1}^n x_j e_j : |x_j| < a, j \in \underline{n}\}$. We have that $W \subset V \subset U$. Meanwhile, $W = \varphi(\{y \in \mathbb{K}^n : \|y\| < a\})$. By the definition of $\mathcal{T}$, $W \in \mathcal{N}_{0,\mathcal{T}}$. With $W \subset U$, this shows that $\mathcal{T}$ is finer than σ, and the claim is proved.

Claim 3.2.2. *The topology σ is finer than $\mathcal{T}$.*

Proof of Claim 3.2.2. This time, let $U \in \mathcal{N}_{0,\mathcal{T}}$ be given. The goal, of course, is to find a subset of U that is a zero neighborhood with respect to σ. As images under φ, $\mathcal{N}_{0,\mathcal{T}}$ is generated by sets of the form

$$U = \left\{ \sum_{j=1}^n x_j e_j : |x_j| < 1, j \in \underline{n} \right\}. \tag{3.9}$$

Observe that the construction of $\mathcal{T}$ by using φ implies that φ is continuous. Next, if $C = \{x = (x_1, x_2, \ldots, x_n) : |x_j| \leq 1, j \in \underline{n}\}$, then $\varphi(C) = \overline{U}$ (the closure being taken with respect to $\mathcal{T}$). The set $\overline{U}$ is compact because $\mathbb{K}^n$ satisfies the Heine–Borel theorem and by virtue of $\overline{U}$ being the continuous image of a compact set.

Up to now, we have that $\mathcal{T}$ and σ are both Hausdorff topologies on $\overline{U}$, and additionally, $\mathcal{T}$ is finer than σ by Claim 3.2.1. That $\overline{U}$ is compact under the finer topology $\mathcal{T}$ implies $\overline{U}$ is also $\mathcal{T}$-compact. Oh, so we have two comparable compact Hausdorff topologies on $\overline{U}$; those topologies must be equivalent on $\overline{U}$ (see Theorem A.3.3). Of course, the hope is that these topologies are equivalent on all of E.

Now, $U \in \mathcal{N}_{0,\mathcal{T}}$ tells us that U is also a zero neighborhood with respect to $\mathcal{T}$ restricted to $\overline{U}$. Hence, there is a $V \in \mathcal{N}_{0,\mathcal{T}}$ for which $V \cap \overline{U} = U$, by the definition of the relative topology on $\overline{U}$. Meanwhile, U being a zero neighborhood with respect to $\mathcal{T}|_{\overline{U}}$ also implies that U is a zero neighborhood with respect to $\sigma|_{\overline{U}}$ from the previous paragraph. Apply the property of the relative topology again, to get $V \in \mathcal{N}_{0,\sigma}$ and $V \cap \overline{U} \subset U$. We then find a balanced σ-zero neighborhood W, where $W \subset V$. Thus, $W \cap \overline{U} \subset V \cap \overline{U} \subset U$. What we really need to prove is that $W \subset U$ outright, because that will complete the proof. For this, let $x \in W$ be arbitrary. Then x is of the form $x = \sum_{j=1}^{n} x_j e_j$, for some $x_j \in \mathbb{K}$. Put $a = \max\{|x_j| : j \in \underline{n}\}$, and we check cases of a to determine if we get $W \subset U$.

<u>Case 1</u>: $a < 1$. Then $W \subset U$ right away, by the definition of U.

<u>Case 2</u>: $a \geq 1$. In this case, $x \notin U$. Moreover, $\frac{1}{a}x = \frac{1}{a}\sum_{j=1}^{n} x_j e_j$, which implies that for some particular $j \in \underline{n}$,

$$\left|\frac{x_j}{a}\right| = 1 \Rightarrow \frac{1}{a}x \notin U. \tag{3.10}$$

On the other hand, you can easily check that $\frac{1}{a}x \in \overline{U}$. We have that $\frac{1}{a} \leq 1$ and we assumed W is balanced, which brings us to the observation that $\frac{1}{a}x \in W$. We observe that

$$\frac{1}{a}x \in W \cap \overline{U} \subset U \Rightarrow \frac{1}{a}x \in U. \tag{3.11}$$

This contradicts Equation (3.10). Therefore, $\frac{1}{a}x \notin U \Rightarrow \frac{1}{a}x \notin W$, which indicates that $x \notin U \Rightarrow x \notin W$, and this leads to the conclusion $W \subset U$, as desired. We conclude that $U \in \mathcal{N}_{0,\sigma}$, so that σ really is finer that $\mathcal{T}$, and we have that the two topologies are equivalent. $\qquad\square$

The result above leads to some important observations in the form of corollaries.

Corollary 3.2.1. *Every finite-dimensional Hausdorff TVS is locally convex.*

Proof. As a normed space, $\mathbb{K}^n$ is locally convex. $\qquad\square$

Corollary 3.2.2. *The topology of any finite-dimensional Hausdorff TVS is equivalent to that of $\mathbb{K}^n$ with any norm, for some $n \in \mathbb{N}$.*

Proof. By Theorem A.5.2(a), all norms on a finite-dimensional normed space over $\mathbb{K}$ are equivalent, and hence, all such normed topologies are equivalent as well. $\qquad\square$

Corollary 3.2.3. *Every finite-dimensional Hausdorff TVS is a Banach space.*

Proof. This result follows by Theorem A.5.2(b). $\qquad\square$

Corollary 3.2.4. *The topology of any finite-dimensional Hausdorff TVS satisfies the Heine–Borel property, namely that every closed, bounded subset is compact.*

Proof. This follows by Theorem A.5.2(c). $\qquad\square$

Corollary 3.2.5. *A finite-dimensional subspace of a Hausdorff TVS is closed.*

Proof. Suppose M is a finite-dimensional subspace of a Hausdorff TVS. Theorem 3.2.1 implies that M is isomorphic to the Banach space $\mathbb{K}^n$ for some $n \in \mathbb{N}$. As a complete subset of a Hausdorff TVS, Proposition 2.5.5(a) applies, and M is closed. $\qquad\square$

A result about finite-dimensional spaces that uses the ideas from the proof of Theorem 3.2.1 is next.

Proposition 3.2.1. *Any linear map from a finite-dimensional Hausdorff TVS to any other TVS is continuous. In particular, on a finite-dimensional TVS, the continuous dual is equal to the algebraic dual.*

Proof. Assume E is a TVS of dimension n, for some $n \in \mathbb{N}$. Let F be any TVS and let $T : E \to F$ be any linear map. It suffices to prove that T is continuous at 0. First, we choose a Hamel basis $\{e_j : j \in \underline{n}\}$ of E. Now let $V \in \mathcal{N}_{0,F}$ be arbitrary. There exists $W \in \mathcal{N}_{0,F}$ such that $\underbrace{W + W + \cdots + W}_{n\,\text{times}} \subset V$. We have that for each $j \in \underline{n}$, $T(e_j) \in F$, and the absorbing nature of W implies that there exists, for each $j \in \underline{n}$, $a_j > 0$ for which $x_j T(e_j) \in W$, whenever $|x_j| \le a_j$. Furthermore, the collection of sets U of the form

$$U = \left\{ \sum_{j=1}^{n} x_j e_j : |x_j| \le a_j, j \in \underline{n} \right\}, \tag{3.12}$$

is a subbase for $\mathcal{N}_{0,E}$. To finish, we note that $T(U) \subset \underbrace{W + W + \cdots + W}_{n\,\text{times}} \subset V$, and we have established that T is continuous. For the second statement, it is always the case that $E' \subset E^*$, and for finite-dimensional spaces the reverse inclusion holds. $\qquad\square$

Here is a seminorm version continuity on finite-dimensional TVS.

Proposition 3.2.2. *Every seminorm on a finite-dimensional Hausdorff LCS is continuous.*

Proof. Exercise 3.2.2. $\qquad\square$

Kolomogorov's Theorem 2.4.1 tells us that if a LCS has a bounded zero neighborhood, then it is a normed space. This result raises the topologically related question of what, if anything, happens if a TVS has a (relatively) compact zero neighborhood. It turns out that for Hausdorff TVS, having a compact zero neighborhood implies that the space must be finite-dimensional (and conversely). In light of Theorem 3.2.1, the consequence of the next result is that a compact zero neighborhood in a TVS is "too small" for the TVS to be interesting. Recall that a topological space is locally compact (see Definition A.3.21) if every point in the space is contained in a neighborhood that is compact. The details are next.

Theorem 3.2.2 (Riesz). *A Hausdorff TVS is locally compact if and only if it is finite-dimensional.*

Proof. ($\Rightarrow$): As usual, it suffices to consider zero neighborhoods. Assume the Hausdorff TVS E is locally compact, with U being a compact zero neighborhood. We construct the following open cover of U: Let $C = \{x + \frac{1}{2}\mathring{U} : x \in U\}$. The compactness assumption on U leads to a finite subcover of the form $C_0 = \{x_j + \frac{1}{2}\mathring{U} : j \in \underline{n}\}$, for some $n \in \mathbb{N}$ and $x_j \in U$. Put $M = \mathrm{span}\{x_j : j \in \underline{n}\}$, creating a subspace of E. We are going to prove that $M = E$ by way of the quotient E/M. First, because $\dim(M) < \infty$, Corollary 3.2.5 implies M is closed, and hence, E/M is Hausdorff. Meanwhile, as the canonical surjection $\varphi : E \to E/M$ is continuous, $\varphi(U)$ is compact in E/M. Making use of the subcover C_0, we observe that

$$U \subset \bigcup_{j=1}^{n}\left(x_j + \frac{1}{2}\mathring{U}\right) \subset \bigcup_{j=1}^{n}\left(x_j + \frac{1}{2}U\right) \subset M + \frac{1}{2}U. \tag{3.13}$$

Applying φ, we obtain

$$\varphi(U) \subset \varphi\left(M + \frac{1}{2}U\right) \subset \{0\} + \frac{1}{2}\varphi(U) \subset \frac{1}{2}\varphi(U), \tag{3.14}$$

and this implies that $2\varphi(U) = \varphi(2U) \subset \varphi(U)$. Thus, for all $k \in N$, $\varphi(2^k U) \subset \varphi(U)$. As U is balanced and absorbing, $E = \cup\{2^k U : k \in \mathbb{N}\} \Rightarrow \varphi(E) = E/M \subset \varphi(U)$. The continuity of φ ensures that $\varphi(U)$ is compact in E/M. An application of Proposition 2.5.10 reveals that $\varphi(U) = E/M$ is bounded, and this means E/M is the closure of the zero vector, in light of Proposition 2.4.2. The word "closure" in the last sentence can be removed because Corollary 3.2.5 notifies us that M is closed. The consequence of the closure of M is that E/M is Hausdorff. Hence, $E/M = \{\overset{\bullet}{0}\}$. We conclude that $M = E$, and E is finite-dimensional.

($\Leftarrow$): If E is finite-dimensional, then Theorem 3.2.1 applies, and E is isomorphic to some $\mathbb{K}^n$, which is a locally compact TVS. $\qquad\square$

An easy consequence of Riesz's theorem is the important fact below that closed bounded subsets of normed spaces need not be compact if the space is infinite-dimensional.

Corollary 3.2.6. *In every infinite-dimensional normed space, the Heine–Borel theorem does not hold.*

Proof. Exercise 3.2.5. □

Example 3.2.1. The closed unit ball of an infinite-dimensional Banach space is an example of a bounded set that is not totally bounded (see Definition 2.5.7).

You can prove this statement in Exercise 3.2.6.

The conclusion we can draw from these results is that finite-dimensional Hausdorff TVS are "uninspiring" in the sense that topologically they are essentially the same as the well-known Euclidean spaces $\mathbb{K}^n$.

Exercises

3.2.1. Suppose A is an absolutely convex, absorbing set in a vector space E. Prove that the Minkowski functional of A is a norm on E if and only if A does not contain a one-dimensional subspace.

3.2.2. Prove Proposition 3.2.2.

3.2.3. Describe a linear map from a normed space to a finite-dimensional space that is discontinuous. *Hint:* Consider the normed space c_{00} (see Definition A.5.1).

3.2.4. Using results of this section, give another proof of the second part of Proposition 3.2.1, namely, that on a finite-dimensional TVS, all linear functionals are continuous.

3.2.5. Prove Corollary 3.2.6.

3.2.6. Prove the statement of Example 3.2.1.

3.2.7. Let E be an infinite-dimensional vector space and equip E with the finest locally convex topology of Example 2.1.3. Prove that a subset of E is bounded if and only if it is contained in a finite-dimensional subspace and is bounded in that subspace.

3.2.8. Suppose E is a normed space and consider a sphere $S = \{x : \|x - x_0\| = r\}$ (where x_0 is a fixed element of E and $r > 0$). Prove that if there exists such a sphere S that is totally bounded, then E is finite-dimensional.

3.3 Product spaces

We start with a summary of products of vector spaces.

Summary 3.3.1. Consider a collection $\{E_\alpha : \alpha \in I\}$ of vector spaces, and the corresponding product $\Pi_{\alpha \in I} E_\alpha$. Here, each E_α is called a **factor space**. Elements $x \in \Pi_{\alpha \in I} E_\alpha$ are expressed as $x = (x_\alpha : \alpha \in I) = (x_\alpha)$ and each x_α is called the αth **coordinate**. For each $\alpha \in I$, the **canonical projection** is $\pi_\alpha : E \to E_\alpha$ given by: $(\forall x = (x_\alpha) \in E)\, \pi_\alpha(x) = x_\alpha$. The vocabulary for such projections is the αth **projection**. Routine calculations show that any product $E = \Pi_{\alpha \in I} E_\alpha$ of vector spaces can be equipped with a linear structure given by: $(\forall x = (x_\alpha), y = (y_\alpha) \in E)$, $x + y = (x_\alpha + y_\alpha)$, and $(\forall a \in \mathbb{K})\, a \cdot x = (ax_\alpha)$.

The next step is to equip $E = \Pi_{\alpha \in I} E_\alpha$ with a topology; not just any topology of course, but one that is compatible with the linear structure above. For this, we appeal to the initial topology (see Definition 2.1.9); that is, we have the following.

Definition 3.3.1. On any product $E = \Pi_{\alpha \in I} E_\alpha$ of vector spaces, the **product topology** is defined as the coarsest topology that makes all projections π_α continuous.

The product topology does not need any adjustments for TVS, as indicated next.

Proposition 3.3.1. *The product topology is compatible with the linear structure of a product of TVS.*

Proof. The proof involves routine calculations, as outlined in Exercise 3.3.1. $\qquad\square$

The next item lists basic properties of products. See Theorem A.3.2 in the Appendix. Except for the fact that each projection π_α is linear (which is easy to check), the other properties are valid in general topological spaces. Detailed proofs can be found in [67, pp. 117–119] or [102, Section 6.2].

Theorem 3.3.1. *On any product $E = \Pi_{\alpha \in I} E_\alpha$ of TVS $(E_\alpha, \mathcal{T}_\alpha)$ with projections π_α and equipped with the product topology, the following hold:*
(a) *Each projection π_α is linear, (continuous), and open. Hence, each projection π_α is uniformly continuous by Corollary 2.3.2.*
(b) *The product $E = \Pi_{\alpha \in I} E_\alpha$ is Hausdorff if and only if each TVS $(E_\alpha, \mathcal{T}_\alpha)$ is Hausdorff.*
(c) *A set $U \in \mathcal{N}_0 \Leftrightarrow U = \Pi_{\alpha \in I} U_\alpha$ such that $(\forall \alpha \in I)\, U_\alpha \in \mathcal{T}_\alpha$, and for all but finitely many α, $U_\alpha = E_\alpha$.*
(d) *For any $A \subset \Pi_{\alpha \in I} E_\alpha$, $\overline{A} = \Pi_{\alpha \in I} \overline{A}_\alpha$, where $A_\alpha = \pi_\alpha(A)$ for each $\alpha \in I$; in particular, $A = \Pi_{\alpha \in I} A_\alpha$ is closed in E if and only if A_α is closed in E_α for each $\alpha \in I$.*
(e) *For any TVS G, suppose a function $f : G \to \Pi_{\alpha \in I} E_\alpha$ is given by $f(w) = (f_\alpha(w) : \alpha \in I)$ where $w \in G$ and $f_\alpha : G \to E_\alpha$. Then f is continuous if and only if each f_α is continuous.*

Proposition 3.3.2. *Arbitrary products of LCS are again LCS.*

Proof. To verify the above statement, we simply cite two items. First, in Exercise 3.3.1, you will have proven that an arbitrary product of TVS is again a TVS. Second, in Theorem 1.1.1(h), we observed that arbitrary products of convex sets remain convex. □

The next result does not require a lot of work but is important.

Theorem 3.3.2. *A product of TVS is complete if and only if each factor space is complete.*

Proof. What is needed is to prove that filters in the product are Cauchy if and only if their images under the projections are Cauchy in the factor spaces. Let $E = \Pi_{\alpha \in I} E_\alpha$ be a product of TVS.

($\Rightarrow$): If $\mathfrak{F}$ is a Cauchy filter in E, then for each $\alpha \in I$, $\pi_\alpha(\mathfrak{F})$ is Cauchy in E_α by the uniform continuity of π_α.

($\Leftarrow$): Suppose each $\pi_\alpha(\mathfrak{F})$ is a Cauchy filter in E_α. Let U be any zero neighborhood in the product E. Then $U = \Pi_{\alpha \in I} U_\alpha$ where U_α is a zero neighborhood in E_α and $U_\alpha = E_\alpha$ for all $\alpha \notin I_F$, where I_F is a finite subset of I. We only need to check to see what happens with respect to U_α for $\alpha \in I_F$. In this case, for each $\alpha \in I_F$, let S_α be such that $\pi_\alpha(S_\alpha) - \pi_\alpha(S_\alpha) \subset U_\alpha$. By the finite intersection property of filters, $S = \bigcap_{\alpha \in I_F} S_\alpha \in \mathfrak{F}$, and we conclude that $S - S \subset U$. □

Presumably, we would like to know if products of quasicomplete spaces are quasicomplete. Addressing this question requires us to know how to determine bounded subsets of products. Such determinations are easy, as we see next.

Proposition 3.3.3. *A subset of a product is bounded if and only if its image under each projection is bounded in the corresponding factor space.*

Proof. ($\Rightarrow$): Let $E = \Pi_{\alpha \in I} E_\alpha$ be a product of TVS. If B is bounded in E, then the continuity of each π_α implies that $\pi_\alpha(B)$ is bounded in E_α.

($\Leftarrow$): Now suppose $B \subset E$ such that $\pi_\alpha(B) = B_\alpha$ is bounded in each E_α and let $U = \Pi_{\alpha \in I} U_\alpha$, where U is as in the proof of Theorem 3.3.2 above. Let $I_F = \{a_1, a_2, \ldots, a_k\}$ denote the finite subset of I for which $U_{\alpha_j} \neq E_{\alpha_j}, j \in \underline{k}$. The assumption that each $\pi_\alpha(B) = B_\alpha$ is bounded in E_α implies that for each $j \in \underline{k}$ there exists $a_j > 0$ such that $B_{\alpha_j} \subset a_j U_{\alpha_j}$. With $a = \max\{a_j : j \in \underline{k}\}$, and noting that for all $\alpha \notin I_F$, $U_\alpha = E_\alpha$, we conclude that $B \subset aU$, and B is therefore bounded in E. □

Exercise 3.3.2 asks you to prove that subsets of products are totally bounded if and only if their projections are totally bounded in each factor space. In addition, you can now prove the following.

Proposition 3.3.4. *Any product of quasicomplete TVS is quasicomplete.*

Proof. Exercise 3.3.4. □

The next result includes an outline of how we can create nonnormable and nonmetrizable LCS.

Theorem 3.3.3.
(a) *A finite product of normed spaces is normed but an infinite product of (nontrivial) normed spaces is never normed.*
(b) *A countable product of metrizable TVS is metrizable but an uncountable product of (nontrivial) metrizable TVS is never metrizable.*

Proof. (a) You will discover in Exercise 3.3.5 that there are at least three ways to prove that a finite product of normed spaces is normed. For the infinite product part, let $E = \Pi_{\alpha \in I} E_\alpha$ be a product of normed spaces, where I is infinite. If E is normable, then Kolomogorov's result (Theorem 2.4.1) requires that there be a bounded zero neighborhood in E. Let us look at an arbitrary zero neighborhood, $U = \Pi_{\alpha \in I} U_\alpha$. Let I_F denote the finite subset of I for which $U_\alpha \neq E_\alpha$. Then for any $\alpha \notin I_F$, $\pi_\alpha(U_\alpha) = E_\alpha$, so by Proposition 3.3.3, U is not bounded. As U was arbitrary, we conclude that there is no bounded zero neighborhood in E, and E is not normable.

(b) First we deal with countable products of metrizable TVS. Consider $E = \Pi_{n=1}^{\infty} E_n$, a countable product of metrizable TVS E_n. Corollary 2.2.1 indicates that each E_n has a countable base of zero neighborhoods; call each base $\mathcal{N}_{n,k} = \{U_{n,k} : k \in \mathbb{N}\}$, for each $n \in \mathbb{N}$. A neighborhood base for the product space is given by the collection of all sets $\Pi_{n=1}^{\infty} W_n$ as follows: Define $W = \Pi_{n=1}^{\infty} W_n$, where for finitely many indices $n_1, n_2, \ldots, n_m$, $W_{n_j} = U_{n,k_j}$, for some $k_j \in \mathbb{N}, j \in \underline{m}$, and $W_n = E_n$, for all $n \neq n_j, j \in \underline{m}$. As a countable collection of finitely many sets, this base is countable. Note that the product is Hausdorff because each factor space is Hausdorff. Apply Corollary 2.2.1 again and we are done with this part. To complete the proof, if I is uncountable, then Corollary 2.2.1 immediately implies that E is not metrizable. $\square$

We will now view a few examples of product spaces in the context of LCS. Example 2.4.1 allowed us to step out from the world of normed spaces to that of nonnormed metrizable TVS. The reason Theorem 3.3.3 has the honor of being denoted as a theorem rather than a proposition is that it shows how we can step outside the realm of metrizable TVS and float in the deep space(s) of TVS that are not even metrizable. We will often encounter such spaces; they are an important and often fascinating part of the theory of TVS. The first example below shows that such a space can be easily constructed.

Example 3.3.1. Any uncountable product of metrizable TVS is an example of a TVS that is not metrizable.

Indeed, Theorem 3.3.3(b) instantly implies that such a TVS is not metrizable.

Example 3.3.2. Any countably infinite product of normed spaces is an example of a LCS that is metrizable but not normable.

Theorem 3.3.3(a) immediately implies that such a LCS is metrizable, but not normable.

In Subsection 2.1.4, we saw examples of seminorms that are not norms. The following is another example of a seminorm that is not a norm, in the context of product spaces.

Example 3.3.3. A Minkowski functional that is not a norm.

Consider $E = \Pi_{n=1}^{\infty}\mathbb{R} = \mathbb{R}^{\mathbb{N}}$, a countable product of the real numbers each with the standard topology. Let $U = (-1,1) \times \Pi_{n=2}^{\infty}\mathbb{R}$. Clearly, U is absolutely convex. The definition of the product topology indicates that U is open. Hence, μ_U is a seminorm. Let $x = (0,1,0,0,\ldots) \in E$. Certainly, for every $a > 0$ such that $a \leq 1, x \in aU$, so $\mu_U(x) = 0$. Meanwhile, the point x is not 0. Therefore, μ_U is not a norm.

Example 3.3.4. A subset of a LCS whose closure cannot be determined by sequences.

Back in Section 1.2, the statement was made that the closure of a set cannot always be determined by sequences. That is, a point can be an element of the closure of a set such that no sequence of elements of that set converges to it. We use products of LCS here to describe a specific example of such a set. Note that the LCS in this example cannot be metrizable. For the example, let E be an uncountable product of $\mathbb{R}$; that is, $E = \mathbb{R}^I$, where I is uncountable. Denote by π_α, the projections of E to $\mathbb{R}$, for each $\alpha \in I$. Let

$$A = \{x = (x_\alpha) : x_\alpha = 1, \text{ for all but finitely many } \alpha \in I\}. \tag{3.15}$$

We intend to prove that $0 \in \overline{A}$ but that there is no sequence of elements of A that converges to 0. To verify this statement, first let $U \in \mathcal{N}_0$ in E be arbitrary. Of course, $U = \Pi_\alpha U_\alpha$, where $U_\alpha = \mathbb{R}$ except for indices in a set $I_F = \{\alpha_1, \alpha_2, \ldots, \alpha_k\} \subset I$, for some $k \in N$. Next, let $x = (x_\alpha)$ such that $x_{\alpha_j} = 0, j \in \underline{k}$ and $x_\alpha = 1$ for all other $\alpha \in I$. Observe that $x \in A$ and $x \in U$, and thus, $x \in \overline{A}$. Third, let $(w^{(n)})$ be any sequence of elements of A, where for each $n \in \mathbb{N}, w^{(n)} = (w_\alpha^{(n)} : \alpha \in I)$. Given any $n \in \mathbb{N}$, let I_n denote the finite subset of I for which $w_\alpha^{(n)} \neq 1$. Notice that $\bigcup_{n=1}^{\infty} I_n$ is countable, being a countable union of finite sets. Simultaneously, I is an uncountable set, so there must be an $\alpha_0 \in I$ such that $\alpha_0 \notin I_n$ for every $n \in \mathbb{N}$. In other words, for every $n \in \mathbb{N}, w_{\alpha_0}^{(n)} = 1$. Now let $V_{\alpha_0} = (-1,1)$ and $V = \pi_{\alpha_0}^{-1}(V_{\alpha_0})$. The continuity of π_{α_0} ensures that $V \in \mathcal{N}_0$. However, for every $n \in \mathbb{N}, w^{(n)} \notin V$, which implies that $(w^{(n)})$ does not converge to 0. As $(w^{(n)})$ was arbitrary, we conclude that $\overline{A}$ cannot be determined by sequences.

Example 3.3.5. A proof of Tychonov's theorem.

As stated in A.3.5, we are to prove that arbitrary products of compact topological spaces are compact. So, let $X = \Pi_{\alpha \in I} X_\alpha$ be any product of topological spaces X_α, with corresponding projections $\pi_\alpha : X \to X_\alpha$. Let $\mathfrak{U}$ be any ultrafilter on X. For each $\alpha \in I, \pi_\alpha(\mathfrak{U})$

is an ultrafilter on X_α, which you can verify in Exercise 3.3.7. By assumption, X_α is compact, so by Theorem 2.5.4, $\pi_\alpha(\mathfrak{U})$ converges to some element of X_α, and this holds for every $\alpha \in I$. Thus, $\mathfrak{U}$ converges to an element of X, and the proof is completed.

Incidently, we indirectly used Zorn's Lemma A.1.1 from Theorem 2.5.3 when we assumed the existence of an ultrafilter on X.

Exercises

3.3.1. Prove Proposition 3.3.1 that an arbitrary product of TVS is again a TVS. *Hint*: Recall that a function $f : (X, \omega) \to \prod_{a \in I}(Y_a, \mathcal{T}_a)$ is continuous if, and only if, $(\forall a \in I)\ f_a : X \to Y_a$ is continuous, where (X, ω) is a topological space, and $\prod_{a \in I}(Y_a, \mathcal{T}_a)$ is a product of topological spaces $(Y_a, \mathcal{T}_a)$, with the product topology.

3.3.2. Prove that a totally bounded subset (see Definition 2.5.7) of a product of TVS is totally bounded if and only if its projection image is totally bounded in each factor space.

3.3.3. Let $\mathfrak{F}$ be a filter on a product space $E = \prod_{a \in I} E_a$ of TVS. For each $a \in I$, let $\mathfrak{F}_a = \pi_a(\mathfrak{F})$ in E_a. Prove that $\mathfrak{F}_a$ is the collection of all subsets of E_a that are of the form $\pi_a(S)$, for all $S \in \mathfrak{F}$. Then prove that $\mathfrak{F}$ converges in E if and only if for every $a \in I$, $\mathfrak{F}_a$ converges in E_a.

3.3.4. Prove Proposition 3.3.4.

3.3.5. Let $E = \prod_{j=1}^{n} E_j$ be a finite product of normed spaces $(E_j, \|\cdot\|_j)$, for $j \in \underline{n}$. Prove that each of the following defines a norm on E and that the norm generates the same topology as the product topology: For $x = (x_1, x_2, \ldots, x_n) \in E$,

(a) $\|x\|^{(1)} = \sum_{j=1}^{n} \|x_j\|_j$.

(b) $\|x\|^{(2)} = [\sum_{j=1}^{n} \|x_j\|_j^2]^{1/2}$.

(c) $\|x\|^{(\infty)} = \max\{\|x_j\|_j : j \in \underline{n}\}$.

3.3.6. Prove that a countable product of Fréchet spaces is a Fréchet space.

3.3.7. Prove that the continuous image of an ultrafilter is an ultrafilter.

3.3.8. Describe an example of an infinite product of LCS, each of which is equipped with the finest locally convex topology of Example 2.1.3, but that the product topology is not the finest locally convex topology.

3.3.9. Prove that every Hausdorff LCS is isomorphic to a subspace of a product of Banach spaces. *Hint:* Let $\{p_a : a \in I\}$ be a family of seminorms that generates the topology of a given Hausdorff LCS E. Adjust, by way of letting F_a be the quotient of E by the nullspace of each p_a. Then each F_a is a normed space under the quotient norm given by $\overset{\bullet}{p}_a$, for each $a \in I$. Next, let x_a be the canonical image in F_a of x, for each $x \in E$. Then verify that the map that takes x to (x_a) into the product space $F = \prod_{a \in I} \widehat{F}_a$ is a linear, continuous injection from E onto its image in F, where $\widehat{F}_a$ is the completion of F_a.

3.4 Initial topologies

In this section, we examine a generalization of products. Although we could define the structures here for general TVS, we restrict our attention to the case of LCS. Consider a collection $\{(E_\alpha, \mathcal{T}_\alpha) : \alpha \in I\}$ of LCS and linear maps $\{f_\alpha : E \to E_\alpha : \alpha \in I\}$ from a fixed LCS E to each E_α. We assume the collection separates points as defined below.

Definition 3.4.1. Given a collection $\{(E_\alpha, \mathcal{T}_\alpha) : \alpha \in I\}$ of LCS and linear maps $\{f_\alpha : E \to E_\alpha : \alpha \in I\}$ from a fixed LCS E to each E_α, we say that the collection **separates points** if for any two distinct elements $x, y \in E$ there is at least one $\alpha \in I$ such that $f_\alpha(x) \neq f_\alpha(y)$.

For notational purposes, we describe the collection by way of $\{(E_\alpha, \mathcal{T}_\alpha, f_\alpha) : \alpha \in I\}$. We use this set up to state the following definition, which is that of initial topologies (see Definition 2.1.9), applied to LCS.

Definition 3.4.2. For a collection $\{(E_\alpha, \mathcal{T}_\alpha, f_\alpha) : \alpha \in I\}$ of LCS with linear maps $\{f_\alpha : E \to E_\alpha\}$ from a fixed LCS E to each E_α and that separates points of E, the initial topology on E is referred to as the **initial (locally convex) topology**. This topology is also referred to as the **projective topology**.

It behooves us to verify that the initial topology for a collection of LCS really does yield a LCS.

Proposition 3.4.1. *For a collection $\{(E_\alpha, \mathcal{T}_\alpha, f_\alpha) : \alpha \in I\}$ of LCS and E as in Definition 3.4.2, the initial topology on E is locally convex.*

Proof. With respect to the initial topology for a collection $\{f_\alpha : E \to E_\alpha : \alpha \in I\}$ of linear maps, continuity with respect to said topology is obtained by finite intersections of sets of the form $f_\alpha^{-1}(V_\alpha)$, for V_α being a zero neighborhood in E_α. These finite intersections form a base of zero neighborhoods in E for a locally convex topology. $\qquad\square$

3.4.1 Examples of initial topologies

Example 3.4.1. Subspaces.

If M is a subspace of a LCS E, then the relative topology on M is the initial topology given by the single linear identity map $\mathrm{id} : M \to E$. Indeed, a set V is open in M if and only if there exists an open set U in E for which $V = U \cap V = \mathrm{id}^{-1}(U)$.

Example 3.4.2. Products.

Consider $\{E_\alpha : \alpha \in I\}$ a collection of LCS and $E = \prod_{\alpha \in I} E_\alpha$. The projections $\pi_\alpha : E \to E_\alpha$ are linear and continuous with respect to the product topology, which is in fact, the initial topology corresponding to the collection of the projections.

We now determine when initial topologies on LCS are Hausdorff.

Proposition 3.4.2. *Let E be a LCS with the initial topology corresponding to a collection $\{(E_\alpha, \mathcal{T}_\alpha, f_\alpha) : \alpha \in I\}$ of LCS and linear maps f_α as in Definition 3.4.2. Then E is Hausdorff if and only if for every nonzero $x \in E$ there exists $\alpha \in I$ and $U_\alpha \in \mathcal{T}_\alpha$ such that the function f_α satisfies $f_\alpha(x) \notin U_\alpha$.*

Proof. ($\Rightarrow$): Assume E is Hausdorff with a base $\mathcal{N}_0$ of zero neighborhoods for the assumed initial topology. Let $x \in E, x \neq 0$. Then there is a $U \in \mathcal{N}_0$ such that $x \notin U$. The definition of the initial topology indicates that there are finitely many zero neighborhoods, n of them, such that $U_{\alpha_j} \subset E_{\alpha_j}$, for each $j \in \underline{n}$ and satisfying $\bigcap \{f_{\alpha_j}^{-1}(U_{\alpha_j}) : j \in \underline{n}\} \subset U$. Hence, for some specific $\alpha_j, f_{\alpha_j}(x) \notin U_{\alpha_j}$.

($\Leftarrow$): Consider any nonzero $x \in E$ again. If for some $\alpha \in I, f_\alpha(x) \notin U_\alpha$, then $x \notin f_\alpha^{-1}(U_\alpha)$ and the fact that $f_\alpha^{-1}(U_\alpha) \in \mathcal{N}_0$ allow us to conclude that E is Hausdorff. $\square$

Next, we find that (totally) bounded subsets in an initial topology are easy to describe.

Proposition 3.4.3. *In a LCS that carries the initial topology corresponding to a collection of linear maps, a set is (totally) bounded if and only if its image by each of the corresponding linear maps is (totally) bounded.*

Proof. Assume E carries the initial topology with respect to $\{(E_\alpha, \mathcal{T}_\alpha, f_\alpha) : \alpha \in I\}$ of LCS and linear maps as in Definition 3.4.2. We prove the bounded part here and leave the totally bounded part as Exercise 3.4.2.

($\Rightarrow$): If $A \subset E$ is bounded, then $(\forall \alpha \in I) f_\alpha(A) \subset E_\alpha$ is bounded by the continuity of f_α.

($\Leftarrow$): Suppose $A \subset E$ is such that $(\forall \alpha \in I) f_\alpha(A) \subset E_\alpha$ is bounded. If V_α is any zero neighborhood in E_α, then V_α absorbs $f_\alpha(A)$, and it is easy to see the $f_\alpha^{-1}(V_\alpha)$ absorbs A. Because zero neighborhoods in the initial topology on E are finite intersections of such $f_\alpha^{-1}(V_\alpha)$, we conclude that A is bounded in E. $\square$

Whenever we chance upon a new type of LCS, another question that comes up is that of determining the continuity of linear maps that involve such new spaces. That is the objective of the next result.

Proposition 3.4.4. *Consider a linear map $T : E \to F$ for LCS E and F such that F carries the initial topology via a collection $\{(F_\alpha, \mathcal{T}_\alpha, f_\alpha) : \alpha \in I\}$. Then T is continuous if and only if $(\forall \alpha \in I) f_\alpha \circ T : E \to F_\alpha$ is continuous.*

Proof. Given any $\alpha \in I$ and an arbitrary zero neighborhood $U_\alpha \subset F_\alpha$, notice that $T^{-1}(f_\alpha^{-1}(U_\alpha)) = (f_\alpha \circ T)^{-1}(U_\alpha)$, and we conclude that T is continuous if and only if $(f_\alpha \circ T)^{-1}(U_\alpha) \in \mathcal{N}_{0,E}$. $\square$

3.4.2 Projective limits

A common construction of initial topologies occurs when the index set is ordered. The details of how such ordering fits with the concept of initial topologies is next. That is, in this subsection we assume the index set I is ordered.

Definition 3.4.3. Let $\{(E_\alpha, \mathcal{T}_\alpha) : \alpha \in I\}$ be a collection of LCS. For any $\alpha, \beta \in I$ such that $\alpha \leq \beta$ suppose $g_{\alpha\beta} : E_\beta \to E_\alpha$ is a continuous linear map. Define E to be collection of elements $x = (x_\alpha) \in \Pi_\alpha E_\alpha$ such that $x_\alpha = g_{\alpha\beta}(x_\beta)$, whenever $\alpha \leq \beta$. Then E called the **projective limit** of $\{E_\alpha : \alpha \in I\}$.

Notation 3.4.1. The notation we will use for projective limits is $\mathrm{proj}_\alpha E_\alpha$.

The following proposition reveals the fact that projective limits of complete (quasicomplete) LCS are complete (quasicomplete). In the interest of efficiency (or maybe impatience for moving on to other topics), the proof is deferred to the references.

Proposition 3.4.5. *A projective limit of Hausdorff (quasi)complete LCS is (quasi)complete.*

Proof. Using methods similar to the proof of the (quasi)completeness of products of Hausdorff (quasi)complete LCS is one way to prove this result, as can be seen in [67, 2.11.3, p. 153]. $\square$

Another approach is to use the result of Theorem 3.4.1 below and apply it to the case where each LCS is Hausdorff. In this situation, the (quasi)completeness follows by citing some results from Section 3.3. The details are left as Exercise 3.4.5.

Theorem 3.4.1. *A projective limit of a collection $\{E_\alpha : \alpha \in I\}$ of Hausdorff LCS is a closed subspace of the product $\prod\{E_\alpha : \alpha \in I\}$.*

Proof. See [70, Proposition 2.6.1, p. 38]. $\square$

Some specific constructions of projective limits may help sort out some details of the previous definition. In many applications, the index set is countable, as we will see next.

Example 3.4.3. A countable projective limit.

Consider a countably infinite collection $\{E_n : n \in \mathbb{N}\}$ such that $E_1 \supset E_2 \supset E_3 \supset \cdots$. A projective limit of $\{E_n : n \in \mathbb{N}\}$ can be formed by using identity maps. Specifically, for $n, k \in \mathbb{N}$, we use $\mathrm{id}_{n,k} : E_n \to E_k$, for $n \geq k$, such that each $\mathrm{id}_{n,k}$ is continuous. Next, put $E = \bigcap_{n=1}^{\infty} E_n$ as a vector space. Then the projective topology on E is the projective limit of $\{E_n : n \in \mathbb{N}\}$. This can be written as $E = \mathrm{proj}_n E_n$.

We now look at a specific example of a countable projective limit.

Example 3.4.4. The LCS $\mathcal{E}(\Omega)$ as a projective limit.

Think back to Example 2.1.8 of the LCS $\mathcal{E}(\Omega)$, where $\Omega \subset \mathbb{R}$ is open. Now define, for each $m \in \mathbb{N}$, the vector space $\mathcal{E}_m(\Omega)$, as the set of all functions $f : \Omega \to \mathbb{R}$ such that $\partial^p f$ exists and is continuous for all multiindices p such that $|p| \leq m$. For each such multiindex p and each compact subset K of Ω, define $q_{K,p}$ by $q_{K,p}(f) = \max\{|\partial^p f(x)| : x \in K\}$. Routine calculations show that each $q_{K,p}$ is a seminorm. Thus, each $\mathcal{E}_m(\Omega)$ is a metrizable LCS (in fact, a Fréchet space). Note that we can define $\mathcal{E}_0(\Omega) = C(\Omega)$, the set of continuous functions on Ω. Next, we notice that for $|p| \leq m + 1$ the collection of seminorms for $\mathcal{E}_{m+1}(\Omega)$ contains all of the seminorms for which $|p| \leq m$ on $\mathcal{E}_m(\Omega)$. This means each injection id $: \mathcal{E}_{m+1}(\Omega) \to \mathcal{E}_m(\Omega)$, for $m \in \mathbb{N}$, is continuous. This is a fancy way of stating that the topology of $\mathcal{E}_{m+1}(\Omega)$ is finer than that of $\mathcal{E}_m(\Omega)$. Moreover, we have that $\mathcal{E}_1(\Omega) \supset \mathcal{E}_2(\Omega) \supset \mathcal{E}_3(\Omega) \supset \cdots$. Additionally, $\mathcal{E}(\Omega) = \bigcap_{m=1}^{\infty} \mathcal{E}_m(\Omega)$. In Exercise 3.4.3, you will have the opportunity to verify the facts claimed here. We conclude that $\mathcal{E}(\Omega) = \text{proj}_m \, \mathcal{E}_m(\Omega)$.

Exercises

3.4.1. For the collections $\{(E_a, \mathcal{T}_a, f_a) : a \in I\}$ of LCS from this section, prove that the following are equivalent:

(a) $\{(E_a, \mathcal{T}_a, f_a) : a \in I\}$ separates points.

(b) $(\forall x \neq 0)(\exists a_0 \in I)$ such that $f_{a_0}(x) \neq 0$.

(c) $\bigcap_{a \in I} f_a^{-1}(\{0\}) = \{0\}$.

3.4.2. Prove the totally bounded part of Proposition 3.4.3.

3.4.3. Prove that each space $\mathcal{E}_m(\Omega)$ of Example 3.4.4 is a Fréchet space and that $\mathcal{E}(\Omega) = \bigcap_{m=1}^{\infty} \mathcal{E}_m(\Omega)$.

3.4.4. Prove that every Hausdorff LCS carries the initial topology with respect to a collection of normed spaces. *Hint:* Consider a family $\mathcal{P}$ of continuous seminorms that define the topology of a Hausdorff LCS. Prove that $\{E_p = E/p^{-1}(\{0\}) : p \in \mathcal{P}\}$ is a collection of normed spaces that forms the desired initial topology on E. Use the Hausdorff assumption to verify that the linear maps corresponding to each $p \in \mathcal{P}$ separate points.

3.4.5. Prove that any projective limit of (quasi)complete Hausdorff LCS is (quasi)complete by applying Theorem 3.4.1 and making use of relevant results from Section 3.3.

3.5 Final topologies

We now consider a construction of LCS that is more or less the opposite of the construction of initial topologies on vector spaces. That is, we now look at collections $\{(E_\alpha, \mathcal{T}_\alpha) : \alpha \in I\}$ and E, for linear maps $f_\alpha : E_\alpha \to E$ and equip E with an appropriate locally convex topology. We later consider ordered index sets like we did for projective limits. Without further ado, let us get started.

Definition 3.5.1. Consider a collection $\{(E_\alpha, \mathcal{T}_\alpha, f_\alpha) : \alpha \in I\}$ of LCS, where for a fixed LCS E, each function $f_\alpha : E_\alpha \to E$ is linear and such that E is the linear span of $\bigcup_{\alpha \in I} f_\alpha(E_\alpha)$. Let $\mathcal{T}_f$ be the finest locally convex topology on E such that every f_α is continuous. Then $\mathcal{T}_f$ is called the **final (locally convex) topology on** E, with respect to $\{(E_\alpha, \mathcal{T}_\alpha, f_\alpha) : \alpha \in I\}$.

Before we start examining spaces with a final topology, we need to verify that such a topology exists. For starters, there is always at least one candidate for being the final topology, namely the trivial locally convex topology under which every f_α is continuous. Proceeding, we check all finer locally convex topologies under which every f_α is continuous. That upper bound of topologies is the finest and, by default, locally convex. The next task is to describe a base of zero neighborhoods for the final topology on a LCS. Fortunately, this can be done as clarified in the next two propositions.

Proposition 3.5.1. *A filterbase for the final topology $\mathcal{T}_f$ on a LCS E with respect to the collection $\{(E_\alpha, \mathcal{T}_\alpha, f_\alpha) : \alpha \in I\}$ is given by all absolutely convex absorbing sets $V \subset E$ for which $(\forall \alpha \in I) f_\alpha^{-1}(V) \in \mathcal{N}_{0,\mathcal{T}_\alpha}$. That is, V is a zero neighborhood with respect to $\mathcal{T}_f$ if and only if for every $\alpha \in I, f_\alpha^{-1}(V) \cap E_\alpha \in \mathcal{T}_\alpha$.*

Proof. Let $\mathfrak{B}$ denote the collection described in the statement. Routine calculations show that $\mathfrak{B}$ is indeed a filterbase. By Theorem 2.1.2, $\mathfrak{B}$ generates a locally convex topology $\mathcal{T}_\mathfrak{B}$ on E. Our task is to show that $\mathcal{T}_\mathfrak{B}$ is the final topology, $\mathcal{T}_f$. To prove $\mathcal{T}_f \cong \mathcal{T}_\mathfrak{B}$, we only need to prove that $\mathcal{T}_\mathfrak{B}$ is finer than $\mathcal{T}_f$, because $\mathcal{T}_f$ is already finer than $\mathcal{T}_\mathfrak{B}$ by definition of the final topology. So, let $V \in \mathcal{T}_f$. Then $(\forall \alpha \in I) f_\alpha^{-1}(V) \in \mathcal{N}_{0,\mathcal{T}_\alpha}$ implies that there is a finite collection of locally convex topologies $\mathcal{T}_j, j \in \underline{n}$ for which $(\forall j \in \underline{n})(\exists W_j \in \mathcal{N}_{0,\mathcal{T}_j})$ such that $\bigcap_{j=1}^n W_j \subset V$, where each W_j is absolutely convex. Put $W = \bigcap_{j=1}^n W_j$, and observe that $W \in \mathcal{T}_f$ because W is absolutely convex, absorbing, and $(\forall \alpha \in I) \bigcap_{j=1}^n f_\alpha^{-1}(W_j) \in \mathcal{N}_{0,\mathcal{T}_\alpha}$. That $W \subset V$ tells us that the proof is completed. $\square$

The next proposition will be useful in terms of describing a final topology using unions of images of sets.

Proposition 3.5.2. *Any zero neighborhood for the final locally convex topology on E with respect to a collection $\{(E_\alpha, \mathcal{T}_\alpha, f_\alpha) : \alpha \in I\}$ is given by all absolutely convex, absorbing sets $U \subset E$ of the form*

$$U = \operatorname{convbal}\left(\bigcup_{\alpha \in I} f_\alpha(U_\alpha)\right), \tag{3.16}$$

where $U_\alpha \in \mathcal{N}_{0,\mathcal{T}_\alpha}$, $\alpha \in I$.

Proof. Any such set U is absolutely convex and $f_\alpha^{-1}(U) \supset U_\alpha$, for $\alpha \in I$, by Equation (3.16). Moreover, it can be quickly established that U is absorbing. Thus $U \in \mathcal{T}_{\mathfrak{B}}$, and by Proposition 3.5.1, $U \in \mathcal{T}_f$. Conversely, if $U \in \mathcal{T}_{\mathfrak{B}}$, then $(\forall \alpha \in I)(\exists U_\alpha \in \mathcal{N}_{0,\mathcal{T}_\alpha})$ such that $U_\alpha \subset f_\alpha^{-1}(U)$. Hence, $f(U_\alpha) \subset U$. $\square$

At this point, we look at three examples of final topologies.

Example 3.5.1. Final topologies of finite-dimensional subspaces.

Consider a vector space E and let $\{E_\alpha : \alpha \in I\}$ be the collection of all finite dimensional subspaces of E. Per Riesz's Theorem 3.2.2, there is but one compatible topology (up to isomorphism) on each E_α, call it $\mathcal{T}_\alpha$. As for the collection of linear maps, we use the canonical injections $\mathrm{id}_\alpha : E_\alpha \to E$. The final topology with respect to this collection creates a LCS. This final topology is in fact the finest locally convex topology, as you can verify in Exercise 3.5.1.

Example 3.5.2. Every quotient of a LCS carries the final topology.

To verify this claim, we merrily state what a quotient topology is. On E/M, the quotient topology is the finest locally convex topology that guarantees the continuity of the canonical quotient map $\varphi : E \to E/M$. Indeed, the quotient topology is the final topology on E/M with respect to the single item $\{(E, \mathcal{T}, \varphi)\}$, where $\mathcal{T}$ is the topology of E.

Example 3.5.3. Locally convex direct sums.

In a product $E = \Pi_{\alpha \in I} E_\alpha$ of vector spaces, consider the vector space spanned by $\bigcup_{\alpha \in I} \mathrm{id}_\alpha(E_\alpha)$, where for each $\alpha \in I$, $\mathrm{id}_\alpha : E_\alpha \to E$ is the canonical injection. This is a vector space and a locally convex topology on it is defined next.

Definition 3.5.2. The vector space formed by $\operatorname{span}\{\bigcup_{\alpha \in I} \mathrm{id}_\alpha(E_\alpha)\}$ in $\Pi_{\alpha \in I} E_\alpha$ is called the **external direct sum**, denoted by $\bigoplus_{\alpha \in I} E_\alpha$. Equipped with the finest locally convex topology such that each injection $\mathrm{id}_\alpha : E_\alpha \to \bigoplus_{\alpha \in I} E_\alpha$ is continuous, the resulting LCS is called the **(locally convex) direct sum.**

Thus, the locally convex direct sum represents an example of a final topology. A detailed discussion of locally convex direct sums and some of their properties, as listed below, can be found in [103, Section V.6, pp. 89–95].

Proposition 3.5.3. *The following hold for a direct sum of LCS:*

(a) *A direct sum consists of the elements of a product that have only finitely many nonzero coordinates. Each element is the sum of the finite number of its nonzero coordinates, except for the zero vector.*
(b) *The topology of a direct sum is finer than the topology induced by the corresponding product. The direct sum can be expressed as an inductive limit (see Subsection 3.5.1) of finite products of LCS.*
(c) *A subset of a direct sum is bounded (resp., totally bounded) if and only if it is contained in and bounded (resp., totally bounded) in a finite sum of subsets of the spaces comprising the direct sum.*

Proof. Proofs of parts (a) and (b) can be seen in [103, Section 5.6, pp. 89–95]. You can verify part (c) in Exercise 3.5.3. $\qquad\square$

In Subsection 3.5.1, we will see some more exciting examples of final topologies. Meanwhile, by way of typical behavior, we determine when linear maps from a LCS with the final topology to an arbitrary LCS are continuous. You will surely notice a comparison with the corresponding case for the initial topology of the previous section.

Proposition 3.5.4. *Suppose E is a LCS equipped with the final topology with respect to $\{(E_\alpha, \mathcal{T}_\alpha, f_\alpha) : \alpha \in I\}$. Let F be any LCS and let $T : E \to F$ be a linear map. Then T is continuous if and only if $(\forall \alpha \in I)\, T \circ f_\alpha : E_\alpha \to F$ is continuous.*

Proof. ($\Rightarrow$): Clear, by the composition of continuous functions.

($\Leftarrow$): Assume each function $T \circ f_\alpha : E_\alpha \to F$ is continuous. Let $V \in \mathcal{N}_{0,F}$ be arbitrary, where we instinctively assume that V is absolutely convex. Observe that $(T \circ f_\alpha)^{-1}(V) = f_\alpha^{-1}(T^{-1}(V)) \in \mathcal{N}_{0,\mathcal{T}_\alpha}$ for each $\alpha \in I$. This absolutely convex absorbing set $T^{-1}(V)$ satisfies the property that $T^{-1}(V) \in \mathcal{N}_{0,E}$. $\qquad\square$

3.5.1 Inductive limits

Just as we did for initial topologies of Section 3.4.3, we consider topologies in which the index set is ordered and write a definition of the final topology of this context. Unlike the case of projective limits, we skip directly to the case of an ordered, countable index set. The main reason for this restriction is that uncountable locally convex inductive limits have little to offer mathematically, at least for our purposes. As Bierstedt [15] indicates, "...it is practically impossible to develop any deep and widely applicable general theory for uncountable l. c. [locally convex] inductive limits" (43–44). That this reduction to the countable case seems to be a significant simplification of the case of arbitrary ordered index sets is deceptive in that we will discover significant complexity (hence, fun) in the context of countable inductive limits. The definition is next.

Definition 3.5.3. Let $\{(E_n, \mathcal{T}_n, f_n) : n \in \mathbb{N}\}$ be a collection of LCS such that for each $n \in \mathbb{N}$, $E_n \subset E_{n+1}$, and $f_n : E_n \to E_{n+1}$ is a continuous linear function. Let $E = \bigcup_{n=1}^{\infty} E_n$ and equip

E with the finest locally convex topology such that each function f_n is continuous. Then E is called the **(locally convex) inductive limit** of the sequence (E_n) of LCS with respect to the collection $\{(E_n, \mathcal{T}_n, f_n) : n \in \mathbb{N}\}$. Each E_n is called a **step** of the inductive limit. If each LCS E_n is a proper subset of E_{n+1}, then the inductive limit is **proper**.

Notation 3.5.1. The notation for an inductive limit are: $E = \operatorname{ind}_n E_n$, and $(E, \mathcal{T}_{\text{ind}}) = \operatorname{ind}_n(E_n, \mathcal{T}_n)$ when details of the topologies are needed.

For most of the inductive limits we will see, the linear functions f_n will be the identity maps $\operatorname{id}_n : E_n \to E_{n+1}$. In this context, the basic idea of the inductive limit topology is that the topologies of the spaces E_n get coarser with each $n \in \mathbb{N}$ and then on the union of the steps we put the finest locally convex topology that when restricted to each step, is still coarser than the topology of that step. Next, some comments about vocabulary are relevant here. In Definition 3.5.3, to state that we equip E with the final topology would not be precise enough. It is because such wording raises the question of "Which final topology"? There are three possibilities for a topology such that each identity map is continuous: The finest *locally convex* topology, the finest compatible (TVS) topology, which need not be locally convex, and the finest of any topology, which need not even be compatible with the linear structure on E. In the context of general inductive limits, it is possible for all three of these topologies to be distinct on a collection of LCS. In fact, an example of such an inductive limit for which these three topologies are distinct can be seen in Bogachev and Smolyanov [18, 2.3.3, pp. 111–112]. Hence, unless otherwise indicated, the phrase *"inductive limit"* refers to a locally convex inductive limit. More detailed discussions of inductive limits can be found in Bierstedt [15] and in [95, Chapter 8]. An example of recent research on inductive limits can be seen in [2].

Let us view two examples of inductive limits.

Example 3.5.4. An example of an inductive limit of Banach spaces.

Consider $E_n = \{f \in C^\infty(\mathbb{R}) : \operatorname{supp}(f) \subset [-n, n], n \in \mathbb{N}\}$, where we equip each E_n with the sup norm: $(\forall f \in E_n) \|f\|_\infty = \sup\{|f(x)| : x \in [-n, n]\}$. Each E_n is a Banach space, and we denote the normed topology of each E_n by $\mathcal{T}_n$. Observe two features of this set up:
(i) $(\forall n \in \mathbb{N}) E_n \subsetneq E_{n+1}$.
(ii) $(\forall n \in \mathbb{N}) \mathcal{T}_{n+1}|_{E_n} = \mathcal{T}_n$. Indeed, by norm properties, we simply restrict the norm of E_{n+1} to E_n.

Thus, for each $n \in \mathbb{N}$, the topology $\mathcal{T}_{n+1}$ being coarser than (actually equivalent to) $\mathcal{T}_n$ implies that $\operatorname{id}_n : E_n \to E_{n+1}$ is continuous. Now put $E = \bigcup_{n=1}^\infty E_n$ to obtain the vector space $C_c^\infty(\mathbb{R})$, which is the set of all infinitely differentiable functions on $\mathbb{R}$ that have compact support. Next, equip E with the inductive limit topology and denote it by $E = \operatorname{ind}_n E_n$. What we end up with is a proper inductive limit of Banach spaces.

Example 3.5.5. An example of an inductive limit of Fréchet spaces.

We now reexamine the important Example 2.1.7 of an inductive limit. Our construction here is over $\mathbb{R}$. Descriptions over general $\mathbb{K}^n$ of this inductive limit can be easily deduced from this version and consist mostly of extra writing of terms and notation; see [67, Examples 2.3.4, p. 83, 2.12.6, p. 165]. Consider $\mathbb{R}$ as the union of compact sets K_n such that $K_n \subset K_{n+1}^{\circ}$, for each $n \in \mathbb{N}$. Define $\mathcal{D}(K_n) = \{f \in C^{\infty}(\mathbb{R}) : \text{supp}(f) \subset K_n\}$, as we have seen previously. On each $\mathcal{D}(K_n)$ define, for each $m \in \mathbb{N}$, seminorms q_m by $q_m(f) = \sup\{|f^{(k)}(x)| : x \in K_n\}$, for $k = 0, 1, 2, \ldots, m$. Here, $f^{(k)}$ represents the kth derivative of f. Let $\mathcal{T}_n$ denote the locally convex topology generated by $\{q_m : m \in \mathbb{N}\}$ on $\mathcal{D}(K_n)$. The countably infinite property of the collection of seminorms q_m allows us to conclude that each $(\mathcal{D}(K_n), \mathcal{T}_n)$ is metrizable. We proceed to prove a variety of claims about $(\mathcal{D}(K_n), \mathcal{T}_n)$ by way of the next proposition.

Proposition 3.5.5. *Each $(\mathcal{D}(K_n), \mathcal{T}_n)$ satisfies the following properties:*
(a) *$(\mathcal{D}(K_n), \mathcal{T}_n)$ is a Fréchet space.*
(b) *For each $n \in \mathbb{N}$, $\mathcal{T}_{n+1}|_{\mathcal{D}(K_n)} = \mathcal{T}_n$ as topologies.*
(c) *For each $n \in \mathbb{N}$, $\mathcal{D}(K_n)$ is a closed subset of $\mathcal{D}(K_{n+1})$.*

Proof. (a) Fix $n \in \mathbb{N}$ and suppose (f_j) is a Cauchy sequence in $(\mathcal{D}(K_n), \mathcal{T}_n)$. Then for each $m \in \mathbb{N}$, $(q_m(f_j))$ is a Cauchy sequence in $\mathbb{R}$. Our use of the suprema in the definition of each q_m, and the fact that K_n is compact implies that $(f_j^{(k)})$ converges uniformly to a continuous function $g^{(k)}$, for each $k \in \mathbb{N}$. That is, for some function $g \in C^{\infty}(\mathbb{R})$, the sequence (f_j) converges to g with respect to $\mathcal{T}_n$. For every $j \in \mathbb{N}$, $\text{supp}(f_j) \subset K_n$, so $\text{supp}(g) \subset K_n$ as well. That is, $g \in \mathcal{D}(K_n)$.

(b) If $f \in \mathcal{D}(K_n)$, then $\sup\{|f^{(k)}(x)| : x \in K_{n+1}\} = \sup\{|f^{(k)}(x)| : x \in K_n\}$, because $\text{supp}(f) \subset K_n \subset K_{n+1}$.

(c) Having a topology generated by a countable number of seminorms, each $\mathcal{D}(K_n)$ is Hausdorff by way of being metrizable due to an application of Corollary 2.2.1. Part (a) additionally allows us to observe that $\mathcal{D}(K_n)$ is a complete, and hence, closed subset of $\mathcal{D}(K_{n+1})$. $\qquad\square$

Now comes the inductive limit part. Put $\mathcal{D} = \bigcup_{n=1}^{\infty} \mathcal{D}(K_n)$ and equip it with the inductive limit topology. Thus, $\mathcal{D} = \text{ind}_n \mathcal{D}(K_n)$. As $\mathcal{D}$ is the LCS that is essential in the construction of distributions, we explicitly define it as such. The comparable definition for the case of $\mathbb{K}^n$ is easily formed.

Definition 3.5.4. $\mathcal{D} = \text{ind}_n \mathcal{D}(K_n)$ is called the **space of test functions**.

Symbolically, this all looks good, but there is an important observation that needs to be addressed so as to ensure that the LCS $\mathcal{D}$ has some legitimacy: Verify that $\mathcal{D}$ is a nontrivial vector space. Said verification is next.

Example 3.5.6. For a fixed positive number a, the function

$$f_a(x) = \begin{cases} e^{1/(x^2-a^2)}, & \text{if } |x| < a \\ 0, & \text{if } |x| \geq a \end{cases}, \tag{3.17}$$

is an element of $\mathcal{D}$.

Some observations from calculus about the definition of f_a reveal that derivatives of all orders of f_a exist and that $\mathrm{supp}(f_a) \subset [-a, a]$. Moreover, linear combinations of functions f_a for various values of a, will also belong to $\mathcal{D}$. That is, $\mathcal{D}$ is a nontrivial in fact, infinite-dimensional, vector space.

Examples 3.5.4 and 3.5.5 motivate the following definition of several types of inductive limits.

Definition 3.5.5. For inductive limits of LCS, we have the following terminology:
(a) An inductive limit of normed spaces is called an (LN) **space**.
(b) An inductive limit of Banach spaces is called an (LB) **space**.
(c) An inductive limit of metrizable spaces is called an (LM) **space**.
(d) An inductive limit of Fréchet spaces is called an (LF) **space**.

We now look at some properties of inductive limits. A common strategy in this context is to try to make use of information we have about the steps E_n in order to deduce properties of the resulting inductive limit. To start with, we adapt Proposition 3.5.2 to the particular case of an inductive limit.

Proposition 3.5.6. *For an inductive limit $E = \mathrm{ind}_n E_n$ of LCS $\{(E_n, \mathcal{T}_n) : n \in \mathbb{N}\}$, a base of zero neighborhoods is given by the collection of sets from either of the forms below:*
(a) *The locally convex topology generated by the filterbase*

$$\mathcal{B} = \{\text{All absolutely convex } V \subset E : (\forall n \in \mathbb{N}) \; \mathrm{id}_n^{-1}(V) \in \mathcal{N}_{0,E_n}\}. \tag{3.18}$$

That is, V is a zero neighborhood in $E = \mathrm{ind}_n E_n$ if and only if for every $n \in \mathbb{N}$, $V \cap E_n \in \mathcal{T}_n$:
(b)

$$\left\{ V \subset E : V = \mathrm{conv\,bal}\left(\bigcup_{n=1}^{\infty} U^{(n)} \right) \right\}, \tag{3.19}$$

where $(\forall n \in \mathbb{N}) \; U^{(n)} \in \mathcal{N}_{0,E_n}$.

Proof. (a) First, if V is an absolutely convex zero neighborhood for the inductive limit topology on E, then the continuity of $\mathrm{id}_n : E_n \to E$ ensures that $(\forall n \in \mathbb{N}) \; \mathrm{id}_n^{-1}(V) \in \mathcal{N}_{0,E_n}$. Second, suppose $V \in \mathcal{B}$. Then V is absolutely convex. The condition that $\mathrm{id}_n^{-1}(V) \in \mathcal{N}_{E_n}$ implies that for each $n \in \mathbb{N}$, $\mathrm{id}_n^{-1}(V)$ is absorbing in E_n. In addition, as $\bigcup_{n=1}^{\infty} \mathrm{id}_n(E_n) = E$,

V is absorbing in E as well. Theorem 1.3.3 now applies and we observe that $\mathcal{B}$ generates a unique locally convex topology on E. Because $\mathcal{B}$ consists of all such V, the corresponding topology on E is the inductive limit topology.

(b) A set V in Equation (3.19) is absorbing because for any $x \in E$, x must belong to some E_n, and then x can be absorbed by any $U^{(n)}$ of the corresponding zero neighborhood base. Secondly, the sets V are, by definition, convex. To finish, we must verify that the base consisting of such sets V generate the inductive limit topology on E. To this end, we first note that, given any V as in Equation (3.19) and given any $n \in \mathbb{N}$, $\mathrm{id}^{-1}(V)$ is a zero neighborhood in E_n because it contains a zero neighborhood in E_n. Conversely, the construction of the sets V indicates immediately that for every $n \in \mathbb{N}$ and any zero neighborhood $U^{(n)}$ in E_n, $\mathrm{id}(U^{(n)}) \subset V$ so that $\bigcup_{n=1}^{\infty} U^{(n)} \subset \mathrm{conv}\,\mathrm{bal}(\bigcup_{n=1}^{\infty} U^{(n)}) = V$. $\quad\square$

Though we will not focus on it here, ordinarily the next step would be to describe how to determine collections of continuous seminorms that define inductive limit topologies. This part of inductive limits does not get as much attention compared to using topological ideas. Nonetheless, in [67, 170–171] you can read about one example of constructing such seminorms. In addition, Bierstedt [15] explains that constructing collections of continuous seminorms on inductive limits is not that difficult, and provides examples to support that claim.

The next result shows how to determine the continuity of a linear map when the domain space is an inductive limit. It is an adaptation of Proposition 3.5.4.

Theorem 3.5.1. *A linear map from an inductive limit to a LCS is continuous if and only if the composition of the map with the injection from each step is continuous.*

Proof. We prove the case of linear functions $f_n : E_n \to E_{n+1}$, with the typical case of $f_n = \mathrm{id}_n$ being an obvious consequence. Thus, let $E = \mathrm{ind}_n E_n$ with respect to the collection $\{(E_n, \mathcal{T}_n, f_n) : n \in \mathbb{N}\}$, and suppose $T : E \to F$ is linear, where F is a LCS. The proof follows from the equivalences described below.

$$
\begin{aligned}
T \text{ is continuous} \;\Leftrightarrow\;& (\forall \text{ absolutely convex } V \in \mathcal{N}_{0,F})\, T^{-1}(V) \in \mathcal{N}_{0,E} \\
\Leftrightarrow\;& (\forall \text{ absolutely convex } V \in \mathcal{N}_{0,F})(\forall n \in \mathbb{N}) \\
& f_n^{-1}(T^{-1}(V)) \in \mathcal{N}_{0,E} \\
\Leftrightarrow\;& (\forall n \in \mathbb{N})\, (T \circ f_n)^{-1}(V) \in \mathcal{N}_{0,E} \\
\Leftrightarrow\;& (\forall n \in \mathbb{N})\, (T \circ f_n) \text{ is continuous on } E_n. \qquad (3.20)
\end{aligned}
$$

$\square$

To illustrate Theorem 3.5.1, we have the following.

Proposition 3.5.7. *A hyperplane in an inductive limit of Hausdorff LCS is closed if and only if its intersection with each step is closed in that step.*

Proof. ($\Rightarrow$): If H is a closed hyperplane in $E = \operatorname{ind}_n E_n$, then by Proposition 3.1.8, $H = T^{-1}(\{0\})$ for some continuous linear functional $T : E \to \mathbb{K}$. Theorem 3.5.1 then indicates that for each $n \in \mathbb{N}$, $T \circ \operatorname{id} : E_n \to \mathbb{K}$ is continuous on E_n. Hence, $H \cap E_n$ must be closed.

($\Leftarrow$): Suppose $H \cap E_n$ is closed in E_n for each $n \in \mathbb{N}$. As a hyperplane on E, $H = T^{-1}(\{0\})$ for some linear functional T. On the other hand, that $H \cap E_n$ is closed for each $n \in \mathbb{N}$ implies that each $T \circ \operatorname{id}$ is continuous, where $\operatorname{id}_n : E_n \to E$ is the canonical injection. Theorem 3.5.1 implies that T is continuous on all of E and we apply Proposition 3.1.8 once more to conclude that H is closed in E. $\qquad\square$

In order to prove some fundamental properties of inductive limits, a geometric property of the zero neighborhoods of a LCS will be useful. That property is proved next.

Lemma 3.5.1 (Neighborhood lemma). *Suppose M is a subspace of a LCS E and U is an absolutely convex zero neighborhood in M. Then:*
(a) *There exists an absolutely convex $V \in \mathcal{N}_{0,E}$ such that $V \cap M = U$.*
(b) *If M is closed, then $(\forall x \notin M)(\exists W \in \mathcal{N}_{0,E})$ such that W is convex and $W \cap M = U$. Moreover, $x \notin W$.*

Proof. Let M and U be as in the statement. (a) We are assuming that M carries the relative topology. In this context, U is a zero neighborhood in M if and only if there exists $W \in \mathcal{N}_{0,E}$ such that $U = W \cap M$. Deciding that W is the neighborhood we seek is premature because W need not be convex. In spite of this, the local convexity of E allows us to find a set $W_1 \subset W$ that is in fact absolutely convex, and we hope W_1 will allow us to complete the proof. Notice that $W_1 \cap M \subset W \cap M = U$. The proof of this part will be completed if we can prove that letting $V = \operatorname{conv}(W_1 \cup U)$ will satisfy $V \cap M = U$. Certainly, $V \in \mathcal{N}_{0,E}$, and we proceed to prove the desired set equality. First, to show $V \cap M \subset U$, let $z \in V \cap M$ be arbitrary. Because W_1 and U are both absolutely convex and $z \in V$, there exist $w \in W_1$ and $u \in U$ for which $z = tw + (1 - t)u$ for some $0 \le t \le 1$. If $t = 0$, then $z \in U$ and the containment is true. Otherwise, $0 < t \le 1$. Solve for w, to get $w = \frac{1}{t}(z - (1 - t)u) \in M - M$, so $w \in M$. We have $w \in W_1 \subset W$ and with $w \in M$, $w \in W \cap M \subset U$. Next, $z = tw + (1 - t)u$ and $w \in U$ tell us that $z \in U$ by the convexity of U. Hence, $V \cap M \subset U$. The proof that $U \subset V \cap M$ follows immediately from $U \subset V$ and $U \subset M$. We have established part (a).

(b) Assume $x \notin M$. Let M and U be as in the statement. If M is closed, then by Proposition 3.1.3 the quotient E/M is Hausdorff. Thus, for the given $x \notin M$ there exists $V \in \mathcal{N}_{0,E}$ for which $V \cap M \subset U$ and $V \cap (x + M) = \emptyset$. Put $W = \operatorname{conv}(V \cup U)$, and note that $W \in \mathcal{N}_{0,E}$. As in part (a), $W \cap M = U$. We aim to prove that $x \notin W$. Assume that $x \in W$ and we will see where that leads us. Observe that $x = tu + (1 - t)v$, for $0 \le t \le 1$ and some $u \in U$, $v \in V$. Then $(1 - t)v = x - tu$. We assumed $x \in W$, and we also have $-tu \in U$ from $u \in U$. A quick calculation shows that $(1 - t)v \in V \cap (x + M)$. This last statement is pleasantly contradictory regarding the earlier observation that $V \cap (x + M) = \emptyset$. $\qquad\square$

For the next part of our discussion, it will be necessary to re*strict* ourselves to a particular type of inductive limit, as defined next.

Definition 3.5.6. An inductive limit $E = \text{ind}_n E_n$ of LCS $(E_n, \mathcal{T}_n)$ is called **strict** if for every $n \in \mathbb{N}$, the topology $\mathcal{T}_{n+1}$ is equivalent to $\mathcal{T}_n$ when restricted to E_n. If in addition each E_n is a closed subset of E_{n+1}, then E is called **hyperstrict**.

It is easy to check that the completeness of each step of the inductive limits of Examples 3.5.4 and 3.5.5 shows that they are hyperstrict.

By its definition, the inductive limit topology of a union of LCS is always coarser than the topology of each step. For strict inductive limits, we achieve equivalence of the inductive limit topology and the step topology. The Neighborhood Lemma is a key tool that gets used in the proof. The next result is attributed to Dieudonné and Schwartz.

Theorem 3.5.2 (Dieudonné–Schwartz-1). *Let* $(E, \mathcal{T}_{\text{ind}}) = \text{ind}_n(E_n, \mathcal{T}_n)$ *be a strict inductive limit. Then for each* $n \in \mathbb{N}$, $\mathcal{T}_{\text{ind}}$ *is equivalent to* $\mathcal{T}_n$ *when restricted to* E_n.

Proof. For each $n \in \mathbb{N}$, let $\omega_n = \mathcal{T}_{\text{ind}}|_{E_n}$ be the restriction of $\mathcal{T}_{\text{ind}}$ to each E_n. Let $\mathcal{N}_{\mathcal{T}_n}$ denote a zero neighborhood base in E_n, for each $n \in \mathbb{N}$. By the definition of E, ω_n is coarser than $\mathcal{T}_n$ for each $n \in \mathbb{N}$. The goal is prove that ω_n is finer than $\mathcal{T}_n$ for each $n \in \mathbb{N}$. Given an arbitrary absolutely convex $V_n \in \mathcal{N}_{\mathcal{T}_n}$, the process of the proof proceeds by constructing a sequence of zero neighborhoods V_{n+k}, for each $k \in \mathbb{N}$ that will be used to find a set from ω_n that is contained in V_n. Let us get started.

Apply the Neighborhood Lemma 3.5.1 to E_n as a subspace of E_{n+1}, to find an absolutely convex $V_{n+1} \in \mathcal{N}_{\mathcal{T}_{n+1}}$ such that $V_{n+1} \cap E_n = V_n$. Apply the lemma again, to obtain an absolutely convex $V_{n+2} \in \mathcal{N}_{\mathcal{T}_{n+2}}$ such that $V_{n+2} \cap E_{n+1} = V_{n+1}$. Observe that $V_n = V_{n+1} \cap E_n = V_{n+2} \cap E_{n+1} \cap E_n = V_{n+2} \cap E_n$. Continuing in this way, we have for each $k \in \mathbb{N}$, an absolutely convex $V_{n+k} \in \mathcal{N}_{\mathcal{T}_{n+k}}$ such that $V_{n+k} \cap E_n = V_n$. Now let $V = \bigcup_{k=1}^{\infty} V_{n+k}$. As $E_n \subset E_{n+1}$ for each $n \in \mathbb{N}$, $V_{n+k} \subset V_{n+k+1}$, for each $k \in \mathbb{N}$. This last statement implies that V is absolutely convex. In addition, V is a zero neighborhood in $\mathcal{T}_{\text{ind}}$, courtesy of Equation (3.19). Meanwhile, $V \cap E_n = V_n$. Given that V_n is a zero neighborhood for ω_n, we end this proof by concluding that ω_n is finer than $\mathcal{T}_n$. $\qquad\square$

An important consequence follows.

Corollary 3.5.1. *A strict inductive limit of a sequence of Hausdorff LCS is Hausdorff.*

Proof. If $E = \text{ind}_n E_n$ is strict and $x \in E$, then there must be some $n \in \mathbb{N}$ such that $x \in E_n$. As E_n is Hausdorff, if $x \neq 0$, then there is a zero neighborhood U in E_n for which $x \notin U$ as deduced from Corollary 1.3.3. Hence, there is a zero neighborhood V in E such that $V \cap E_n \subset U$, and $x \notin V$. We have proven that E is, in fact, Hausdorff. $\qquad\square$

Using induction, one can prove the following as well.

Corollary 3.5.2. *If in the previous theorem, the inductive limit is hyperstrict, then for each* $n \in \mathbb{N}$, E_n *is closed in* E.

Proof. Exercise 3.5.10. □

Applying Theorem 3.5.2 to $\mathcal{D}$ of Definition 3.5.4, we obtain the following.

Corollary 3.5.3. *The LCS $\mathcal{D} = \mathrm{ind}_n \mathcal{D}(K_n)$ satisfies the property that for each $n \in \mathbb{N}$, the inductive limit topology, restricted to $\mathcal{D}(K_n)$, is equivalent to the topology of $\mathcal{D}(K_n)$.*

Proof. $\mathcal{D} = \mathrm{ind}_n \mathcal{D}(K_n)$ is a strict inductive limit. □

One useful property of a TVS is completeness. In Propositions 3.3.2 and 3.4.5, we saw that the completeness of products and projective limits of complete LCS are straightforward to establish. Completeness of a strict inductive limit of complete spaces can also be obtained, but the proof is not as short as those for products and projective limits. The extra complexity of the next result foretells the reality that inductive limits are generally more complicated than projective limits.

Theorem 3.5.3. *A strict inductive limit of complete Hausdorff LCS is complete.*

Proof. We begin a path to contradiction by assuming that the strict inductive limit $E = \mathrm{ind}_n E_n$ of complete, Hausdorff spaces E_n is not complete. The path to be taken goes through the completion $\hat{E}$ of E (see Definition 2.5.4). If E is not complete, then there exists an element $y \in \hat{E} \setminus E$, which then implies that $(\forall n \in \mathbb{N}) y \notin E_n$. Meanwhile, the completeness of the Hausdorff spaces E_n tells us that each E_n is closed in $\hat{E}$. Hence, there are absolutely convex zero neighborhoods W_n of $\hat{E}$ such that

$$(\forall n \in \mathbb{N})\,(y + W_n) \cap E_n = \emptyset. \tag{3.21}$$

Without loss of generality, we assume that for each $n \in \mathbb{N}$, $W_{n+1} \subset W_n$ as neighborhoods in the LCS $\hat{E}$. Next, define

$$U = \mathrm{convbal}\left[\bigcup_{n=1}^{\infty}\left(\frac{1}{2}W_n \cap E_n\right)\right]. \tag{3.22}$$

By its construction, U is a zero neighborhood of $\hat{E}$ and, therefore, so is $\overline{U}$. A property of the completion is that E is dense in $\hat{E}$, and this means that by Equation (1.8), for some $n_0 \in \mathbb{N}$,

$$(y + \overline{U}) \cap E_{n_0} \neq \emptyset. \tag{3.23}$$

Showing that $\overline{U} \subset W_{n_0} + E_{n_0}$ will lead to the desired contradiction. To this end, we start with $\overline{U} \subset U + \frac{1}{2}W_{n_0}$. Let $x \in U$ be arbitrary. Then x is of the form $x = \sum_{j=1}^{k} a_j x_j$, for some $k \in \mathbb{N}$ and such that $x_j \in \frac{1}{2}W_j \cap E_j$ and $\sum_{j=1}^{k} |a_j| = 1$ via absolute convexity. Observe that we can split $\sum_{j=1}^{k} a_j x_j$ into two parts: First, $\sum_{j=1}^{n_0} a_j x_j \in \sum_{j=1}^{n_0} a_j E_j \subset E_{n_0}$ because $E_n \subset E_{n+1}$ for all $n \in \mathbb{N}$. Second, $\sum_{j>n_0} a_j x_j \in \sum_{j>n_0} a_j \frac{1}{2}W_j \subset \frac{1}{2}W_{n_0}$, where we have applied the assumption of $W_{n+1} \subset W_n$, for $n \in \mathbb{N}$. Up to now, we have that $x \in \frac{1}{2}W_{n_0} + E_{n_0}$, and the arbitrariness of x gives us $\overline{U} \subset (\frac{1}{2}W_{n_0} + E_{n_0}) + \frac{1}{2}W_{n_0} = W_{n_0} + E_{n_0}$. In Equation (3.23), the last statement implies that $(y + W_{n_0}) \cap E_{n_0} \neq \emptyset$, contradicting Equation (3.21). The conclusion: E is complete. $\qquad\square$

There is one more item in this subsection and that item concerns bounded sets. The result below is also attributed to Dieudonné and Schwartz.

Theorem 3.5.4 (Dieudonné–Schwartz-2). *A subset of a hyperstrict inductive limit is bounded if and only if it is contained in and bounded in one of the steps.*

Proof. Let $(E, \mathcal{T}_{\mathrm{ind}}) = \mathrm{ind}_n(E_n, \mathcal{T}_n)$.

($\Rightarrow$): Suppose $B \subset E$ is bounded but is not contained in any step E_n. Without loss of generality, assume B is absolutely convex. We will first reach a contradiction to prove that B must be contained in some step E_m, then apply Proposition 2.2.5 to show that B is bounded in that step. Construct a sequence (x_n) as follows: B not being contained in E_1 means there is at least one point in B that cannot be absorbed by any zero neighborhood of E_1; that is, there is $x_1 \in B$ such that $x_1 \notin E_1$. On the other hand, $B \subset E$, so there is a $j_2 \in \mathbb{N}$ such that $x_1 \in E_{j_2}$. We may assume that j_2 is the smallest positive integer that satisfies the last condition. Next, B is not contained in E_{j_2} so there is an $x_2 \in B \setminus E_{j_2}$. The point x_2 must be somewhere in E, so as in the previous case, we find the smallest positive integer j_3 for which $x_2 \in E_{j_3}$. Continuing this way, we have an increasing sequence (j_k) in $\mathbb{N}$ and a corresponding sequence (x_{n_k}) of points of B; call it (x_k), such that $x_k \in E_{j_{k+1}} \setminus E_{j_k}$, for each $k \in \mathbb{N}$.

These calculations now allow further constructions, this time of neighborhoods. First, $x_1 \notin E_1$ means there is some zero neighborhood $W_1 \in \mathcal{T}_1$ such that $x_1 \notin W_1$. The "hyper" part of the assumption that the inductive limit is hyperstrict allows you to prove in Exercise 3.5.11 that E_n is a closed subspace of E_m for any $m \geq n$ and this allows us to apply part (b) of the Neighborhood Lemma 3.5.1. There is a zero neighborhood $W_2 \in \mathcal{T}_{j_2}$ such that $W_1 = W_2 \cap E_1$, and for which $x_1 \notin W_2$. Next, there exists $x_2 \notin E_{j_2}$ such that $\frac{1}{2}x_2 \notin W_2$. In other words, neither x_1 nor $\frac{1}{2}x_2$ belong to W_2. The "strict" part of the assumption that $\mathcal{T}_{n+1}$ restricted to $\mathcal{T}_n$ is equivalent to $\mathcal{T}_n$ implies there are absolutely convex zero neighborhoods V_3 and V_3' in E_{j_3} such that $W_2 = V_3 \cap E_{j_2}$ with $x_1 \notin V_3$ and $W_2 = V_3' \cap E_{j_2}$, with $\frac{1}{2}x_2 \notin V_3'$. Put $W_3 = V_3 \cap V_3'$. Then we have $W_2 = W_3 \cap E_{j_2}$, with $x_1, \frac{1}{2}x_2 \notin W_3$. We continue this pattern, obtaining zero neighborhoods W_k in E_{j_k} such that $W_{k-1} = W_k \cap E_{j_{k-1}}$ and for which

$$x_1, \frac{1}{2}x_2, \frac{1}{3}x_3, \ldots, \frac{1}{k}x_k \notin W_{k+1}, \tag{3.24}$$

for every $k \in \mathbb{N}$. Put $W = \bigcup_{k=1}^{\infty} W_k$. Then W is a zero neighborhood in E by Proposition 3.5.2. Hence, there is some $N \in \mathbb{N}$ such that for all $k \geq N$,

$$\left\{ \frac{x_k}{k}, \frac{x_{k+1}}{k+1}, \ldots \right\} \subset W. \tag{3.25}$$

In particular, $\frac{1}{N}x_N \in W \cap E_{j_{N+1}}$, which by the absolute convexity assumption of B contradicts Equation (3.24). Therefore, B must be contained in some step E_m. Showing that B is in fact bounded in E_m will complete the proof. Indeed, Equation (3.25) indicates that $\frac{1}{k} \to 0$ in $\mathbb{K}$, but $\frac{1}{k}x_k \nrightarrow 0$ in E for the sequence (x_k) in B. That is, B is not bounded in E by Proposition 2.2.5. This contradiction completes the proof of this implication.

($\Leftarrow$): If B is a bounded subset of some E_m, then as a continuous linear image of $\mathrm{id}_m : E_m \to E$, B is bounded in E. $\qquad\square$

The last few results motivate the following list of some types of inductive limits.

Definition 3.5.7. An inductive limit $E = \mathrm{ind}_n E_n$ of a sequence of LCS $(E_n, \mathcal{T}_n)$ is
(a) α-**regular** if every bounded subset of E is contained in one of the steps.
(b) **Regular** if every bounded subset of E is contained in one of the steps and is bounded in that step.
(c) **Boundedly retractive** if every bounded subset of E is contained in one of the steps and on that set the inductive limit topology agrees with the topology of the step.
(d) **Sequentially retractive** if every convergent sequence in E is contained in one of the steps and converges in the topology of that step.
(e) **Compactly regular** if every compact subset of E is contained in one of the steps and is compact in the topology of that step.

In a moment, we will see some relations and properties regarding these types of inductive limits. There exist examples that distinguish the various properties as well as conditions under which some are equivalent. Most of those examples and conditions are beyond the scope of this book. Such deeper details and additional exciting information about inductive limits can be found in [15], [70, Chapter 4], and [95, Chapter 8] as well as in the references cited in those works. Someone probably needs to write an up-to-date work that focuses on inductive limits and their important role within the context of topological vector spaces. The theorem that follows provides a summary of some properties connected to Definition 3.5.7. We assume the steps in the inductive limits are Hausdorff, though some of the results are valid in more general contexts. In fact, one of the most important properties that we desire is that an inductive limit Hausdorff LCS be Hausdorff, which is addressed in the next result.

Theorem 3.5.5. *For inductive limits of Hausdorff LCS, the following properties hold:*
(a) *A hyperstrict inductive limit is regular and Hausdorff. In fact, any regular inductive limit of Hausdorff LCS is Hausdorff.*
(b) *A hyperstrict inductive limit is quasicomplete if and only if each step is quasicomplete.*
(c) *Boundedly retractive inductive limits are compactly regular, which in turn are sequentially retractive, which are regular, and hence, such inductive limits are Hausdorff.*

Proof. Let $E = \mathrm{ind}_n E_n$ be an inductive limit of Hausdorff LCS.

(a) The conclusion of regularity is nothing more than a restatement of the conclusion of the Dieudonné–Schwartz-2 Theorem, 3.5.4. For the Hausdorff part, consider $\overline{\{0\}}$ in E, where E is any regular inductive limit of Hausdorff LCS. Then $\overline{\{0\}}$ is a bounded subspace of E, and hence, is contained in some E_m and is bounded there. E_m is Hausdorff, so $\overline{\{0\}} = \{0\}$ in E_m. Then Dieudonné–Schwartz-1, Theorem 3.5.2 implies that $\overline{\{0\}} = \{0\}$ in E as well. Thus, E is Hausdorff by Proposition 1.3.3.

(b) First, suppose the hyperstrict inductive limit $E = \mathrm{ind}_n E_n$ is quasicomplete. If B is any bounded subset of a step E_n, then the continuity of the injection from E_n to E implies that B is bounded in E. Hence, the closure of B in E is complete. Next, the Dieudonné–Schwartz-1 Theorem 3.5.2 indicates that the closure of B in E_n coincides with the closure of B in E. Thus, the closure of B is complete in E_n and E_n is quasicomplete. The reverse implication is similar and is left as Exercise 3.5.13.

(c) Certainly, compact subsets of Hausdorff LCS being bounded tells us that boundedly retractive implies compactly regular. That compactly regular implies sequentially retractive is left for you as Exercise 3.5.12. The proof that sequentially retractive inductive limits are regular can be found in [46]. Regular inductive limits of Hausdorff LCS are Hausdorff by part (a). $\qquad\square$

Corollary 3.5.4. *A strict inductive limit of complete LCS is hyperstrict, hence is regular, complete, and Hausdorff.*

Proof. This follows instantly from Theorem 3.5.5 and the fact that complete subsets of Hausdorff LCS are closed. $\qquad\square$

Part (c) of Theorem 3.5.5 above is summarized succinctly in Equation (3.26) below. Conditions under which some of these properties are equivalent as well as counterexamples that distinguish the properties can be found in the references cited before Theorem 3.5.5:

$$\text{Boundedly retractive} \Rightarrow \text{compactly regular} \Rightarrow$$
$$\text{sequentially retractive} \Rightarrow \text{regular} \Rightarrow \text{Hausdorff.} \tag{3.26}$$

For more results about regularity properties of inductive limits, see [15], Chapter 8 of [95], and [61], along with the references therein.

3.5.2 Some intriguing complications of inductive limits

It appears that projective limits and inductive limits (at least in the case of countable sets of indices) are inversely related. There is some truth to this suspicion. There is a connection between inductive and projective limits in the theory of duality of LCS that can be seen in [111, IV.4]. In fact, it turns out that inductive limits, and more generally locally convex final topologies, tend to be more complicated and unpredictable creatures than projective limits. Below are a few of those complications.

Example 3.5.7. An example of an inductive limit of a Hausdorff LCS of that is not Hausdorff.

Indeed, Example 3.1.2 shows an example a discontinuous linear functional $f : E \to \mathbb{K}$, where E is Hausdorff. With M being the nullspace of f, the situation is that E is Hausdorff, but M is not closed by Proposition 3.1.8. We invoke Proposition 3.1.3 to conclude that the quotient E/M, which carries the inductive limit topology, is not Hausdorff.

Second, it turns out that there are examples such that each step is complete, but that the corresponding inductive limit is not complete. That is, completeness is not preserved by general inductive limits. An example will be described once we have discussed the closed graph theorem in Chapter 8. In fact, in [18, 2.10.24, p. 147], there are outlines of examples of (LB) spaces that are not regular or that have some other pathological properties.

Third, the inductive limits of Examples 3.5.4 and 3.5.5 are not metrizable. In fact, a more general statement can be made as done in the next proposition and its corollary.

Proposition 3.5.8. *Proper, hyperstrict inductive limits are never Baire.*

Proof. Consider a proper hyperstrict inductive limit $E = \text{ind}_n E_n$. By Theorem 3.5.2, each E_n is closed in E. Moreover, no E_n is absorbing in E, which means the interior of every E_n with respect to the inductive limit topology is empty. Applying the equivalences of being a Baire space in the form of Theorem A.3.10, we conclude that E simply cannot be a Baire space. □

Corollary 3.5.5. *Proper, hyperstrict inductive limits of complete LCS are never metrizable.*

Proof. By Theorem 3.5.3, such an inductive limit is complete. If it were metrizable, then the Baire Category Theorem A.3.9 would apply and the space would be a Baire space, in contradiction to Proposition 3.5.8 above. □

The following corollary represents a nontrivial motivation for the study of LCS by exhibiting the nonmetrizability of the space of test functions.

Corollary 3.5.6. *The LCS $\mathcal{D}$ is a proper, nonmetrizable (LF) space.*

It turns out that the metrizability of (LB) and (LF) spaces is more complicated than what we just saw above. First, Köthe [79, 31.6, pp. 434–435] constructed the following result that indicates some of said complication.

Theorem 3.5.6. *There exists an (LB) space that is not regular and not complete.*

Proof. See [79, 31.6, pp. 434–435]. □

The example of Theorem 3.5.6 is quite important in the context of inductive limits. It is sometimes referred to as Köthe's example. In Example 8.2.2, we will see that this LCS is incomplete in a significantly more general way.

Proposition 3.5.9. *There exist (LF) spaces that are metrizable and even normable.*

Proof. A detailed discussion of such spaces can be found in Pérez Carreras and Bonet [95, Section 8.7, pp. 309–315]. □

Proposition 3.5.10. *Proper (LB) spaces are never metrizable.*

Proof. A proof of this result can also be found in [95, Proposition 8.5.18, p. 288]. □

If you are hoping to learn more about the topic of inductive limits, we will return to it several times going forward. If you would like to delve deeper into this topic, two places to start would be Bierstedt [15] and [95, Chapter 8].

Exercises

3.5.1. Prove that the topology of Example 3.5.1 is, in fact, the finest locally convex topology.

3.5.2. Prove the following regarding the function $f_a \in \mathcal{D}_{K_1} = C^\infty([-1,1])$ of Example 3.5.6, restricted here to $K = [-1,1]$:
  (a) If $g_b(x) = \sin(bx)$, for any $a > 0$, then $g_b f_a \in C^\infty([-1,1])$ whenever $f_a \in \mathcal{D}_{K_1}$.
  (b) Describe a function $g \in C([-1,1])$ such that $gf_a \notin C^\infty([-1,1])$. Justify your claim.

3.5.3. Prove part (c) of Proposition 3.5.3.

3.5.4. Let E be an (LB) space. Prove that an absolutely convex set $U \subset E$ is a zero neighborhood in $E = \mathrm{ind}_n E_n$ if and only if for every $n \in \mathbb{N}$, the set U absorbs $B_{n,1}$, where $B_{n,1}$ is the unit ball of E_n.

3.5.5. Prove that no proper regular inductive limit is Mackey first countable.

3.5.6. Prove that the (LB) space of Example 3.5.4 is hyperstrict.

3.5.7. Prove that any strict (LN) space has a fundamental sequence of bounded sets.

3.5.8. Prove that a proper, strict inductive limit of infinite-dimensional Banach spaces is not docile (Definition 2.2.9).

3.5.9. Prove directly that any hyperstrict inductive limit is sequentially retractive.

3.5.10. Prove Corollary 3.5.2.

3.5.11. From the proof of the Dieudonné–Schwartz-2 Theorem 3.5.4, prove that in a hyperstrict inductive limit, for any $n \in \mathbb{N}$, the LCS E_n is closed in E_m for any $m \geq n$.

3.5.12. Prove compactly regular implies sequentially retractive in Theorem 3.5.5.

3.5.13. Prove the second implication of part (c) of Theorem 3.5.5.

3.5.14. Give a direct proof that the LCS $\mathcal{D}$ is not metrizable. *Hint:* Consider any nonzero element $\varphi \in \mathcal{D}$ and examine the collection of functions $\varphi_{j,k} \in \mathcal{D}$ given by $\varphi_{j,k}(x) = \frac{1}{k}\varphi(x/j)$ and prove that $\varphi_{j,k}$ converges to 0 as $k \to \infty$ but the sequence (φ_{j,k_j}) does not converge whenever $k_j \to \infty$.

3.5.15. Let E be a Hausdorff inductive limit of an increasing sequence of Hausdorff LCS E_n such that $(\forall n \in \mathbb{N})(\exists U_n \in \mathcal{N}_{E_n})$ such that $U_n \subset U_{n+1}$. Prove that if A is bounded in E, then there exists $n \in \mathbb{N}$ such that $A \subset n\overline{U_n}$ where the closure is taken in E.

4 Duality

In this chapter, we find ourselves in front of a door. It is a colorful, beautiful door behind which are numerous mathematical treasures. The key that unlocks the door is the Helly–Hahn–Banach Theorem 4.1.2, which is one of the most fundamental results in analysis. The consequences of this result include an entirely new (at least for us) direction in the realm of LCS, called duality. Duality is based on pairing elements of a LCS with continuous linear functionals, and such pairings lead to the creation of a variety of topologies such as the weak topology (Definition 4.2.3), which turns out to be rather important. Polars (Definition 4.2.4) can be used to define a variety of topologies, some of which preserve the continuous dual, called consistent topologies (Definition 4.2.5). We discover in Theorem 4.3.2 that the closures of convex sets are the same in any consistent topology. Theorem 4.3.3 tells us that *every* locally convex topology is generated by the polars of some collection. The Alaoğlu–Bourbaki Theorem 4.4.1 shows us that the polar of any zero neighborhood is weakly compact. The Mackey–Arens Theorem 4.4.2 reveals that there is a specific range of topologies consistent with duality on a LCS. The chapter ends with another deep result about convex sets known as the Kreĭn–Mil'man Theorem 4.6.4. Let us get to it!

4.1 The Helly–Hahn–Banach theorem

First, a note on the title of this section and theorem. Traditionally, the name of this theorem is listed as the *Hahn–Banach theorem*, and you will find it listed this way in most earlier books on TVS. On the other hand, Narici and Beckenstein in [91, 7.13, pp. 214–216] give a compelling argument for including Helly as one of the main architects of this result. Next, in the previous paragraph, I wrote some words that imply that the Helly–Hahn–Banach theorem is a profound result, so some justification of that is needed. To qualify as a deep result in mathematics, a theorem (although sometimes such items are listed otherwise, such as a lemma) should have at least some of the following qualities: Its proof either relies on an earlier fundamental result or involves fundamentally new techniques. Another quality is that the applications of such a theorem open a pathway to significant developments or even spark the creation of a new subtopic within mathematics. The words "fundamental" and "significant" are of course, open to interpretation. Nevertheless, there is typically wide agreement of whether a result deserves such adjectives. In the case of the Helly–Hahn–Banach theorem, we will address these adjectives regarding the Helly–Hahn–Banach theorem after going through the proofs, so that the statements have some context.

Recall Definition 1.1.13 of the algebraic dual of a vector space, namely the set of all linear functionals defined on the vector space. If the vector space has a compatible topology (see Definition 1.3.2), then we recall that the continuous dual of a TVS is the set of all continuous linear functionals defined on that TVS (Definition 2.3.2(d)).

https://doi.org/10.1515/9783111392868-004

Definition 4.1.1. The **continuous dual** of a TVS is the set of all continuous linear functionals on that TVS.

Notation 4.1.1. The notation for the continuous dual of a TVS E is E'.

At the risk of whining, here is another reminder (see the comment after Definition 1.1.13) that the notation typically used for the continuous dual of a general TVS is the notation that is typically used for the algebraic dual in the context of normed spaces and vice versa.

Because linear combinations of continuous linear functions are continuous, E' is a vector space. Equipping E' with a locally convex topology will be the focus of much of this chapter. In fact, we will find that there are numerous locally convex topologies that can be put on E'. Continuous functions always get our attention, however, before we start studying the continuous dual of a LCS, we need to know if there are any nontrivial elements of such a dual. Certainly the zero function belongs to the continuous dual of any TVS. In the case of finite-dimensional TVS, the answer is "yes" the continuous dual is nontrivial, by Proposition 3.2.1. On the other hand, we have seen (Corollary 3.2.2) that finite-dimensional TVS are, well, mostly insignificant. This means the question about whether there are nonzero elements of the continuous dual of a TVS focuses mainly on this question for infinite-dimensional vector spaces. It turns out, as you can see in references, such as [73] and [106], that for general TVS it can happen that the *only* continuous linear functional is the zero function. It is precisely the Helly-Hahn–Banach theorem that will guarantee that for LCS, the continuous dual always has nonzero elements, and even has enough nonzero elements to separate points of the original LCS, where the phrase "separate points" is defined next. Note that it is necessary to define the precise meaning of the phrase "separates points" in this context, considering that this phrase has already appeared in the context of initial topologies (see Definition 3.4.1). Linearity simplifies the following definition.

Definition 4.1.2. A collection of linear functionals on a vector space E **separates points** if for any nonzero element $x \in E$ there is at least one functional f in the collection such that $f(x) \neq 0$.

For our first result, we show that the *algebraic* dual always, in fact, separates points. It is a nontrivial result because its proof uses the deep result that every vector space has a Hamel basis.

Theorem 4.1.1. *The algebraic dual of any nontrivial vector space separates points.*

Proof. Let $E \neq \{0\}$ be a vector space, and let $x_0 \in E \setminus \{0\}$. The goal is to find some $f \in E^*$ such that $f(x_0) \neq 0$ in $\mathbb{K}$. Because $x_0 \neq 0$, the set $\{x_0\}$ can be extended to a Hamel basis $\mathcal{B}$ for all of E. Define $f : E \rightarrow \mathbb{K}$ by setting $f(x_0) = 1$ and extend f to a linear functional defined on all of E. The resulting element of the algebraic dual satisfies Definition 4.1.2. $\qquad\square$

The function f in the proof above turns out to be a hyperplane, as you can verify in Exercise 4.1.2. Before diving into the next big result, it will be helpful to use our imagination, as well as to state a few definitions of geometric concepts. To start, imagine the set $z = \pi$ in $\mathbb{R}^3$. It is a horizontal plane (using the standard orientation of the xyz coordinates) of height π. In fact, this set is like a hyperplane but is in fact a linear translation of a hyperplane. Such a set needs a few correspondingly modified terms as defined next.

Definition 4.1.3. A translate of a linear subspace is called a **linear manifold** (also called an **affine subspace**). A translate of a hyperplane is called an **(affine) hyperplane**.

Some authors include in their general definition of hyperplane what is parenthetically listed here as an affine hyperplane. We will follow this pattern and use the term "hyperplane" to include translates of hyperplanes that were defined in Definition 3.1.4. Meanwhile, observe that a hyperplane is of the form $H = H_0 + x_0$ for some possibly nonzero $x_0 \in E \setminus H_0$, and where H_0 is a hyperplane of Definition 3.1.4. If H_0 is defined by the equation $f(x) = 0$ for a linear functional f, then for $a = f(x_0)$ in $\mathbb{K}$, we have that $a \neq 0$ and the equation for H is expressed by $H = \{x : f(x) = a\}$. In addition, we define the following.

Definition 4.1.4. For a hyperplane H with equation $H = \{x : f(x) = a\}$, the sets $\{x : f(x) \leq a\}$ and $\{x : f(x) \geq a\}$ are called **(algebraically) closed half spaces** determined by H; the sets $\{x : f(x) < a\}$ and $\{x : f(x) > a\}$ are called **(algebraically) open half spaces** determined by H.

From the example of the hyperplane given by $z = \pi$ in $\mathbb{R}^3$, we have $H = H_0 + x_0$, where $x_0 = (0, 0, \pi)$ and $H_0 = \{(x, y, 0) : x, y \in \mathbb{R}\}$.

We are now ready to study the Helly–Hahn–Banach theorem. As a prelude to what will be done, it turns out that we will need three statements and proofs of this theorem: A geometric version where $\mathbb{K} = \mathbb{R}$, an analytic version for $\mathbb{K} = \mathbb{R}$, and a general version that includes the case of $\mathbb{K} = \mathbb{C}$.

Theorem 4.1.2 (Helly–Hahn–Banach theorem—geometric version). *Let E be a TVS over the field $\mathbb{R}$ and let A be an open, convex subset of E. Let M be an affine subspace that does not intersect A. There exists a closed hyperplane that contains M but does not intersect A.*

Before the proof, we need a lemma. The lemma gives a clue as to how the proof of the Helly–Hahn–Banach is nontrivial even for moving up by one dimension. The details are next.

Lemma 4.1.1. *Assume $(E, \mathcal{T})$ is a TVS over the field $\mathbb{R}$, and that $A \subset E$ is open and convex. Suppose H is a subspace of E such that $H \cap A = \emptyset$. Then either H is a hyperplane or there exists $x_0 \in E$ for which $x_0 \notin H$ and $\mathrm{span}\{x_0, H\} \cap A = \emptyset$.*

Proof. Define $C = H + \bigcup\{aA : a < 0\}$. This set is open because A is open and in a TVS the sum of an open set with any other set is open. Observe that $-C = H + \bigcup\{aA : a < 0\}$, because $-H = H$. These observations allow us to prove that $C \cap (-C) = \emptyset$, as follows.

Suppose there exists $z \in C \cap (-C)$. Then $z = h + ax = k - by$ for some $h, k \in H, a, b > 0$, and $x, y \in A$. We calculate $h + ax = k - by \Rightarrow ax + by \in H$. As A is convex, $aA + bA = (a+b)A$, by Proposition 1.1.1. This implies that $(a + b)A \cap H = \emptyset$, contradicting $ax + by \in H$. Now consider $H \cup C \cup (-C)$. Of course, either $H \cup C \cup (-C) = E$, or $H \cup C \cup (-C) \neq E$. Suppose first that $H \cup C \cup (-C) \neq E$. In this case, there is an element x_0 in E for which $x_0 \notin H$ and $x_0 \notin C \cup (-C)$, and hence, x_0 is such that $\mathrm{span}\{x_0, H\} \cap A = \emptyset$. We move on to the other possibility, namely that $H \cup C \cup (-C) = E$. We have the following.

Claim 4.1.1. *If $H \cup C \cup (-C) = E$, then H is a hyperplane.*

Proof of Claim 4.1.1. Suppose H is not a hyperplane. By the definition of a hyperplane, there exists some $x_0 \in E$ such that $\mathrm{span}\{x_0, H\} \neq E$. Obviously, $x_0 \notin H$, so without loss of generality, $x_0 \in C$. We have that $C \cap (-C) = \emptyset$, so there must be some point $y_0 \in -C$ such that $y_0 \notin \mathrm{span}\{x_0, H\}$. Define $f(t) = ty_0 + (1 - t)x_0$ for $0 \leq t \leq 1$. As a line segment between two fixed elements, f is continuous. Hence, if $P = f^{-1}(C)$ and $Q = f^{-1}(-C)$, then P and Q are open subsets of $[0, 1]$ in $\mathbb{R}$ (with $0 \in P, 1 \in Q$). In addition, $P \cap Q = \emptyset$. Define $s = \sup\{p : p \in P\}$. Then $s \in \overline{P}$, and in fact, s is in the boundary of P. Thus,

$$s \in \overline{P} \cap \overline{P^c} \subset \overline{Q^c} \cap \overline{P^c} = P^c \cap Q^c = (P \cup Q)^c, \tag{4.1}$$

by P and Q being open. Using this property, it follows that $f(s) \in H$. Hence, $f(s) = sy_0 + (1 - s)x_0 \in H$ implies that $sy_0 \in (s-1)x_0 + H$. Observe that $s \neq 0$, and this means $y_0 \in \frac{s-1}{s}x_0 + H$. However, this last statement implies that $y_0 \in \mathrm{span}\{x_0, H\}$, a contradiction. We have simultaneously completed the proof of the claim and the proof of the lemma. $\qquad\square$

We can now prove the geometric version of the Helly–Hahn–Banach theorem.

Proof of Theorem 4.1.2. What remains to show is that the hyperplane H can be constructed and that it is, in fact, closed. To this end, define $\mathcal{E}$ to be the collection of subspaces of E that contain M and do not intersect A. We order $\mathcal{E}$ by set inclusion. Then $\mathcal{E}$ satisfies the hypotheses of Zorn's Lemma A.1.1. That is, $\mathcal{E}$ contains a maximal element $\mathcal{M}$. Define H to be the union of all subspaces in $\mathcal{M}$. By the ordering of $\mathcal{E}$, H is a subspace of E. Moreover, $H \cap A = \emptyset$ by the definition of $\mathcal{E}$. Applying Lemma 4.1.1, we conclude that either H is a hyperplane or $(\exists x_0 \notin H)$ $\mathrm{span}\{x_0, H\} \cap A = \emptyset$. This second condition is impossible because if it occurred, then $\mathcal{M}' = \mathrm{span}\{x_0, H\}$ would be a subspace of E that strictly contains the maximal subspace $\mathcal{M}$. Therefore, H is some kind of hyperplane. That H is closed is left as Exercise 4.1.4. $\qquad\square$

For the next theorem, we need a definition.

Definition 4.1.5. In a vector space E, consider a hyperplane $H = \{x : f(x) = a\}$ for $f : E \to \mathbb{R}$ linear and $a \in \mathbb{R}$. Suppose $A, B \subset E$. If $A \subset \{x : f(x) \leq a\}$ and $B \subset \{x : f(x) \geq a\}$, then H **separates** A and B. If $A \subset \{x : f(x) < a\}$ and $B \subset \{x : f(x) > a\}$, then H **strictly separates** A and B.

The result that follows describes how open, convex sets can be (strictly) separated by a hyperplane in a TVS. It is a consequence of the geometric version of the Helly–Hahn–Banach theorem.

Theorem 4.1.3. *Let E be a TVS, and suppose A and B are both nonempty convex subsets such that $A \cap B = \emptyset$. If A is open, then there exists a closed hyperplane that separates A and B. If B is also open, the separation is strict.*

Proof. We examine the set difference $A - B$ (see Definition 1.1.1). From Chapter 1 material, we know that $A - B$ is convex as a set difference of convex sets, and that $A - B = \bigcup \{A - y : y \in B\}$ is open by the assumption that A is open. We will use the subspace $M = \{0\}$ of E. The first step is to observe that $0 \notin A - B$. Indeed, if 0 were an element of $A - B$, then there would exist $x \in A, y \in B$ such that $x - y = 0$, but then $x = y$ contradicts $A \cap B = \emptyset$. That $0 \notin A - B$ implies that $(A - B) \cap M = \emptyset$. This is our opportunity to apply the geometric version of the Helly–Hahn–Banach Theorem 4.1.2 to conclude that there exists a closed hyperplane H such that $(A - B) \cap H = \emptyset$, and $M \subset H$. Proposition 3.1.6 indicates that there is some linear functional f such that $H = f^{-1}(\{0\})$. Next, we observe that either $A - B \subset \{w : f(w) > 0\}$, or $A - B \subset \{w : f(w) < 0\}$ in E. With no loss of generality, we may assume $A - B \subset \{w : f(w) > 0\}$, in which case $f(A - B)$ is a set of positive numbers; that is, for all $x \in A$ and $y \in B, f(x - y) = f(x) - f(y) > 0$, hence, $f(x) > f(y)$.

In order to determine the value of a of Definition 4.1.5, we will spend some time in a quotient space. Consider the canonical quotient map $\varphi : E \to E/H$ and $h : \mathbb{R} \to E/H$ given by $h(b) = b\varphi(w)$ where $b \in \mathbb{R}$, and $w \in E$ such that $f(w) = 1$. Proposition 3.1.2 reveals that φ and h are open maps (Definition A.3.15), and are additionally continuous. The result of this observation is that $f = h^{-1} \circ \varphi$ is an open map, as a composition of open maps. Moreover, because f is real valued as well as the fact that A and B are convex, it follows that $f(A)$ and $f(B)$ are both convex, with $f(A)$ being open. If B is open, then $f(B)$ is also open. Another observation is that $f(A) \cap f(B) = \emptyset$ by $f(x) > f(y)$. Therefore, $f(A)$ and $f(B)$ are disjoint intervals in $\mathbb{R}$, with $f(A)$ an open interval and $f(B)$ also open if B is assumed open. The disjoint nature of $f(A)$ and $f(B)$ implies that for some $a \in \mathbb{R}$, $A \subset \{x : f(x) < a\}$ and $B \subset \{x : f(x) \geq a\}$, with a strict inequality regarding if B is assumed to be open. $\square$

Here comes a significant consequence of Theorem 4.1.3 above.

Corollary 4.1.1. *Consider nonempty convex subsets K and C of a LCS such that K is compact and C is closed, with $K \cap C = \emptyset$. Then K and C can be strictly separated.*

Proof. Let us go back to Theorem 1.3.4, an early example of how topology and linear algebra mingle. In that result, Claim 1.3.1 was made (and verified) to the effect that for a compact subset K and a closed subset C of a TVS for which $K \cap C = \emptyset$, it is possible to find $U \in \mathcal{N}_0$ such that $(K + U) \cap (C + U) = \emptyset$. In this context, we assume that the set U is open and convex. Letting $A = K + U$ and $B = C + U$ and recalling that both sets are open, we apply Theorem 4.1.3 to reach the conclusion. $\square$

Next is another significant consequence of Theorem 4.1.3.

Corollary 4.1.2. *The continuous dual of any LCS separates points of the LCS; that is, if E is a LCS, then E' separates points of E.*

Proof. Exercise 4.1.6. $\qquad\qquad\square$

The next goal is to prove what is referred to as the analytic version of the Helly–Hahn–Banach theorem. The focus is on extending linear functionals. The proof uses the concept of a sublinear functional, which is defined below.

Definition 4.1.6. A **sublinear functional** on a vector space E is a function $p : E \to \mathbb{R}$ such that for all $x, y \in E$ and all $t \geq 0$:
(a) $p(x + y) \leq p(x) + p(y)$.
(b) $p(tx) \leq tp(x)$.

Before we state and prove the next gem of a theorem, a definition of relevant buzzwords is in order.

Definition 4.1.7. Suppose a function f is defined on a set A. An **extension** of f to a superset B of A is a function $\tilde{f}$ that is defined on B such that $\tilde{f}|_A = f$. If the functions f and $\tilde{f}$ are linear, then $\tilde{f}$ is a **linear extension** of f.

Definition 4.1.8. The expression $f(x) \leq p(x)$ for all $x \in M$ is described by saying that f is **majorized by** the sublinear functional p.

Theorem 4.1.4 (Helly–Hahn–Banach theorem—analytic version). *If on a TVS over $\mathbb{R}$, a linear functional is majorized on a subspace by a sublinear functional, then there is a linear extension of that functional to the entire TVS that is still majorized by the same sublinear functional. That is, let E be a TVS over the field $\mathbb{R}$. Suppose M is a subspace of E and that $p : E \to \mathbb{R}$ is a sublinear functional. If $f : M \to \mathbb{R}$ is linear and satisfies $f(x) \leq p(x)$ for all $x \in M$, then there exists a linear functional $\tilde{f}$, defined on all of E such that the following hold:*
(a) *$\tilde{f}(x) = f(x)$ for all $x \in M$.*
(b) *$\tilde{f}(x) \leq p(x)$ on all of E.*
(c) *For every $x \in E$, $-p(-x) \leq \tilde{f}(x)$.*

Proof. Assume $M \neq E$. Choose $x_1 \notin M$ and define $M_1 = \{x + tx_1 : x \in M, t \in \mathbb{R}\}$. Obviously, M_1 is an affine subspace of E. By the assumption that $f(x) \leq p(x)$ for all $x \in M$, we observe that for all $x, y \in M$,

$$f(x) + f(y) = f(x + y) \leq p(x + y) = p(x - x_1 + x_1 + y) \leq p(x - x_1) + p(x_1 + y), \qquad (4.2)$$

which leads to $f(x) - p(x - x_1) \leq p(x_1 + y) - f(y)$. If $s = \sup\{f(x) - p(x - x_1) : x \in M\}$, in Exercise 4.1.5(a) you can verify that $f(x) - s \leq p(x - x_1)$ and $f(y) + s \leq p(y + x_1)$. Now define f_1 on M_1 by: for each $x + x_1 \in M_1, f_1(x + x_1) = f(x) + st$, where $x \in M, t \in \mathbb{R}$. Certainly, f_1 is a linear functional on M_1 and letting $t = 0$ reveals that $f_1 = f$ on M. In Exercise 4.1.5(b), you can verify that for all $z \in M_1, f_1(z) \le p(z)$. Meanwhile, we now welcome Zorn's Lemma A.1.1 to this proof.

Let $\mathcal{P}$ represent all ordered pairs of the form $(M^\sharp, f^\sharp)$ such that $M^\sharp$ is a subspace of E that contains M and $f^\sharp$ is a linear extension of f from M for which $f^\sharp \le p$ on $M^\sharp$. We equip $\mathcal{P}$ with a partial order $\preceq$ as follows: Given ordered pairs $(M^\sharp, f^\sharp), (L^\sharp, g^\sharp) \in \mathcal{P}$, define

$$(M^\sharp, f^\sharp) \preceq (L^\sharp, g^\sharp) \iff M^\sharp \subset L^\sharp, \quad \text{and} \quad f^\sharp = g^\sharp \text{ on } M^\sharp. \tag{4.3}$$

Let P be any totally ordered subset of $\mathcal{P}$. An upper bound of P can be obtained by taking the union of an increasing collection of subspaces and applying the conditions of membership in $\mathcal{P}$. Now apply Zorn's Lemma A.1.1 to conclude that there exists a maximal element $\tilde{M}$ on $\mathcal{P}$. By its construction, $\tilde{M}$ is a subspace of E. We define on $\tilde{M}$ a functional $\tilde{f}$ in the following way. First, note that, given any $x \in \tilde{M}$, there exists $(M^\sharp, f^\sharp) \in \mathcal{P}$ such that $x \in M^\sharp$. Define $\tilde{f}(x) = f^\sharp(x)$. Before continuing, we need to check to see if $\tilde{f}$ is well-defined. Suppose $x \in M^\sharp$ and $x \in L^\sharp$. By the total ordering of $\mathcal{P}$, $x \in N^\sharp$ where $M^\sharp \cup L^\sharp \subset N^\sharp$. In addition, $\tilde{f}$ is defined on $N^\sharp$.

Part (a) now follows directly from the definition of $f^\sharp$. There are but three items left in order to finish the proof. The first item is to show that $\tilde{M} = E$. If this is not the case, then there is another subspace that strictly contains $\tilde{M}$, contradicting the maximality of $\tilde{M}$. The other two items are to verify the calculations (b) and (c) of the statement of Theorem 4.1.4. Those two verifications can be done as Exercise 4.1.5(c). We can now relax, knowing that the claims of this theorem have been proven (some of them by you). $\quad\square$

As with the geometric version of the Helly–Hahn–Banach theorem, it will probably come as no surprise that the analytic version of the Helly–Hahn–Banach theorem leads to some significant consequences. The first corollary is quite significant because of the presence of the mathematically significant word *continuous*.

Corollary 4.1.3. *In a LCS E, if f is a linear functional that is continuous on a subspace M, then there exists a continuous linear extension $\tilde{f}$ of f to all of E.*

Proof. If f is continuous on M, then $V = f^{-1}([0,1])$ is a zero neighborhood in M. There exists an absolutely convex $U \in \mathcal{N}_{0,E}$ such that $U \cap M \subset V$. If $p = \mu_U$ (the Minkowski functional of U), then p is a continuous seminorm on E by Proposition 2.1.8 and, moreover, $|f| \le p$ on M because $U \cap M \subset V$. By Theorem 4.1.4, there is an extension $\tilde{f}$ of f for which $|\tilde{f}| \le p$ on all of E. We finish by noting that $\tilde{f}$ is continuous because p is continuous. $\quad\square$

Corollary 4.1.4. *In any nontrivial LCS E, the continuous dual is nontrivial; that is, $E' \ne \{0\}$.*

Proof. Let x_0 be a nonzero element of E and define $M = \text{span}\{x_0\}$. Naturally, M is a nontrivial subspace of E. Define $f : M \to \mathbb{R}$ by: $f(ax_0) = a$ for all $a \in \mathbb{R}$. Evidently, f is a nonzero linear functional on M. Even more so, f is continuous as a linear functional on a finite-dimensional TVS (via Theorem 3.2.1). By Corollary 4.1.3 above, there exists a continuous linear extension $\tilde{f}$ of f to all of E. This means the continuous dual E' contains at least one nonzero element. $\qquad\square$

So, the continuous dual E' of any LCS is nontrivial. On the other hand, it would be even better if E' separates points, per Definition 4.1.2. In fact, it does. Part (b) of the next corollary establishes that fact.

Corollary 4.1.5. *For any nontrivial Hausdorff LCS E, the following hold:*
(a) *Given a point $x_0 \in E$ and a seminorm p on E, there exists a linear functional f on E such that for every $x \in E$, $|f(x)| \leq p(x)$, and $f(x_0) = p(x_0)$.*
(b) *The continuous dual of E separates points; that is, given any $x_0 \neq 0$ in E, there exists $f \in E'$ such that $f(x_0) \neq 0$ in $\mathbb{K}$.*

Proof. (a) Let $M = \text{span}\{x_0\}$ and define f on M by letting $f(ax_0) = ap(x_0)$ for every $a \in \mathbb{K}$, where $ax_0 \in M$. Apply Theorem 4.1.4 and extend f from M to $\tilde{f}$ on all of E.

(b) Recall that the topology of E can be generated by a collection of continuous seminorms by way of Theorem 2.1.4. The assumption that E is Hausdorff allows us to apply Proposition 2.1.12 and seek out a continuous seminorm on E to use in part (a) for which $p(x_0) \neq 0$. Take f as in part (a) such that $f(x_0) = p(x_0)$. Its extension is in E'. We have discovered a continuous linear functional f on E such that $f(x_0) \neq 0$, as desired. $\qquad\square$

The Helly–Hahn–Banach theorem can be proved in a version in which sublinear functionals are replaced by seminorms, as stated below.

Theorem 4.1.5 (Helly–Hahn–Banach theorem—seminorm version). *Assume p is a seminorm on a TVS E over the field $\mathbb{R}$. Suppose M is a subspace of E and f is a linear functional on M that satisfies the condition $|f(x)| \leq p(x)$ for each $x \in M$. There exists a linear extension $\tilde{f}$ of f such that*
(a) $\tilde{f}(x) = f(x)$ *for all $x \in M$.*
(b) $|\tilde{f}(x)| \leq p(x)$ *on all of E.*

Proof. Exercise 4.1.18. $\qquad\square$

The third and last major issue in this section with which to contend is that of the case where $\mathbb{K} = \mathbb{C}$, the obstacle being that $\mathbb{C}$ is not ordered. To deal with this involves some definitions and observations. You may wish to consult an introductory source on complex variables if necessary.

Definition 4.1.9. On a vector space E over the field $\mathbb{C}$:
(a) A functional f is **real linear** if for every $x \in E$ and every $a \in \mathbb{R}$, $f(ax) = af(x)$.
(b) A functional f is **complex linear** if for every $x \in E$ and every $a \in \mathbb{C}$, $f(ax) = af(x)$.

For the next item, recall that for a complex number $z = a + bi$, where $a, b \in \mathbb{R}$, the real part of z is $\mathrm{Re}(z) = a$. An elementary fact about complex numbers is that for any $z \in \mathbb{C}$, $z = \mathrm{Re}(z) - i\,\mathrm{Re}(iz)$ (just observe that $\mathrm{Re}(iz) = -b$). With this, we have a few calculations of complex numbers that are needed for proving the Helly–Hahn–Banach theorem when $\mathbb{K} = \mathbb{C}$. In the lemma below, we effectively get an equivalent expression for a linear functional that is defined over the field $\mathbb{C}$.

Lemma 4.1.2. *Assume E is a vector space over $\mathbb{C}$ and $f : E \to \mathbb{C}$ is linear. Then the following hold:*
(a) *If u is the function defined by $u = \mathrm{Re}(f)$, then $f(x) = u(x) - iu(ix)$ for each $x \in E$.*
(b) *If $u : E|_{\mathbb{R}} \to \mathbb{R}$ is linear, then the function $f : E \to \mathbb{C}$ defined by $f(x) = u(x) - iu(ix)$ for each $x \in E$ is linear over $\mathbb{C}$.*

Proof. What follows amounts to the general ideas of the claims, with the remaining minor details left as Exercise 4.1.7. (a) Let $z = a + bi \in \mathbb{C}$ be arbitrary. From the comments above, $z = \mathrm{Re}(z) - i\,\mathrm{Re}(iz)$. With the assumption that f is linear, we have that for every $x \in E$, $if(x) = f(ix)$, so $\mathrm{Re}(if(x)) = \mathrm{Re}(f(ix)) = u(ix)$. In other words, $f(x) = u(x) - iu(-x)$.

(b) Notice that if u is linear over $\mathbb{R}$, then f is also linear over $\mathbb{R}$. This means $f(ix) = u(ix) - iu(-x) = if(x)$. Thus, $f(ix) = if(x)$ allows us to conclude that f is linear over $\mathbb{C}$, too. $\qquad\square$

Theorem 4.1.6. *The Helly–Hahn–Banach Theorem 4.1.4 is valid over the field $\mathbb{C}$.*

Proof. Consider the statement of the Helly–Hahn–Banach theorem as written in Theorem 4.1.4. Let $u = \mathrm{Re}(f)$. Then $|\mathrm{Re}(f)| \le |f|$, so $|u(x)| \le p(x)$, for all $x \in M$. By Theorem 4.1.4, there exists a linear $g : E|_{\mathbb{R}} \to \mathbb{R}$ such that for all $x \in M$, $g(x) = u(x)$, and $|g(x)| \le p(x)$ for all $x \in E$. Put $\tilde{f} = g - ig$. Then $\tilde{f}$ is linear and $\tilde{f}|_M = f$. To show that $|\tilde{f}| \le p$, let $z \in \mathbb{C}$ be arbitrary. There exists a scalar $a \in \mathbb{C}$ such that $|z| = az$, with $|a| = 1$ (as a complex absolute value), as you can verify as Exercise 4.1.8. Hence, we obtain $|\tilde{f}(x)| = a\tilde{f}(x) = \tilde{f}(ax)$, by linearity. Furthermore, as $|\tilde{f}(x)| \in \mathbb{R}$, we have $|\tilde{f}(x)| = g(x) \le p(ax) = |a|p(x) = p(x)$, and the proof is completed. $\qquad\square$

To finish this section, let us contemplate the Helly–Hahn–Banach theorem and its various versions, in terms of what it means to be a profoundly deep result. First, does the theorem rely on some deep result? Yes, it makes use of Zorn's Lemma A.1.1. Second, does the theorem lead to a pathway to some significant development? Yes indeed, namely, for example, its application to LCS showing that they all have nontrivial, point separating, continuous duals leads to the development of duality theory. We have only seen the very beginnings of duality theory, however, it is profound enough that it will be part of almost every topic we discuss going forward.

Exercises

4.1.1. Verify the details of Equation (4.1).

4.1.2. Prove that the function f of Theorem 4.1.1 is a hyperplane.

4.1.3. Explain why the collection $\mathcal{E}$ of the proof of Theorem 4.1.2 satisfies the hypotheses of Zorn's Lemma A.1.1.

4.1.4. Prove that the hyperplane of the proof of the geometric form of the Helly–Hahn–Banach Theorem 4.1.2, is in fact, closed.

4.1.5. From the proof of the analytic version of the Helly–Hahn–Banach Theorem 4.1.4, verify the following:
   (a) $f(x) - s \le p(x - x_1)$ and $f(y) + s \le p(y + x_1)$.
   (b) $(\forall z \in M_1) f_1(z) \le p(z)$.
   (c) Verify claims (b) and (c), namely (b): $\tilde{f}(x) \le p(x)$ on all of E, and (c): For every $x \in E$, $-p(-x) \le \tilde{f}(x)$.

4.1.6. Prove 4.1.2. *Hint:* You can make short order of this proof by way of Corollary 4.1.1.

4.1.7. Fill in the missing details of the proof of Lemma 4.1.2.

4.1.8. From a line in the proof of Theorem 4.1.6, verify that if $z \in \mathbb{C}$ is arbitrary then there exists a scalar $a \in \mathbb{C}$ such that $|z| = az$, with $|a| = 1$. *Hint:* Consider the polar form $z = re^{i\theta}$.

4.1.9. Prove that in a Hausdorff LCS a closed, convex set is equal to the intersection of all closed half spaces that contain it.

4.1.10. Suppose H is a hyperplane in E.
   (a) Prove that $(\forall y \in E)\, y = x + ak$, for some $x \in H$, $a \in \mathbb{K}$, $k \notin H$.
   (b) Use (a) to prove that $(\exists f \in E^* \setminus \{0\})\, H = f^{-1}(\{0\})$.

4.1.11. Given the inequality $f(x) - p(x - x_1) \le f(y + x_1) - f(y)$, and $a = \sup\{f(x) - p(x - x_1) : x \in M\}$ prove that $(\forall x, y \in M)$, we have the following:
   (a) $f(x) - a \le p(x - x_1)$.
   (b) $f(y) + a \le p(y + x_1)$.

4.1.12. Consider the definition of p as in the proof of Theorem 4.1.4, that is, $p(x + y) \le p(x) + p(y)$, and $p(tx) = tp(x)$, $(\forall t \ge 0)$. Prove that p is a linear function if and only if $(\forall x, y \in E)\, p(x) + p(-x) = 0$.

4.1.13. Let E be a Hausdorff LCS over $\mathbb{R}$. Suppose M is a linear subspace of E. Prove that the following are equivalent:
   (a) M is dense in E.
   (b) $(\forall f \in E')$, if $f(x) = 0$ for every $x \in M$, then $f = 0$ on all of E.

4.1.14. Consider the space of square Lebesgue integrable functions $L_2 = L_2([-1, 1])$, with respect to the norm $\|f\|_2 = [\int_{-1}^{1} |f|^2 d\lambda]^{\frac{1}{2}}$, where λ is Lebesgue measure. Define, for each $a \in \mathbb{K}$, $E_a = \{f \in C([-1, 1]) : f(0) = a\}$.
   (a) Prove that E_a is convex for any a.
   (b) Is E_a balanced?
   (c) Prove that for any a, E_a is dense in L_2, and deduce that no two distinct E_a, E_b can be separated by a continuous linear functional in L_2. *Hint:* The set of continuous functions with compact support is dense in L_2.

4.1.15. Let L be a closed subspace of a LCS E. A subset A of L is **total** if L is the smallest closed subspace that contains A. Use the Helly–Hahn–Banach theorem to prove that A is total in L if and only if whenever $f \in E'$ is such that $f(x) = 0$ on A, then $f(x) = 0$ on all of L.

4.1.16. Let A and B be disjoint, open convex subsets of a LCS E. Let

$$\mathcal{F} = \big\{ \text{subspaces } L \text{ of } E : L \cap (A - B) = \emptyset \big\}. \tag{4.4}$$

It can be shown that $\mathcal{F}$ has a maximal element H. Prove that H is a hyperplane.

4.1.17. In Theorem 4.1.3, describe examples requested below to show that the hypotheses are necessary. For brevity, let $\Omega = A - B$ be a general open, convex set.

(a)   The subspace M must be nonempty.

(b)   The set Ω must be nonempty.

(c)   The set Ω must be open. *Hint:* Consider the xy plane and let Ω be the open upper half plane union with the origin; that is, $\Omega = \{(x, y) : y > 0\} \cup \{(0, 0)\}$. Let M be the affine subspace consisting of the point $\{(1, 0)\}$.

(d)   The set Ω must be convex. *Hint:* Consider the xy plane and the washer shape defined by $\Omega = \{(r\cos(\theta), r\sin(\theta)) : 0 \le \theta \le 2\pi, 1/2 < r < 1\}$, and $M = \{(0, 0)\}$.

(e)   If B is not open in the expression $A - B$, then the separation need not be strict. *Hint:* Consider the xy plane with A being the open upper half plane and B being the lower closed half plane. Let H be the x axis.

4.1.18.   Prove Theorem 4.1.5. *Hint:* The steps are similar to those of the analytic version. See [67, pp. 179–180] for more hints.

4.2 Pairings and polars

Unless otherwise noted, we assume that all spaces are LCS because we want to make sure that the dual of a space is sufficiently useful.

In the last section, we began to see that a LCS E and its dual E' seem to be connected. Such a connection is only the beginning of the story. In the rest of this chapter, and indeed in most of the rest of this book, we will see that a LCS and its dual are in fact intimately related in the sense that information about one of these often tells us important and sometimes astounding information about the other, or about the relationship between the two. This section focuses on some formalities of the connections between a LCS and its dual. First, there needs to be something that connects a space with its dual. That something is defined next.

4.2.1 Pairings

Definition 4.2.1. For two vector spaces F and G over the same field, we have the following:

(a) A **bilinear form** between the vector spaces is a function from the product of the vector spaces to the field that is linear in both variables.

(b) For two vector spaces over the field $\mathbb{K}$ and a bilinear form B defined on $F \times G$, the expression $(F, G; B)$ is called a **pairing**.

(c) A bilinear form B as defined above **separates points** of F if for any nonzero $x \in F$, there exists $y \in G$ such that $B(x, y) \neq 0$. Similarly, B separates points of G if for any nonzero $y \in G$, there exists $x \in F$ such that $B(x, y) \neq 0$.

(d) If B separates points of both F and G, then the pairing is called **separated**. Other names for this include that (F, G) is a **dual pair** or a **dual system** with respect to B.

Example 4.2.1. Given a vector space E over $\mathbb{K}$ and its algebraic dual E^*, consider any $f \in E^*$. Define $B : E \times E^* \to \mathbb{K}$ by $B(x, f) = f(x)$ for all $x \in E$.

This function is clearly bilinear. Moreover, the typical notation for this function is $\langle x, f \rangle$, and has a specific name.

Definition 4.2.2. The function $\langle \cdot, \cdot \rangle : F \times G \to \mathbb{K}$ given by: For a fixed $x \in F$ and for $f \in G$ $\langle x, f \rangle = f(x)$ is called the **evaluation map**.

The evaluation map is sometimes called the **canonical bilinear form**.

Example 4.2.2. A nontrivial vector space E and its algebraic dual E^* form a dual pair with respect to the evaluation map.

Indeed, using the evaluation map of Definition 4.2.2 and Exercise 4.2.1, we deduce that $\langle \cdot, \cdot \rangle$ separates points of E. The definition of the zero function tells us that $\langle \cdot, \cdot \rangle$ separates points of E^*. That is, (E, E^*) is always a dual pair.

The next example represents one of the most important examples of this book.

Example 4.2.3. A Hausdorff LCS E and its continuous dual E' form a dual pair with respect to the evaluation map.

Using the evaluation map of Definition 4.2.2 and Corollary 4.1.5, we deduce that $\langle \cdot, \cdot \rangle$, restricted to E', separates points of E and E'. It is the fact that E is a LCS that allows us to always be able to form the dual pair (E, E').

Example 4.2.4. Suppose $(E, \| \cdot \|)$ is a normed space and E' is the dual space equipped with the topology of the operator norm. Then (E, E') forms a dual pair, as is well known in the world of normed spaces (see [89], for example). Of course, normed spaces are all LCS.

The symmetric nature of defining a pairing indicates that for vector spaces paired as (F, G) we can interchange F and G and consider the pairing (G, F). The evaluation map also works two ways. It can be described by $\langle x, f \rangle = f(x)$ or $\langle f, x \rangle = f(x)$. In other words, from the dual pair (F, G), a duality can also be formed as (G, F), where we consider x to be a linear functional on G. Another way to express this idea is that F can be identified with a subspace of $(G)^*$ via $\langle x, f \rangle = f(x)$ and G can be identified with a subspace of F^* via $\langle f, x \rangle = f(x)$. On the other hand, it is worth noting that this symmetry does not preserve all properties. For example, if E is any incomplete normed space, its strong dual is always complete (see [89, Theorem 1.10.7, p. 87]).

If we find a vector space among the trees of a park, our typical response is to want to equip that vector space with a locally convex topology, with some expectation that the chosen topology would result in a LCS of some interest. Specifically, the idea here is to equip either F or G with a locally convex topology that involves some connection between F and G. The activity then amounts to studying properties that do or do not hold under such topologies. The first such topology comes about by defining a collection of seminorms and then using the initial topology that depends on those seminorms. The details are next.

Consider a dual pair (F, G) with a bilinear form B. Then we have the following.

Proposition 4.2.1. *For $y \in G$, the function $p_y : F \to \mathbb{K}$ given by $p_y(x) = |B(x, y)|$ for all $x \in F$, is a seminorm on F.*

Proof. Exercise 4.2.2. $\qquad\qquad\square$

With this construction, the following topology is defined.

Definition 4.2.3. For a dual pair (F, G), define the topology $\sigma(F, G)$ on F to be the coarsest locally convex topology that satisfies: For each $y \in G$, the seminorm p_y defined on F by letting $p_y(x) = |B(x, y)|$ for every $x \in F$, is continuous. The topology $\sigma(F, G)$ is called the **weak topology** with respect to the duality (F, G). Similarly, we may define, for each $x \in F$, a seminorm q_x on G given by: $q_x(y) = |B(x, y)|$ for all $y \in G$, and the corresponding coarsest locally convex topology on G such that each q_x is continuous is called the **weak∗ topology** (pronounced "**weak star**").

Observe that if $\mathcal{B}$ is a Hamel basis for the vector space G, then we may express $y \in G$ as $y = \sum_{j=1}^{n} a_j e_j$, for some $n \in \mathbb{N}$, where $e_j \in \mathcal{B}, j \in \underline{n}$. We may subsequently calculate p by using $e \in \mathcal{B}$ so that $p_e(x) = |B(x, e)|$. Furthermore, Theorem 2.1.4 applies to conclude that the topology of the definition above is in fact a locally convex topology. Some clarifying vocabulary remarks are relevant here, too.

Remark 4.2.1. The majority of publications that include some aspect of duality for LCS use the phrase *weak topology* as indicated in Definition 4.2.3 to denote the *coarseness* of the topology. However, in general topological contexts, the term *weak* or *weaker* does not always imply *coarse* or *coarser*.

Next, in what follows, we verify that if F is equipped with the weak topology then the dual of F with respect to that weak topology is the same as the dual with respect to the original topology of F. A preliminary lemma comes first.

Lemma 4.2.1. *Consider a vector space E and let $u_1, u_2, \ldots, u_n \in E^*$. If $g \in E^*$ is such that $g^{-1}(\{0\}) \supset \bigcap_{j=1}^{n} u_j^{-1}(\{0\})$, then g is a linear combination of $u_1, u_2, \ldots, u_n$.*

Proof. Without loss of generality, we may assume that all the functions u_j are nonzero, for $j \in \underline{n}$. Notice that as a nullspace of a linear functional each $u_j^{-1}(\{0\})$ is a hyperplane. We proceed by induction. For $n = 1$, if $g = 0$, then $g = 0 \cdot u_1$, and if $g \neq 0$, then for some $a_1 \in \mathbb{K}, a_1 \neq 0, g = a_1 u_1$. Now assume the conclusion of the lemma holds for $n - 1$; that is, there are $n - 1$ scalars a_j such that $g(x) - \sum_{j=1}^{n-1} a_j u_j = 0$, for all $x \in u_n^{-1}(\{0\})$. Applying the case of $n = 1$, we deduce that there is a scalar a_n such that $g - \sum_{j=1}^{n-1} a_j u_j = a_n u_n$, and the proof is completed. $\qquad\square$

Proposition 4.2.2. *If (F, G) is a dual pair with respect to the bilinear form B, then $(F, \sigma(F, G))' = G$.*

Proof. Let f be a linear functional on F such that f is $\sigma(F, G)$-continuous. By properties of seminorms, we have the following construction. There exist $y_1, y_2, \ldots, y_n \in G$ such that $|f(x)| \leq \max\{|B(x, y_j)| : j \in \underline{n}\}$, and this holds for every $x \in F$. Note that if $\langle x, y_j \rangle = 0$ for every $j \in \underline{n}$, then $f(x) = 0$ for every $x \in F$. The lemma above applies and we conclude that there exist $a_1, a_2, a_3, \ldots, a_n \in \mathbb{K}$ such that $f = \sum_{j=1}^{n} a_j \tilde{y}_j$, where $\tilde{y}_j$ is a linear functional on F for each j, because F^* can be identified with a subspace of G. Put $y = \sum_{j=1}^{n} a_j \tilde{y}_j$. Hence, $f(x) = B(x, y)$ for all $x \in F$. This implies that f is continuous with respect to $\sigma(F, G)$. We have shown that for every $f \in F'$ there is a $y \in G$ such that $f(x) = B(x,y)$. In other words, $(F, \sigma(F, G))' = G$. $\qquad\square$

There is a direct connection between when a pairing separates points and the weak topology, as shown next.

Proposition 4.2.3. *A pair (F, G) separates points if and only if the corresponding weak topology is Hausdorff.*

Proof. ($\Rightarrow$): Assume (F, G) is a dual pair with respect to the bilinear form B. By way of seminorms, if the dual pair separates points of F, then for any $x \neq 0$ in F, there is a $y \in G$ such that $p_y(x) = |B(x,y)| \neq 0$. This statement enjoins us to apply Proposition 2.1.12, which tells us that $\sigma(F, G)$ is Hausdorff.

($\Leftarrow$): Suppose $\sigma(F, G)$ is Hausdorff and let x be any nonzero element of F. Applying Proposition 2.1.12 again, we find that there is a seminorm p on F such that $p(x) \neq 0$. Because the topology $\sigma(F, G)$ is generated by the bilinear form B, we deduce that there is a $y \in G$ such that $p(x) = B(x,y) \neq 0$. Thus, $\sigma(F, G)$ separates points. $\qquad\square$

In the case of LCS, Proposition 4.2.2 tells us that there is a coarsest locally convex topology on a LCS E that leaves the continuous dual E' of E unchanged. The natural question that comes up is whether there are other topologies that leave the dual space unchanged. In particular, we can wonder if there is a finest locally convex topology for which E' remains the same. There are such topologies, and we will be able to determine them by way of the next concept.

4.2.2 Polars and polar topologies

The following definition is fundamental in the realm of duality.

Definition 4.2.4. Suppose (F, G) is a dual pair of LCS with regard to the bilinear form $B(\cdot, \cdot)$. The **(absolute) polar** of a subset A of F is

$$A^\circ = \{y \in G : |B(x,y)| \leq 1, \text{ for all } x \in A\}. \tag{4.5}$$

Notation 4.2.1. Observe the difference in notation between the interior $\mathring{A}$ of a set and the polar A° of a set.

Next, we derive some basic properties of polars.

Proposition 4.2.4. *For (F, G) a dual pair of LCS with regard to the bilinear form B, the following properties hold: Given $A, B \subset F$,*
(a) *If $A \subset B$, then $A^\circ \supset B^\circ$.*
(b) *$A^\circ = [\mathrm{bal}(A)]^\circ$.*
(c) *$A \subset (A^\circ)^\circ = A^{\circ\circ}$.*

(d) $A^\circ = A^{\circ\circ\circ}$.

(e) A° *is absolutely convex and* $\sigma(G, F)$-*closed.*

(f) *For any scalar* $a \neq 0$, $(aA)^\circ = \frac{1}{a}A^\circ$.

(g) *If* $\{A_\alpha : \alpha \in I\}$ *is a collection of subsets of* F, *then*

$$\left(\bigcup_{\alpha \in I} A_\alpha \right)^\circ = \bigcap_{\alpha \in I} A_\alpha^\circ. \tag{4.6}$$

Proof. Parts (a) and (b) follow directly from the definition of polars.

(c) If $x \in A$, then $|B(x, y)| \leq 1$ for each $y \in A^\circ$, which implies that $x \in A^{\circ\circ}$.

(d) Exercise 4.2.4.

(e) First let $x, z \in A^\circ$ and $t \in [0, 1]$ be arbitrary. For $y \in G$ being linear, we calculate $|B(tx + (1 - t)z, y)| = |tB(x, y) + (1 - t)B(z, y)| \leq t|B(x, y)| + (1 - t)|B(z, y)| \leq t + (1 - t) \leq 1$. We already knew that A° is balanced; hence, A° is absolutely convex. To prove that A° is $\sigma(G, F)$-closed, observe that by Definition 4.2.3 the linear map y_x given by $y_x = B(x, y)$ for all $x \in F$ is continuous. If $A_x = \{y \in G : |B(x, y)| \leq 1\} = y_x^{-1}([-1, 1])$ for $x \in A$, then A_x is closed in G, as a continuous inverse image of a closed set. Then $A^\circ = \bigcap_{x \in A} A_x$ is $\sigma(G, F)$-closed.

(f), (g) Exercise 4.2.4. $\qquad\square$

The next result represents the first indication of the importance of bounded sets in the context of polars.

Proposition 4.2.5. *For a dual pair* (F, G) *with a bilinear form* B, *if* A *is a subset of* F, *then* A° *is absorbing in* G *if and only if* A *is* $\sigma(F, G)$-*bounded.*

Proof. ($\Rightarrow$): Suppose A° is absorbing in G, and let $y \in G$ be arbitrary. Then there exists a scalar $a > 0$ such that $ay \in A^\circ$; that is, for all $x \in A$, $|B(x, ay)| = |aB(x, y)| = |B(ax, y)| \leq 1$. Applying Proposition 4.2.4(f), we have $y \in (aA)^\circ = \frac{1}{a}A^\circ$. Recall from Definition 4.2.3 that the weak topology on the pair (F, G) is generated by the seminorms p_y, defined as $p_y(x) = |B(x, y)|$ for each $x \in F$. In particular, we have that for all $x \in A$, $p_y(x) = |B(x, y)| \leq \frac{1}{a}$. As $y \in G$ was arbitrary, we apply Proposition 2.2.6 and conclude that A is $\sigma(F, G)$-bounded.

($\Leftarrow$): Exercise 4.2.6. $\qquad\square$

Speaking of polars, here comes a deep theorem. It is deep because the proof relies on the Helly–Hahn–Banach Theorem 4.1.2.

Theorem 4.2.1 (Bipolar theorem). *If* A *is a nonempty subset of* F *of the dual pair* (F, G), *then*

$$A^{\circ\circ} = \overline{\operatorname{conv} \operatorname{bal}(A)}^{\sigma(F, G)}. \tag{4.7}$$

Note that the closure is taken with respect to the $\sigma(F, G)$ topology (see the definition of this notation in A.3.1 of the Appendix).

Proof. We know already that $A^{\circ\circ}$ is absolutely convex and $\sigma(F, G)$-closed. What we need to show is that if D is any absolutely convex $\sigma(F, G)$-closed set that contains A, then $A^{\circ\circ} \subset D$. The pathway to this conclusion goes through the contrapositive. That is, given such a set D, suppose there is $x \in A^{\circ\circ}$ such that $x \notin D$. The set D is convex, so we may separate D from $\{x\}$ by way of the Helly–Hahn–Banach theorem as expressed in Theorem 4.1.3. We first consider the case of F_0 of F over the reals $\mathbb{R}$, and find a hyperplane $f \in F_0$ for which there is a scalar $c > 0$ such that $f(w) < c$ for all $w \in D$ while $f(x) > c$. Because $0 \in D$ by absolute convexity, we may use $c = 1$; that is, $f(w) < 1$ for all $w \in D$ and $f(x) > 1$. We apply the duality between F and G to find $y \in G$ that corresponds to f. The set D is balanced, so it follows that for each $w \in D$, $|f(w)| = |B(w, f)| \le 1$. Naturally, this means $y \in D^{\circ}$. By Proposition 4.2.4(b), we conclude that $D^{\circ} = A^{\circ}$. In other words, $y \in A^{\circ}$. On the other hand for the element $x \in A^{\circ\circ}$, $B(x, f) = f(x) > 1$, which implies that $x \notin A^{\circ\circ}$. The complex case is left as Exercise 4.2.7. $\qquad\square$

An application of the Bipolar Theorem 4.2.1 shows that there is a difference in the order of calculating unions and intersections of polars (compare with Proposition 4.2.4(g)), as seen next.

Proposition 4.2.6. *Suppose $\{A_\alpha : \alpha \in I\}$ is a collection of absolutely convex, nonempty $\sigma(F, G)$-closed subsets of F. Then*

$$\left(\bigcap_{\alpha \in I} A_\alpha \right)^{\circ} = \overline{\operatorname{conv} \operatorname{bal} \bigcup_{\alpha \in I} A_\alpha^{\circ}}^{\,\sigma(F, G)}. \tag{4.8}$$

Proof. We calculate as follows:

$$\bigcap_{\alpha \in I} A_\alpha \overset{\text{Bipolar}}{=} \bigcap_{\alpha \in I} A_\alpha^{\circ\circ} = \left(\bigcup_{\alpha \in I} A_\alpha^{\circ} \right)^{\circ}, \tag{4.9}$$

because each A_α is absolutely convex and $\sigma(F, G)$-closed. Applying the Bipolar Theorem 4.2.1 again and in concert with Proposition 4.2.4(g), we obtain

$$\left(\bigcap_{\alpha \in I} A_\alpha \right)^{\circ} = \left(\bigcup_{\alpha \in I} A_\alpha^{\circ} \right)^{\circ\circ} \overset{\text{Bipolar}}{=} \overline{\operatorname{conv} \operatorname{bal} \bigcup_{\alpha \in I} A_\alpha^{\circ}}^{\,\sigma(F, G)}. \tag{4.10}$$

$\square$

Returning to the question of which locally convex topologies other than the weak topology might produce the same continuous dual as that of the original topology of a LCS, we optimistically give a name to such topologies.

Definition 4.2.5. Suppose (F, G) is a dual pair. A topology on F is called **consistent with the duality** of (F, G) if the continuous dual of F under that topology is G. Such a topology is also said to be a **topology of the dual pair**, or called **compatible with the duality**.

Up to now, we know that for any LCS there are at least two topologies that are consistent with the duality, namely, the original topology and the weak topology. These two topologies are generally distinct. Indeed, in Exercise 4.2.9 you can prove that if the weak and original topology of a normed space are equivalent, then the normed space is finite-dimensional. Therefore, in any infinite-dimensional normed space, the normed and weak topologies are distinct. Specifically, we can state the following.

Proposition 4.2.7. *On any infinite-dimensional normed space, the normed topology is strictly finer than the weak topology.*

Exercises

4.2.1.   Prove that if $E \neq \{0\}$, then $E^* \neq \{0\}$ and use this to deduce that the evaluation map $\langle \cdot, \cdot \rangle$ of Example 4.2.2 separates points of E. *Hint:* Extend some nonzero value of an element of a Hamel basis using the result of Exercise 1.1.15.

4.2.2.   Prove Proposition 4.2.1.

4.2.3.   Determine and explain the following polars:

  (a) E°.

  (b) $\{0\}^\circ$.

4.2.4.   Prove parts (d), (f), and (g) of Proposition 4.2.4. Note that the proof of property (d) involves the Bipolar Theorem 4.2.1.

4.2.5.   Denote by A^r the "radial" polar: $\{y \in E' : |\langle x, y \rangle| \leq r\}$, where $r > 0$. Prove that for any set A, we have $A^r = rA^\circ$.

4.2.6.   Prove the implication ($\Leftarrow$) of Proposition 4.2.5.

4.2.7.   Prove the case of F being over the complex numbers $\mathbb{C}$ of the Bipolar Theorem 4.2.1. *Hint:* Use the expression $g(x) = f(x) - i f(ix)$.

4.2.8.   Consider two vector spaces F and G over $\mathbb{K}$ in duality with respect to a bilinear form B. For a subset A of F, we can define the **real polar** A^{real} of A as $A^{\text{real}} = \{y \in G : (\forall x \in A) \; \mathrm{Re}(B(x,y)) \leq 1\}$, where $\mathrm{Re}(B(x,y))$ represents the real part of $B(x,y)$ (of course, if $\mathbb{K} = \mathbb{R}$, then we have just $B(x,y)$). Prove the following properties of A^{real}.

  (a)   If $A \subset F$, then $A^\circ \subset A^{\text{real}}$, and if A is balanced, then $A^\circ = A^{\text{real}}$.

  (b) $A^\circ = (\mathrm{bal}(A))^{\text{real}}$.

  (c)   If $A \subset B$ in F, then $A^{\text{real}} \supset B^{\text{real}}$.

  (d) $A^{\text{real}} \subset A^{\text{real real}}$.

  (e) $A^{\text{real}} = A^{\text{real real real}}$ (really!).

  (f) A^{real} is absolutely convex and $\sigma(G, F)$-closed.

  (g)   For any scalar $a \neq 0$, $(aA)^{\text{real}} = \frac{1}{a} A^{\text{real}}$.

  (h)   If $\{A_\alpha : \alpha \in I\}$ is a collection of subsets of F, then

$$\left(\bigcup_{\alpha \in I} A_\alpha \right)^{\text{real}} = \bigcap_{\alpha \in I} A_\alpha^{\text{real}}. \tag{4.11}$$

  (i)   For a subset A of F, A^{real} is absorbing in G if and only if A is $\sigma(F, G)$-bounded.

  (j)   If A is a nonempty subset of F, then

$$A^{\text{real real}} = \overline{\mathrm{conv}\big(A \cup \{0\}\big)}^{\,\sigma(F,G)}. \tag{4.12}$$

4.2.9.   Suppose E is a normed space. Prove that if $\sigma(E, E')$ coincides with the topology of the norm, then $\dim(E) < \infty$. *Hint:* Apply Lemma 4.2.1.

4.3 Creating topologies on the dual space

In the last section, we began to see a connection between polars and locally convex topologies, namely that the weak topology $\sigma(F, G)$ is related to polars and to bounded sets. In this section, we will delve deeper into this connection, discovering that polars of a variety of collections of bounded sets yield a variety of topologies on a LCS E and its dual E'. The definition that will appear shortly sets up this interesting line of ideas. As a prelude, some explanation is needed in order to be sure that a collection of such sets can generate a topology.

Suppose $\mathcal{S}$ is a collection of $\sigma(F, G)$-bounded subsets of F. Proposition 4.2.5 tells us that for every $A \in \mathcal{S}$, A° is absolutely convex and absorbing (as well as $\sigma(F, G)$-closed). Theorem 2.1.2 indicates that a locally convex topology can be created by way of collections of such sets. The definition below formalizes this observation.

Definition 4.3.1. Let $\mathcal{S}$ be a collection of $\sigma(F, G)$-bounded subsets of F of a dual pair (F, G). The $\mathcal{S}$ **topology** on G is the locally convex topology generated by the collection of polars of the sets of $\mathcal{S}$. When confusion is unlikely, an $\mathcal{S}$ topology is referred to as a **polar topology**.

There is more—we also obtain collections of seminorms for generating such topologies. Indeed, given any $A \subset F$ with $A \in \mathcal{S}$, define for each $y \in G$, $p_A(y) = \sup\{|B(x, y)| : x \in A\}$. In Exercise 4.3.1, you can verify that this statement defines a seminorm on G. Furthermore, we have the following.

Proposition 4.3.1. *The following hold for a dual pair (F, G), where F, G are LCS.*
(a) *A subset A of F is bounded with respect to $\sigma(F, G)$ if and only if for every $f \in G$, $\{|f(x)| : x \in A\}$ is a bounded subset of $\mathbb{K}$.*
(b) *A sequence (x_n) in F converges to 0 with respect to $\sigma(F, G)$ if and only if for every $f \in G$ the sequence $(f(x_n))$ converges to 0 in $\mathbb{K}$.*

Proof. Exercise 4.3.3. $\qquad\qquad\qquad\qquad\qquad\qquad\qquad\qquad\qquad\qquad\qquad\qquad\square$

Remark 4.3.1. When speaking/writing about properties related to weak topologies, there is a tendency to use adverbs. For the weak topology on a LCS E, namely $\sigma(E, E')$, the adverb is **weakly** so, for example, in Proposition 4.3.1(a) above, we say that the set is **weakly bounded**, and in part (b), we say the sequence is **weakly convergent**. There is also the phrase **weakly continuous**. Likewise, for the weak topology on the dual, namely $\sigma(E', E)$, we use the adverb **weak∗** (pronounced "weak star"), where the use of ∗ harkens back to the notation for normed spaces. Thus, terms like **weak∗ bounded**, **weak∗ convergent**, and **weak∗ continuous**. When the context is clear, one can simply say "weakly."

An alternate choice of the phrase "weak∗" is "weakly∗," which comes from Megginson [89, footnote 5, p. 223] and is attributed to Mahlon Day.

Based on the last proposition, we define the following.

Definition 4.3.2. A set is **weakly bounded** if it is bounded with respect to the weak topology. A sequence is **weakly convergent** if it converges with respect to the weak topology.

An unexpectedly surprising fact is that, given a collection S of $\sigma(F, G)$-bounded subsets of F of a dual pair (F, G), the collection S can be replaced by several related collections of $\sigma(F, G)$-bounded subsets of F, *without changing* the corresponding locally convex topology. The details are next.

Theorem 4.3.1. *An S topology on G of a dual pair (F, G) remains unchanged if S is replaced by any of the following collections of subsets of F:*
(a) *All subsets of the sets of S.*
(b) *All finite unions of the sets of S.*
(c) *All scalar multiples of the sets of S.*
(d) *The balanced hulls of each of the sets of S.*
(e) *The $\sigma(F, G)$ closures of each of the sets of S.*
(f) *The $\sigma(F, G)$ closures of the absolutely convex hulls of each of the sets of S.*

Proof. (a) If $A \in S$ and $C \subset A$, then C is also $\sigma(F, G)$-bounded. Moreover, $C^\circ \supset A^\circ$, which indicates that C° is a zero neighborhood with respect to the polar topology generated by S.

(b) Suppose $A_1, A_2, \ldots, A_n \in S$. Proposition 4.2.4(g) alerts us to the fact that $(\bigcup_{j=1}^{n} A_j)^\circ = \bigcap_{j=1}^{n} A_j^\circ$, which as a finite intersection, is a zero neighborhood with respect to the polar topology generated by S.

(c) Exercise 4.3.2.

(d) If $D = \mathrm{bal}(A)$ for $A \in S$, then Proposition 4.2.4(b) implies that $D^\circ = A^\circ$, the latter of which is a zero neighborhood with respect to the polar topology generated by S.

(e), (f) Exercise 4.3.2. $\qquad\qquad\Box$

Corollary 4.3.1. *If (F, G) is a dual pair and S is a collection of $\sigma(F, G)$-bounded subsets of F, then $\{A^\circ : A \in S\}$ forms a base of zero neighborhoods for a locally convex topology on G, if the two properties below are satisfied:*
(a) *If $A, B \in S$, then there is a set $C \in S$ such that $A \cup B \subset C$.*
(b) *If $A \in S$ and $a \in \mathbb{K}$ is any scalar, then $aA \in S$.*

Proof. Exercise 4.3.4. $\qquad\qquad\Box$

Additionally, when necessary in order to ensure that S topologies are Hausdorff, we will assume the following.

Assumption 4.3.1. If (F, G) is a dual pair and S is a collection of $\sigma(F, G)$-bounded subsets of F, then assume $\mathrm{span}(\cup\{A : A \in S\}) = F$.

A slight generalization of this can be obtained as in Exercise 4.3.5.

Some examples of polar topologies are worth viewing.

Example 4.3.1. The weak topology $\sigma(G, F)$.

In this case, we choose $\mathcal{S}$ to be the collection of all finite subsets of F. Moreover, Theorem 4.3.1 indicates that $\sigma(G, F)$ is the same polar topology on G with respect to any of the following collections of subsets of F, among other possibilities:

(i) The collection of all singleton sets in F.

(ii) The collection of all sets of the form $\sum_{j=1}^{n} a_j \{x_j\}$ for which $\sum_{j=1}^{n} |a_j| \leq 1$ in $\mathbb{K}$, and $x_1, x_2, \ldots, x_n$ are arbitrary elements of F.

(iii) The collection of all $\sigma(F, G)$-bounded, finite-dimensional subsets of F.

Definition 4.3.3. The Mackey topology $\mu(G, F)$ is the finest locally convex topology that is consistent with the duality (Definition 4.2.5). A LCS that carries the Mackey topology is called a **Mackey space**.

Here, we consider the collection $\mathcal{S}$ of all absolutely convex $\sigma(F, G)$-compact subsets of F. This topology has the specific name *Mackey topology* because Mackey was one of the early pioneers in developing the topic of topological vector spaces. References [85] and [86] contain many fundamental results of TVS, and specifically of LCS. Also, note that we cannot just use the collection of $\sigma(F, G)$-compact subsets of F because the convex hull of a compact set might fail to be compact, as can be seen in [91, Example 4.8.8, p. 94].

Example 4.3.2. The strong topology $\beta(G, F)$.

Remark 4.3.2. As with the weak topology (see Remark 4.2.1), the word "strong" is intended to imply "fine" but that in general topological vocabulary that is not always the intended meaning of "strong." A similar statement holds for the term "stronger."

This time, to define $\beta(G, F)$ we use the collection of *all $\sigma(F, G)$-bounded subsets* of F. This is the finest possible $\mathcal{S}$ topology. When the dual space carries the strong topology, it has a formal name, stated next.

Definition 4.3.4. The dual of a LCS, equipped with the strong topology $\beta(E', E)$, is called the **strong dual** of the LCS E.

The choice of the letter beta is to indicate the collection of all $\sigma(F, G)$-**b**ounded sets. Moreover, in Example 4.5.1 you can see that for a nonreflexive Banach space, the strong topology is not always consistent with the duality (see Definition 4.2.5). That is, in a nonreflexive infinite-dimensional Banach space E, the strong topology of the dual E' is stronger than the Mackey topology $\mu(E', E)$.

If we choose S to be the collection of all finite subsets of F, then for a particular $x \in F$, $\{x\}^\circ = \{y \in G : |B(x,y)| \le 1\}$ indicates that the polars of the finite subsets of F generate the weak topology $\sigma(G,F)$ on G, as we have just seen. This is the coarsest, labeled as the *weakest*, possible consistent topology. Thus, we could call $\sigma(G,F)$ the topology of pointwise convergence. Similar reasoning shows that, for example, the Mackey topology $\mu(G,F)$ represents uniform convergence on the absolutely convex, $\sigma(F,G)$-compact subsets of F. In other words, we can label S polar topologies as stated next.

Definition 4.3.5. If S is a collection of $\sigma(F,G)$-bounded subsets of F of a dual pair (F,G), then the polar topology generated by S is also called the **topology of uniform convergence** on the sets of S. In particular, if S is the collection of all finite subsets of F, then the corresponding polar topology is called the **topology of pointwise convergence**.

Starting at this point, we restrict our focus to the dual pair (E,E') of LCS over the field $\mathbb{K}$, using the evaluation map $\langle \cdot, \cdot \rangle$ as the bilinear form. Up to now, we have seen that there are up to four possible topologies that are consistent with the pairing (E,E'), namely the original topology on E, as well as the weak, Mackey, and strong topologies. In general, these topologies can all be distinct.

From general topology, we know that two distinct topologies need not have the same collections of closed sets. The next result shows a surprising situation in which certain closed sets are in fact the same for distinct topologies.

Theorem 4.3.2. *For a dual pair, the closed convex sets are the same for all consistent topologies.*

Proof. It is easy to see that a set is convex in E over $\mathbb{K}$ if and only if it is convex over the field $\mathbb{R}$. Also, by Exercise 4.1.9, we know that a closed convex set is equal to the intersection of all closed half spaces that contain the set. Such half spaces may be denoted by $\{x \in E : f(x) \le a\}$, where $f \in E'$, $a \in \mathbb{R}$. By the definition of consistent topologies, f is continuous with respect to all such topologies, which in turn indicates that the closed half spaces containing a set are the same for all compatible topologies. $\square$

The next goal is to instill in our minds that polar topologies are important by way of proving that *every* locally convex topology is in fact a polar topology. A useful tool for such instillation is that of equicontinuity, which will now be defined, for general TVS.

Definition 4.3.6. Let E and F be TVS. A set $\mathcal{H}$ of linear maps from E to F is **equicontinuous** if for every $V \in \mathcal{N}_{0,F}$ there is a $U \in \mathcal{N}_{0,E}$ satisfying the condition that for all $f \in \mathcal{H}$, $f(U) \subset V$.

It is clear from the linearity of the topologies and the mappings that an equicontinuous collection $\mathcal{H}$ is in fact *uniformly* equicontinuous. In addition, we may define a collection $\mathcal{H}$ to be (uniformly) equicontinuous on a set $A \subset E$ with obvious modifications of the definition. Moreover, it is easy to prove, which you can do as Exercise 4.3.8, that a collection $\mathcal{H}$ is equicontinuous on E if and only if $\mathcal{H}$ is equicontinuous at zero.

As usual, in the case of LCS there is a seminorm version of equicontinuity, as follows.

Proposition 4.3.2. *Let E and F be LCS and $\mathcal{H}$ a collection of linear maps from E to F. Suppose $\{p_\alpha : \alpha \in I\}$ and $\{q_\beta : \beta \in J\}$ are collections of seminorms that generate the topologies of E and F, respectively. Then $\mathcal{H}$ is equicontinuous if and only if for each $\beta \in J$ there exist finitely many $\alpha \in I$ given by $\alpha_1, \alpha_2, \ldots, \alpha_n$, and there is a scalar $a > 0$ for which $q_\beta(f(x)) \leq a \max\{p_{\alpha_j}(x) : j \in \underline{n}\}$ for every $f \in \mathcal{H}$ and $x \in E$. In particular, if $F = \mathbb{K}$, then $\mathcal{H} \subset E'$ is equicontinuous on a subset A of E if and only if there is a continuous seminorm p on E such that for each $f \in \mathcal{H}$ and for all $x \in A$, $|f(x)| \leq p(x)$.*

Proof. The proof of this is similar to that of Proposition 2.1.11 and is left as Exercise 4.3.9. $\qquad\square$

Here is a connection between polars and equicontinuous sets of linear functions.

Proposition 4.3.3. *A subset $\mathcal{H}$ of E' is equicontinuous if and only if $\mathcal{H} \subset V^\circ$, for some $V \in \mathcal{N}_{0,E}$.*

Proof. ($\Rightarrow$): Suppose $\mathcal{H}$ is equicontinuous. Then there is a $V \in \mathcal{N}_{0,E}$ such that for each $x \in V$ and for all $f \in \mathcal{H}$, $|\langle x, f \rangle| \leq 1$; that is, $\mathcal{H} \subset V^\circ$.

($\Leftarrow$): Now suppose there exists $V \in \mathcal{N}_{0,E}$, such that $\mathcal{H} \subset V^\circ$. This means that whenever $\varepsilon > 0$ and $x \in \varepsilon V$, we have that for all $f \in \mathcal{H}$, $|\langle x, f \rangle| \leq \varepsilon$. Thus, $\mathcal{H}$ is equicontinuous. $\qquad\square$

With these last few items, we can now prove that every locally convex topology is a polar topology. You will notice that we invoke not one, but two significant results during the proof.

Theorem 4.3.3. *The topology of any LCS is a polar topology.*

Proof. Consider any LCS $(E, \mathcal{T})$. We intend to prove that $\mathcal{T}$ is the polar topology given by $\mathcal{S}$, where $\mathcal{S}$ is the set of all equicontinuous subsets of E'. Indeed, a base of $\mathcal{N}_0$ in E is given by the collection of all absolutely convex, closed zero neighborhoods. The collection of absolutely convex, closed subsets is the same for any consistent topology, by Theorem 4.3.2. In particular, all such sets U are $\sigma(E, E')$-closed. Next, by the Bipolar Theorem 4.2.1, we have $U = U^{\circ\circ}$, and this means U° is equicontinuous in E' by Proposition 4.3.3. We conclude that the polar topology generated by $\mathcal{S}$ is finer than the original topology $\mathcal{T}$ on E. On the other hand, $\mathcal{H} \subset E'$ being equicontinuous implies, according to Proposition 4.3.3 again, that $\mathcal{H} \subset V^\circ$ for some absolutely convex $\sigma(E, E')$-closed set that belongs to $\mathcal{N}_{0,E}$. Hence, $\mathcal{H}^\circ \supset V^{\circ\circ} = V$, which tells us that $\mathcal{T}$ is finer than the polar topology generated by $\mathcal{S}$. This fun with polars allows us to chill out, because the proof is completed. $\qquad\square$

Exercises

4.3.1. Given a dual pair (F, G) generated by a collection $\mathcal{S}$ of subsets of F, with the bilinear form B, prove that the following defines a seminorm on G: $(\forall A \in \mathcal{S})\, p_A(y) = \sup\{|B(x,y)| : x \in A\}, y \in G$.

4.3.2. Prove parts (c), (e), and (f) of Theorem 4.3.1.

4.3.3. Prove Proposition 4.3.1.

4.3.4. Prove Corollary 4.3.1.

4.3.5. Prove that an $\mathcal{S}$ topology on a dual pair (F, G) is Hausdorff if and only if the span of the union $\bigcup_{A \in \mathcal{S}} A$ is dense in F with respect to the $\sigma(F, G)$ topology.

4.3.6. Suppose E is a finite-dimensional LCS. Prove that all polar topologies (including the strong topology) are consistent with the duality on E.

4.3.7. Prove that every absolutely convex subset of E' that is $\sigma(E', E)$-compact, must be $\beta(E', E)$-bounded.

4.3.8. Prove that a collection $\mathcal{H}$ of linear maps from a TVS E to a TVS F is equicontinuous on E if and only if $\mathcal{H}$ is equicontinuous at zero.

4.3.9. Prove Proposition 4.3.2.

4.3.10. Consider an inductive limit $E = \mathrm{ind}_n E_n$ of a sequence of LCS of the form $\{(E_n, \mathcal{T}_n, f_n) : n \in \mathbb{N}\}$, where for each $n \in \mathbb{N}$, $E_n \subset E_{n+1}$ and $f_n : E_n \to E_{n+1}$ is linear and continuous (See Definition 3.5.3). Let $\mathcal{H}$ be a collection of linear maps from E to a LCS F. Prove that $\mathcal{H}$ is equicontinuous if and only if, $(\forall n \in \mathbb{N})$ the collection $\mathcal{H} \circ f_n = \{h \circ f_n : h \in \mathcal{H}\}$ of linear maps from E_n to F is equicontinuous.

4.3.11. Suppose (F, G) is a dual pair. A collection $\mathcal{S}$ of $\sigma(f, G)$-bounded subsets of F is **saturated** if:

    (i)   Every subset of a member $A \in \mathcal{S}$ belongs again to $\mathcal{S}$.

    (ii)   Finite unions of members of $\mathcal{S}$ belong to $\mathcal{S}$.

    (iii) If $A \in \mathcal{S}$, then $\alpha A \in \mathcal{S}(\forall \alpha \in \mathbb{K}\setminus\{0\})$.

    (iv)

$$\overline{\mathrm{conv\, bal}(A)}^{\,\sigma(F,G)} \in \mathcal{S}, \quad (\forall A \in \mathcal{S}). \tag{4.13}$$

Suppose $\mathcal{S}_1$ is a saturated collection of $\sigma(F, G)$ bounded subsets of F, and $\mathcal{S}_2$ is another such collection for which $\mathcal{S}_1 \subsetneq \mathcal{S}_2$. Prove that the corresponding $\mathcal{S}_2$ topology is strictly finer than the $\mathcal{S}_1$ topology.

4.4 The theorems of Alaoğlu–Bourbaki, Mackey–Arens, and related results

We continue our study of the properties of polar topologies. In this section, some deep connections between polar topologies and sets that are either bounded or compact will be revealed. We start with a completeness result.

Proposition 4.4.1. *The algebraic dual of a Hausdorff LCS is weakly complete. That is, if E is a LCS, then $(E^*, \sigma(E^*, E))$ is complete.*

Proof. Because a LCS need not be metrizable, we appeal to our old friends, filters. Let $\mathfrak{F}$ be a Cauchy filter on $(E^*, \sigma(E^*, E))$. Given a set $S \in \mathfrak{F}$ in E^*, we write for any $x \in E$, $\langle x, S \rangle = \{\langle x, s \rangle : s \in S\}$ for brevity. Notice that for every $x \in E$ and every $\varepsilon > 0$, there exists $S \in \mathfrak{F}$ such that for every $s \in S$, $|\langle x, s \rangle| < \varepsilon$. This implies that for each $x \in E$, the filter $\{\langle x, S \rangle : S \in \mathfrak{F}\}$ is Cauchy in $\mathbb{K}$ and, therefore, converges to some scalar. We label that scalar $f(x)$. Hence, for each $x \in E$, $f(x)$ exists. As suggestive as the notation $f(x)$ seems, we must still prove that it defines a linear functional. To be sure, $f(x) \in \mathbb{K}$ for each $x \in E$. To prove linearity, first let $x, y \in E$, and $\varepsilon > 0$ be arbitrary. By convergence, there is a set $S_1 \in \mathfrak{F}$ such that for all $s_1 \in S_1$, $|\langle x, s_1 \rangle - f(x)| < \frac{\varepsilon}{2}$. Likewise, there is a set $S_2 \in \mathfrak{F}$ such that for all $s_2 \in S_2$, $|\langle y, s_2 \rangle - f(y)| < \frac{\varepsilon}{2}$. Let $S_3 = S_1 \cap S_2$. By Definition 1.2.1(b), $S_3 \in \mathfrak{F}$. In addition (literally), for every $s \in S_3$, $|\langle x + y, s \rangle - (f(x) - f(y))| < \frac{\varepsilon}{2} + \frac{\varepsilon}{2} = \varepsilon$. Another way to state this is that $\langle x + y, \mathfrak{F} \rangle \to f(x) + f(y)$, and because E is Hausdorff the uniqueness of limits gives us $f(x + y) = f(x) + f(y)$. The scalar multiple part is left as Exercise 4.4.1. $\qquad\square$

It is worth noting that for the continuous dual Proposition 4.4.1 generally does not hold. See Theorem 4.5.5.

Proposition 4.4.2. *For a LCS E, $(E', \sigma(E', E))$, any weakly bounded subset is weakly totally bounded.*

Proof. In Proposition 2.5.15, we use $\mathcal{V}$ in E given by $\mathcal{V} = \{x : |\langle x, y \rangle| \le 1\}$, where $y \in E'$. Now, if $A \subset E$ is $\sigma(E, E')$-bounded, then for all $y \in E'$, $\{\langle x, y \rangle : x \in A\}$ is a bounded set of scalars, which can be covered by finitely many sets of the form $I_j, j \in \underline{n}$ for which the diameter in $\mathbb{K}$ of each I_j is less than 1. Observe that each set $y^{-1}(I_j)$, for $j \in \underline{n}, y \in E'$ forms a covering of A by sets that are small of order V, where each V is of the form $V = \{x : |\langle x, y \rangle| \le 1\}$. $\qquad\square$

Proposition 4.4.3. *Suppose $\mathcal{T}_1$ and $\mathcal{T}_2$ are two topologies on the same set K such that $\mathcal{T}_1$ is finer than $\mathcal{T}_2$. If K is compact with respect to $\mathcal{T}_1$, then K is compact with respect to $\mathcal{T}_2$.*

Proof. Exercise 4.4.2. $\qquad\square$

The previous propositions allow for the proof of the following fundamental result. We refer to it here as the Alaoğlu–Bourbaki theorem, though it also goes by the name Banach–Alaoğlu theorem.

Theorem 4.4.1 (Alaoğlu–Bourbaki theorem). *For a Hausdorff LCS, the polar of any zero neighborhood is weakly compact. That is, if E is a LCS, then the polar of any zero neighborhood in E is $\sigma(E', E)$-compact.*

Proof. We start with E^*, equipping it with the $\sigma(E^*, E)$ topology. Suppose U is a zero neighborhood in E. As an absorbing set, U° is $\sigma(E', E)$-totally bounded by Proposition 4.4.2. Next, we apply Proposition 4.4.1 to observe that because U° is $\sigma(E^*, E)$-closed, U° is $\sigma(E^*, E)$-complete, where we have applied Proposition 2.5.5(c). Thus, U° is $\sigma(E^*, E)$-compact by Theorem 2.5.5. To complete the proof, we need to find our way to E'. This is accomplished by noting that, in fact, $U^\circ \subset E'$. Moreover, $\sigma(E^*, E)|_{E'} = \sigma(E', E)$. Now apply Proposition 4.4.3 to the equivalent (hence technically finer) topology $\sigma(E', E)$, from which we see that U° is $\sigma(E', E)$-complete and totally bounded. The result now follows from another application of Theorem 2.5.5. $\square$

Corollary 4.4.1. *An equicontinuous subset of the dual of a LCS is weakly relatively compact. That is, if $\mathcal{H} \subset E'$ is equicontinuous, then $\mathcal{H}$ is $\sigma(E', E)$ relatively compact.*

Proof. An application of Proposition 4.3.3 reveals that such a set $\mathcal{H}$ must be contained in the polar of a zero neighborhood of E. That polar is, by the Alaoğlu–Bourbaki Theorem 4.4.1, $\sigma(E', E)$-compact. $\square$

Corollary 4.4.2. *The unit ball in the dual of a normed space is weakly compact. That is, if $(E, \|\cdot\|)$ is a normed space, then the unit ball in the continuous dual E' is $\sigma(E', E)$-compact.*

Proof. Indeed, the unit ball in E' is the polar of the unit ball in E. $\square$

For next big theorem, we will again need some preliminary propositions.

Proposition 4.4.4. *Suppose E is a Hausdorff LCS with a base $\mathcal{N}_0$ of zero neighborhoods. Then $E' = \bigcup\{U^\circledast : U \in \mathcal{N}_0\}$, where $U^\circledast$ represents the polar taken in E^*.*

Proof. An element $y \in E^*$ is continuous if and only if there exists $U \in \mathcal{N}_0$ such that for each $x \in U$, $|\langle x, y \rangle| \le 1$. $\square$

Proposition 4.4.5. *If (E, E') is a dual pair, then the dual of E' under an $\mathcal{S}$ topology is given by $\bigcup\{A^{\circ\circledast} : A \in \mathcal{S}\}$, where the polars are taken in E'^*.*

Proof. This result follows from Proposition 4.4.4 above. $\square$

Proposition 4.4.6. *Suppose E is a Hausdorff LCS. Then the dual of E' is E whenever E' carries the topology of uniform convergence on a set $\mathcal{S}$ of absolutely convex compact subsets of E.*

Proof. According to Proposition 4.4.5, the dual of E' is given by $\bigcup_{A \in \mathcal{S}} A^{\circ\circledast}$. Next, we prove that for an absolutely convex compact subset A of E, $A^{\circ\circledast} = A$, a subset of E'^*. To see this, note that $A^{\circ\circledast}$ is the $\sigma(E'^*, E')$ closure of A. By hypothesis, A is $\mathcal{T}$-compact. As $\sigma(E'^*, E')|_E$ is equivalent to $\sigma(E, E')$ and $\sigma(E, E')$ is coarser than $\mathcal{T}$, we have that A is $\sigma(E'^*, E')$-compact, too, by Proposition 4.4.3. Furthermore, A is $\sigma(E'^*, E')$-closed by the Hausdorff assumption. We have established that $A^{\circ\circledast} = A$. Finally, by Assumption 4.3.1, we conclude that $\bigcup\{A^{\circ\circledast} : A \in \mathcal{S}\} = \bigcup\{A : A \in \mathcal{S}\}$, and the proof is done. $\qquad\square$

Up next is another big theorem.

Theorem 4.4.2 (Mackey–Arens theorem). *Consider a dual pair (E, E') for a Hausdorff LCS E. A locally convex topology $\mathcal{T}$ on E is consistent with the duality (E, E') if and only if $\mathcal{T}$ is a topology of uniform convergence with respect to a collection $\mathcal{S}$ of absolutely convex $\sigma(E', E)$-compact subsets of E'.*

Proof. ($\Rightarrow$): Assume $\mathcal{T}$ is consistent with the duality (E, E'). We seek a collection $\mathcal{S}$ of absolutely convex $\sigma(E', E)$-compact subsets of E'. To this end, consider the collection of all absolutely convex $\mathcal{T}$-closed zero neighborhoods in E. By Theorem 4.3.2, these sets are all $\sigma(E, E')$-closed as well. We can then use $\mathcal{S}$ to be the set of all polars U°, where U is a zero neighborhood in E. It follows that $\mathcal{T}$ is the topology of uniform convergence on the set of all U°, because $U^{\circ\circ} = U$. Applying the Alaoğlu–Bourbaki Theorem 4.4.1 in the context of roll reversal, we have that the sets U° are $\sigma(E', E)$-compact, and are of course absolutely convex.

($\Leftarrow$): Now assume that for $(E, \mathcal{T})$, the topology $\mathcal{T}$ is a topology of uniform convergence on a collection of absolutely convex, $\sigma(E', E)$-compact subsets of E'. Apply Proposition 4.4.6 again with the roles of E and E' reversed, to conclude that the dual of E is in fact E'. $\qquad\square$

It is worth summarizing the Mackey–Arens Theorem 4.4.2 above for the case of the original space E. That is, for the dual pair (E, E') involving the Hausdorff LCS E, denoting the weak topology by $\sigma(E, E')$, any consistent topology on E by $\mathcal{T}$, the Mackey topology by $\mu(E, E')$ and letting "$\mathcal{T}_1 \subset \mathcal{T}_2$" mean $\mathcal{T}_1$ is a coarser topology than $\mathcal{T}_2$. We have

$$\sigma(E, E') \subset \mathcal{T} \subset \mu(E, E'). \tag{4.14}$$

Likewise, reversing roles between E and E', and denoting by v any consistent topology on E' we have

$$\sigma(E', E) \subset v \subset \mu(E', E). \tag{4.15}$$

Example 4.4.1. For a nonreflexive Banach space, the strong topology on the dual is strictly finer than the Mackey topology.

Indeed, the strong topology for such spaces is not consistent with the duality (see Example 4.5.1).

We finish this section with a brief look at the finest locally convex topology. The proofs of the following results are left as enlightening exercises.

Proposition 4.4.7. *Consider a vector space E with the finest locally convex topology of Example 2.1.3. Denote this topology by $\mathcal{T}_{\text{finest}}$. Then the following hold:*

(a) *The continuous dual of E is the same as the algebraic dual of E, that is, $E' = E^*$.*
(b) *The topology $\mathcal{T}_{finest}$ on E is the Mackey topology.*
(c) *The only $\mathcal{T}_{finest}$-bounded subsets of E are those that are finite-dimensional.*
(d) *All of the polar topologies, from $\sigma(E',E)$ to $\beta(E',E)$, are equivalent on E^*.*
(e) *If $(E,\mu(E,E'))$ is normable, then E is finite-dimensional.*

Proof. Exercise 4.4.3 $\qquad\qquad\square$

Exercises

4.4.1. Prove the scalar multiple part of Proposition 4.4.1, namely that $(\forall x \in E, a \in \mathbb{K})\, f(ax) = af(x)$.

4.4.2. Prove Proposition 4.4.3.

4.4.3. Prove Proposition 4.4.7. *Hint:* For (c), suppose $\{e_n : n \in \mathbb{N}\}$ is an infinite collection of linearly independent elements of E. Extend this set to a Hamel basis $\mathcal{B}$ of E. Define a linear functional f on E by letting $f(x_n) = n$, and $f(e) = 0$, for all other $e \in \mathcal{B}$. Now prove that $\{x_n : n \in \mathbb{N}\}$ cannot be bounded. For part (d), apply part (c).

4.5 Normed and Banach spaces—some duality properties

As the title insinuates, we will look at a number of results about normed and Banach spaces as viewed with the lens of duality. In particular, infinite-dimensional Banach spaces equipped with their weak topologies provide concrete examples of surprising properties of LCS. Hence, this section represents something of a "Stop and smell the theorems" experience. By the way, many of the results in this section are developed in resources on normed and Banach spaces such as [89].

Below is one of many applications of the Helly–Hahn–Banach theorem to normed and Banach spaces. The result below is a consequence of the analytic form via Corollary 4.1.5. To set up the result, we recall some facts about the norms of linear maps on normed spaces. A note of clarification is to recall that our notation E^* refers to the *algebraic* dual of a normed space while E' represents the *continuous* dual of a normed space.

Definition 4.5.1. Consider a normed space $(E, \| \cdot \|)$. For $f \in E^*$, define

$$\|f\|^* = \sup\left\{ \frac{|f(x)|}{\|x\|} : \|x\| \neq 0 \right\}. \tag{4.16}$$

It is a routine exercise to prove that $\| \cdot \|^*$ does in fact define a norm on E^*. The Helly–Hahn–Banach result is next.

Theorem 4.5.1. *In a normed space $(E, \| \cdot \|)$, let M be a subspace of E, and let f be a linear functional that is continuous on M. Apply Equation* (4.16) *to $\| \cdot \|$ restricted to M and suppose the norm of f on M has the value a. There exists $\tilde{f} \in E'$ such that $\tilde{f}|_M = f$ and $\|\tilde{f}\|^* = a$.*

Proof. According to the statement of the theorem, we have

$$\|f\|_M = \sup\left\{ \frac{|f(x)|}{\|x\|} : \|x\| \neq 0, x \in M \right\}. \tag{4.17}$$

In Corollary 4.1.5(a), use $p = \| \cdot \|^*$, and we have $\|\tilde{f}\|^* = \|f\|_M = a$. That is, the continuous extension $\tilde{f}$ has the same norm on all of E as f has on M. □

Proposition 4.5.1. *In a Banach space $(E, \| \cdot \|)$, the strong topology is the operator norm; that is, the strong topology is generated by the norm given by: For all $f \in E'$, $\|f\| = \sup\{|f(x)| : \|x\| \leq 1, x \in E\}$.*

Recall Definition 4.3.2 of the strong topology.

Proof. A subset A of E is weakly bounded if and only if $f(A)$ is a bounded subset of $\mathbb{K}$ for all $f \in E'$. In this case, using the evaluation map, we have $|f(x)| = |\langle x, f \rangle|$, for all $x \in A$. We can view $A \subset E$ as being equivalent to a subset of E'^* via the map $\langle x, f \rangle$, for $x \in A$ and all $f \in E'$. That is, we consider x to be an element of E'^*. Applying the Banach–Steinhaus theorem for Banach spaces A.5.4, we have that pointwise bounded linear functionals on a Banach space are uniformly bounded; that is, for A being a weakly bounded subset of E, there exists an $M \geq 0$ such that for every $x \in A$ one has $|\langle x, f \rangle| \leq M$, for $f \in E'$. In particular, for $x \in E$, $\|x\| = \sup\{ \frac{|\langle x, f \rangle|}{\|f\|} \leq M : f \in E'\}$, which tells us that A is bounded with respect to the norm. $\qquad \square$

The previous proposition allows us to exhibit a topology on the dual that is not consistent with the duality. Recall that a Banach space E is **reflexive** if its second dual is equal to the original space; that is, if $E'' = E$. See [89, Section 1.11, pp. 97–109] for more details. By the way, the Banach–Steinhaus theorem will be examined in the general case of LCS in Chapter 7 (see Theorem 7.2.6).

Example 4.5.1. For a nonreflexive Banach space, the strong dual is not consistent with the duality.

If E is not reflexive, then the dual of E' is strictly larger than E. Thus, the normed topology on its dual E' cannot be consistent with the duality.

Next, we have a corollary to the Alaoğlu–Bourbaki Theorem 4.4.1, as applied to certain Banach spaces. In the item below, reference is made of reflexive normed spaces. Details on such spaces in the context of LCS can be found in Section 5.3. Meanwhile, in the case of normed spaces, we have the following.

Corollary 4.5.1. *If E is a reflexive Banach space, then the closed unit ball of E is $\sigma(E, E')$-compact.*

By [89, Theorem 2.8.2, p. 245], a normed space is reflexive if and only if its closed unit ball is weakly compact. On the other hand, for $(E, \| \cdot \|)$ where the dimension of E is infinite, if the closed unit ball were compact with respect to the norm $\| \cdot \|$, then by Riesz's Theorem 3.2.2, the dimension of E would have to be finite, a contradiction. Therefore, if the dimension of a reflexive Banach space is infinite, the closed unit ball of that space is weakly compact but is *not* compact with respect to the normed topology. Hence, the Alaoğlu–Bourbaki Theorem 4.4.1 provides a way to conclude that for infinite-dimensional reflexive Banach spaces, the normed and weak topologies are not equivalent. Another observation here is that, although the Heine–Borel property is *never* true for infinite-dimensional normed spaces (by Corollary 3.2.6), it is *always* true with respect to the weak topology on reflexive Banach spaces, as indicated by the previous corollary. Speaking of the Alaoğlu–Bourbaki theorem, the next result makes use of that theorem.

Theorem 4.5.2. *The closed unit ball in the dual of a normed space is metrizable with respect to the weak topology if and only if the normed space is separable.*

Before the proof, recall Definition A.3.12 of separability, if necessary.

Proof. ($\Rightarrow$): See [116, Theorem 16.11, pp. 213–214].

($\Leftarrow$): Assume the normed space $(E, \|\cdot\|)$ is separable. In Exercise 4.5.6, you can verify some of the details. Let $\{x_n\}$ be a countable dense subset of E. Let $D = \{y \in E' : \|y\| \leq 1\}$ denote the closed unit ball of the dual of E. Define the following on D: For any $w, y \in D$,

$$d(w,y) = \sum_{k=1}^{\infty} \frac{|\langle w - y, x_k \rangle|}{2^k (1 + |\langle w - y, x_k \rangle|)}. \tag{4.18}$$

Then d is a metric on D and generates a topology $\mathcal{T}_d$ that is coarser on D than $\sigma(E', E)$. Next, using the Alaoğlu–Bourbaki Theorem 4.4.1, one shows that D is $\sigma(E', E)$-compact. Finally, considering comparable topologies and compactness, it follows that $\mathcal{T}_d$ is equivalent to $\sigma(E', E)$ on D, and hence, D is $\sigma(E', E)$ metrizable. $\qquad\square$

Corollary 4.5.2. *If E is a separable normed space and (y_n) is a bounded sequence with respect to the norm in the dual E', then there is a subsequence of (y_n) that is $\sigma(E', E)$-convergent.*

4.5.1 Examples of infinite-dimensional Banach spaces with their weak topologies

Note that several of the results in this subsection are valid for more general infinite-dimensional normed spaces. First, we look at normability or in this case, the lack thereof.

Proposition 4.5.2. *When equipped with their weak topologies, infinite-dimensional Banach spaces are never normable.*

Proof. Let E be an infinite-dimensional Banach space. By way of Definition 4.2.3, we have that the weak topology for the dual pair (E, E') is generated by the collection $\{p_f : f \in E'\}$ of seminorms defined by $p_f(x) = |\langle x, f \rangle|$ for all $x \in F$. In particular, for any given $f \in E'$, a zero neighborhood with respect to $\sigma(E, E')$ contains a set of the form $U = \{x \in E : |f(x)| \leq 1\}$, which in turn contains the set $f^{-1}(\{0\})$. We recognize $f^{-1}(\{0\})$ as a closed hyperplane by way of Proposition 3.1.8. As the dimension of E is infinite, U contains an infinite-dimensional subspace. In other words, no $\sigma(E, E')$ zero neighborhood of E is bounded. Applying Kolomogorov's Theorem 2.4.1, we deduce that E, with its weak topology, cannot be a normed space. $\qquad\square$

Let that last result float around in our minds. We know that if a LCS is not a normed space, then every zero neighborhood must contain an unbounded set. Focusing on an absolutely convex subset of such a zero neighborhood, we see that this means every zero neighborhood contains a nontrivial linear subspace. In the case of an infinite-dimensional Banach (or even normed) space, every zero neighborhood is not just unboundedly large, it is infinitely dimensionally huge! We can further prove that the weak dual of any infinite-dimensional normed space is not even metrizable. Although the next result could have been done first, leaving Proposition 4.5.2 as a corollary, the proof in the context of metrizable LCS differs from that for the case of normed spaces. At any rate, viewing the results independently may give you some extra insight into these ideas. First, is a preparatory proposition.

Proposition 4.5.3. *The polar of the unit ball of a normed space is the unit ball of its dual, and conversely. That is, if B is the unit ball of a normed space $(E, \|\cdot\|)$, then B° is the unit ball B' in the dual $(E', \|\cdot\|')$, where $\|\cdot\|'$ is the operator norm (see Definition A.5.5). Conversely, $(B')^\circ = B$ (the polar being taken in the dual).*

Proof. Use the notation of the statement of the proposition. We make use of the surface S of the unit ball; that is, $S = \{x \in E : \|x\| = 1\}$. Then with obvious meanings of notation, for each $f \in E'$,

$$\|f\|' = \sup\{|f(x)| : x \in B\} = \sup\{|f(B)|\} = \sup\{|f(S)|\}. \tag{4.19}$$

In the dual, let S' be the surface of the unit ball in E'. Thus, $B^\circ = \{f \in E' : |\langle B, f\rangle| \le 1\} = \{f \in E' : \|f\|' \le 1\} = B'$. In Exercise 4.5.1, you can prove that, correspondingly, $\|x\| = \sup\{|\langle x, B'\rangle|\}$ for all $x \in E$, which leads to

$$(B')^\circ = \{x \in E : \sup\{|\langle x, B'\rangle| \le 1\}\} = \{x \in E : \|x\| \le 1\} = B, \tag{4.20}$$

and the proof is completed. $\qquad\square$

Proposition 4.5.3 creates a pathway to the proof of the following result.

Theorem 4.5.3. *When equipped with their weak topologies, infinite-dimensional normed spaces are never even metrizable.*

Proof. Let $(E, \|\cdot\|)$ be an infinite-dimensional normed space. By a standard result from Banach space theory ([89, Theorem 1.10.7, p. 87]), E', with the norm $\|\cdot\|'$ is a Banach space. By Proposition 4.5.3, E' must be infinite-dimensional and with that information, we invoke a standard result of Banach spaces (see [11]), which states that a Hamel basis of any infinite-dimensional Banach space is uncountable. Thus, we conclude that E' is of uncountable dimension. Apply Proposition 4.2.1 to observe that the topology of $\sigma(E, E')$ is generated by uncountably many seminorms. To finish, we apply Corollary 2.2.1 and conclude that $(E, \sigma(E, E'))$ just cannot be metrizable. $\qquad\square$

Let us now look at the l_p spaces for $1 \le p < \infty$ (see Subsection A.5.1 in the Appendix). We start with a standard example that distinguishes convergence with respect to norm versus the weak topology.

Example 4.5.2. In l_2, the sequence consisting of the unit vectors given by $(x_n^{(k)} : k \in \mathbb{N})$, where $x_n^{(k)} = 1$ exactly when $n = k$, is weakly convergent to zero but not norm convergent to zero.

The verification of this is a routine exercise from normed spaces material and is left as Exercise 4.5.2. Meanwhile, the next result is rather surprising.

Theorem 4.5.4 (Schur's theorem). *In l_1, weakly convergent sequences are also norm convergent. For $1 \le p < \infty$, this is the only l_p space for which this property holds.*

The proof of this theorem is beyond the scope of this book, however, an accessible proof can be found in Megginson [89, Example 2.5.24, pp. 218–219]; see also [37]. We can give a name to this property for which the weak and original topologies have the same convergent sequences for general LCS in the following definition, which is adapted from [89, Definition 2.5.25, p. 220].

Definition 4.5.2. A LCS satisfies **Schur's property** if every weakly convergent sequence is convergent with respect to the original topology of the LCS.

We note here for reference that in the context of LCS Schur's property is equivalent to the original and weak topologies and have the same collections of sets that satisfy a variety of forms of compactness, as proved by Gabriyelyan [48].

There is one more result to mention regarding infinite-dimensional normed spaces, the proof of which can be found in [89, Proposition 2.5.15, p. 215]

Theorem 4.5.5. *Equipped with its weak topology, a normed space is complete if and only if it is finite-dimensional.*

Exercises

4.5.1. In the proof of Proposition 4.5.3, prove that for all $x \in E$,

$$\|x\| = \sup\{|\langle x, B'\rangle|\}. \tag{4.21}$$

4.5.2. Prove the statement of Example 4.5.2.

4.5.3. Prove that none of the spaces l_p, for $p \in (1, \infty)$ as well as c_0, satisfy Schur's property.

4.5.4. Prove that c_0, with its weak topology, is not sequentially complete.

4.5.5. Prove that c_0 cannot be the dual of a Banach space. *Hint:* Construct a sequence $(x^{(n)})$ in c_0 such that for any continuous linear functional f on c_0, $\lim_{n \to \infty} x^{(n)}$ exists, however, there is no element of c_0 that is equal to that limit.

4.5.6. Regarding the proof of Theorem 4.5.2, prove the following:

(a) Prove that the function d is indeed a metric and that the topology $\mathcal{T}_d$ generated by d is coarser than $\sigma(E', E)$.

(b) Prove that D is $\sigma(E', E)$-compact and deduce that $\mathcal{T}_d$ is equivalent to $\sigma(E', E)$ on D.

4.6 The Kreĭn–Mil'man theorem

If you thought the fun of deep theorems about convex and compact sets in a TVS came to an end in the last section, then you will be pleased to know that there are more such results in this section. To set the stage for the main result of this section, imagine a collection of eight points in $\mathbb{R}^3$ that serve as vertices of a rectangular solid. By appropriately drawing line segments to connect the dots including the solid enclosed by those dots, the created figure, including the sides and the eight points, is a convex, compact subset of $\mathbb{R}^3$. The eight vertices are special in the sense that no vertex can be the midpoint of a proper line segment in the figure. That is, such points are *extreme*, a term we will formalize shortly. In this section, we will obtain a result, the Kreĭn–Mil'man Theorem 4.6.4 that reveals how compact convex sets of a LCS can be described, (under certain closure circumstances) using extreme points. As is typical, we first need to define some terms and develop some preliminary results. In this section, we first consider LCS over the field $\mathbb{R}$ of real numbers.

Definition 4.6.1. Let E be a vector space over $\mathbb{R}$ and suppose A is a subset of E. A point $x \in A$ is an **extreme point** if the set A satisfies either of the following equivalent conditions.

(a) The point x is not the midpoint of any proper line segment in A.

(b) $A \setminus \{x\}$ is convex.

Some simple examples follow.

Example 4.6.1. The vertices of any regular polygon in $\mathbb{R}^2$ are extreme points. Similarly, the vertices of any regular polyhedron in $\mathbb{R}^3$ are extreme points.

Example 4.6.2. Consider the sets shown in Figures 4.1.

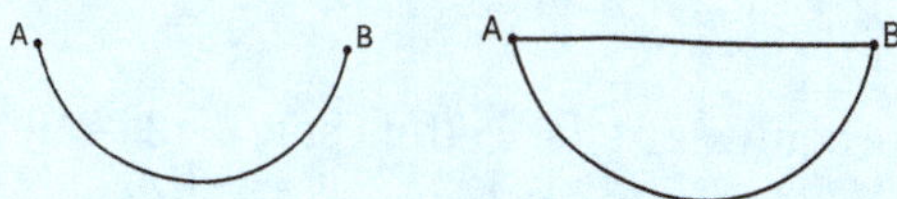

Figure 4.1: Distinct sets with the same set of extreme points.

In the set on the left in Figure 4.1, all the points of the set are extreme points, while in the set on the right, the line segment of the figure (excluding the endpoints) does not contain any extreme points. Nevertheless, we conclude that two distinct sets can have the same set of extreme points.

Example 4.6.3. Consider the sets shown in Figures 4.2.

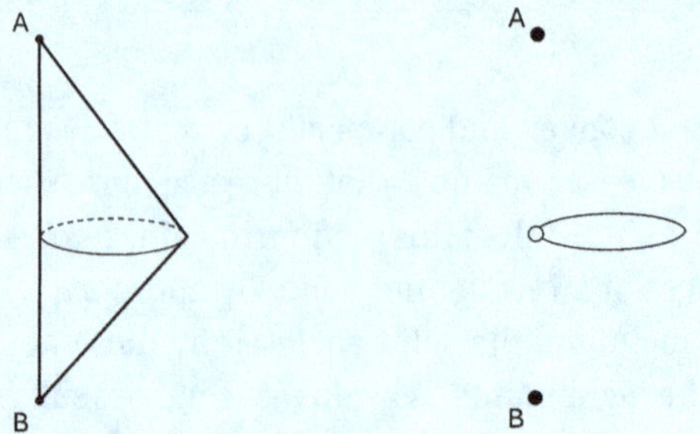

Figure 4.2: Distinct sets with the same set of extreme points.

The sets shown in Figure 4.2 are also two distinct sets that have the same set of extreme points. The set on the left is convex, hence, connected as well as being closed. Meanwhile, the set on the right is neither closed, nor convex, nor connected.

Next is a description of some equivalent expressions for extreme points of convex sets.

Proposition 4.6.1. *Let A be a convex subset of the vector space E over $\mathbb{R}$. For a point $x \in A$, the following are equivalent:*

(a) *If $x = \frac{y+z}{2}$ such that $y, z \in A$, then $x = y = z$.*

(b) *If x is a convex combination of distinct elements y and z of A, then x must be one of the endpoints of the segment between y and z; that is, for $x = ty + (1-t)z$, where $y, z \in A$, $0 \le t \le 1$, and $y \ne z$, then either $t = 0$ or $t = 1$.*

(c) *In part (b) above, if $0 < t < 1$, then $x = y = z$.*

(d) *For a more general convex combination such as in Definition 1.1.7, with each coefficient t_j being positive for $j \in \underline{n}$, we conclude that x is equal to each of the vectors of the convex combination. In other words, if $x_1, x_2, \ldots, x_n, \in A$, and $t_1, t_2, \ldots, t_n \in (0,1)$ such that $\sum_{j=1}^{n} t_j = 1$, and the sum $x = \sum_{j=1}^{n} t_j x_j$, then $x = x_j$ for every $j \in \underline{n}$.*

(e) *The point x is an extreme point of A.*

Proof. First, we prove that (a), (b), and (c) are equivalent. (a) $\Rightarrow$ (b): Let $w = (2t-1)y + 2(1-t)z$, for $\frac{1}{2} \le t \le 1$. Because A is convex, $w \in A$. By the assumption in (a), $x = \frac{y+w}{2} = \frac{y2ty+2z-2tz}{2}$, implies that $x = ty + (1-t)z$. Applying the second part of (a) produces $x = y = w$. In particular, $x = y$ leads to $y = ty + (1-t)z$, which reduces to $(1-t)(z-y) = 0$. In (b), it is assumed that $y \ne z$, and we conclude that $t = 1$.

(b) $\Rightarrow$ (c): The assumptions of (b) and the equation $(1-t)(z-y) = 0$ imply that either $t = 1$ or $y = z$, however, the assumption in (c) is that $0 < t < 1$, which reduces the choice to only one option: $y = z$, and the result follows.

(c) $\Rightarrow$ (a): Exercise 4.6.1.

(c) $\Leftrightarrow$ (d): Exercise 4.6.1.

(c) $\Rightarrow$ (e): Exercise 4.6.1.

(e) $\Rightarrow$ (c): We take the route through the contrapositive. Assume (c) does not hold, then $x = ty + (1 - t)z$, for $0 < t < 1$ and for which $x \neq y$ or $x \neq z$. We examine $A \setminus \{x\}$ and observe that $x \notin A$ however, $x = ty + (1 - t)z$. Thus, $A \setminus \{x\}$ is not convex. $\qquad\square$

The next concept is about what we imagine as surfaces between vertices of a solid.

Definition 4.6.2. Consider a vector space E over $\mathbb{R}$ and a convex subset B of E. Then $A \subset B$ is a **face** of B if given $\{x_1, x_2, \ldots, x_n\} \subset B$ for which the convex combination created from $t_j \in (0,1], j \in \underline{n}, \sum_{j=1}^{n} t_j = 1$ of the form $\sum_{j=1}^{n} t_j x_j$ belongs to A, then all of the points $x_j, j \in \underline{n}$ belong to A.

Example 4.6.4. Consider a rectangular solid in $\mathbb{R}^3$ and a line segment between two of the vertices. That line segment is a face of the rectangular solid. If we exclude one of the vertices, that is, one of the endpoints of the line segment, then the resulting segment is not a face.

The next item is a verification of a sort of transitive property of being a face; in other words, that a face of a face is again a face.

Proposition 4.6.2. *In a vector space E over $\mathbb{R}$, consider convex sets A, B, C such that $A \subset B \subset C$. Then if A is a face of B and B is a face of C, then A is a face of C.*

Proof. Let $\{x_1, x_2, \ldots, x_n\} \subset C$ and consider the convex combination $\sum_{j=1}^{n} t_j x_j$, for $t_j \in (0,1], j \in \underline{n}, \sum_{j=1}^{n} t_j = 1$. Suppose $\sum_{j=1}^{n} t_j x_j \in A$. Then $A \subset B \subset C$ and the assumption that B is a face of C imply that $\sum_{j=1}^{n} t_j x_j \in B$, which in turn implies that $\{x_1, x_2, \ldots, x_n\} \subset B$. To finish, we note that $A \subset B$ implies that $\{x_1, x_2, \ldots, x_n\} \subset A$. $\qquad\square$

Next, we see that any nonempty intersection of faces is again a face.

Proposition 4.6.3. *In a vector space E over $\mathbb{R}$, suppose $\{A_\alpha : \alpha \in I\}$ is a collection of faces of a convex set B. If $C = \bigcap \{A_\alpha : \alpha \in I\}$ is nonempty, then C is a face of B.*

Proof. Let $\{x_1, x_2, \ldots, x_n\} \subset B$ and consider the convex combination $\sum_{j=1}^{n} t_j x_j$, for $t_j \in (0,1], j \in \underline{n}, \sum_{j=1}^{n} t_j = 1$. Because $C \subset A_\alpha$ for all $\alpha \in I$, and each A_α is a face of B, we have that for every $\alpha \in I$ and for every $j \in \underline{n}, x_j \in A_\alpha$. Hence, each x_j satisfies

$$x_j \in \bigcap_{\alpha \in I} A_\alpha = C, \tag{4.22}$$

and we conclude that C is a face of B. $\qquad\square$

Assume the vector space E over $\mathbb{R}$ is equipped with a Hausdorff locally convex topology. Naturally, a topology can be used to answer questions such as whether a set of extreme points is or is not closed. Here comes an example.

Example 4.6.5. In $\mathbb{R}^3$, Figure 4.2(a) shows a set A, and (b) shows the set of extreme points of A.

Thus, the set of extreme points of a closed set need not be closed.

The next result makes use of Zorn's Lemma A.1.1.

Theorem 4.6.1. *Let E be a Hausdorff LCS over* $\mathbb{R}$. *Let C be a convex compact subset of E. Consider the collection*

$$F = \{A \subset C : A \neq \emptyset \text{ and } A \text{ is closed}\}, \tag{4.23}$$

where we define an order $\sqsubseteq$ *on F by:* $A_1 \sqsubseteq A_2$ *if* $A_1 \subset A_2$, *and* A_1 *is a face of* A_2. *Then:*
(a) *The relation* $\sqsubseteq$ *is a partial order on F.*
(b) *F, with the partial order* $\sqsubseteq$ *has a minimal element.*

Proof. Part (a) follows from Proposition 4.6.2. Meanwhile, Proposition 4.6.3 reveals that any chain in F that is decreasing is bounded below. Thus, we apply Zorn's Lemma A.1.1 to conclude that part (b) holds. $\square$

The next proposition is part of the pathway to a larger result that will indicate conditions on which a compact convex subset of a LCS must have at least one extreme point.

Proposition 4.6.4. *Let E be a Hausdorff LCS over* $\mathbb{R}$ *and A a convex compact subset of E. Suppose f is a linear, real valued functional on E that is continuous on A. If*

$$F = \{x \in A : f(x) = \sup\{f(y) : y \in A\}\}, \tag{4.24}$$

then F is a face of A.

Proof. Put $S = \sup\{f(y) : y \in A\}$. The assumption that f is continuous on the compact set A implies that S is finite. To show that F is a face of A, we look at the set of sums $\sum_{j=1}^{n} t_j x_j$, for $t_j \in (0,1]$, $j \in \underline{n}$, $\sum_{j=1}^{n} t_j = 1$, where each x_j belongs to A. The goal is to show that $f(x_j) \in F$, for each $j \in \underline{n}$. Equivalently, we must show that for each $j \in \underline{n}$, $f(x_j) = S = \sup\{f(y) : y \in A\}$. We proceed by way of contradiction; that is, assume that for some $j \in \underline{n}, f(x_j) \neq S$. Because it is always the case that for each $j \in \underline{n}, x_j \in A$, this implies that $f(x_j) \leq S$. Thus, assume that for some $j \in \underline{n}, f(x_j) < S$. This assumption leads to the following calculation:

$$S = \sum_{j=1}^{n} t_j S > \sum_{j=1}^{n} t_j f(x_j) = f\left(\sum_{j=1}^{n} t_j x_j\right) = S, \tag{4.25}$$

where the linearity of f was used. Moreover, the last equality comes from applying the convexity of A by observing that $\sum_{j=1}^{n} t_j x_j \in F$ means $f(\sum_{j=1}^{n} t_j x_j) = S$, which in turn creates a contradiction. In other words, $f(x_j) = S$ for each $j \in \underline{n}$, and hence, F is a face of A. $\square$

For the upcoming theorem and items thereafter, notation for the set of all extreme points of a set will be useful. Said notation is defined next, after which we define what it means for a face to be minimal.

Notation 4.6.1. For a subset A of a vector space, the set of all extreme points of A will be denoted by Ext(A).

Definition 4.6.3. Let E be a Hausdorff LCS over $\mathbb{R}$ and A a convex compact subset of E. A face of A is **minimal** if it contains no proper subset that is also a face.

Theorem 4.6.2. *Let E be a Hausdorff LCS over $\mathbb{R}$ and A a convex compact subset of E. A face of A is minimal if and only if it consists of a singleton from the set of extreme points of A.*

Proof. ($\Rightarrow$): Assume F is a minimal face of A and let $u, v \in F$. Suppose $u \neq v$. Because E is Hausdorff implies E' separates points by way of the duality (E, E'), there exists some $f \in E'$ such that $f(u) \neq f(v)$. Put $Y = \{x \in F : f(x) = \sup\{f(y) : y \in F\}\}$. Then $Y \neq \emptyset$ and by the compactness of F, Y is closed. By Theorem 4.6.1, $Y \sqsubseteq F$. Observe that if $u, v \in Y$, then $f(u) = f(v)$. Because F is already minimal, $Y \neq F$ is impossible; that is, $Y = F$. We conclude that $F = \{u\}$. That $u \in \text{Ext}(A)$ is a short calculation and is left as Exercise 4.6.3.

($\Leftarrow$): Assume $F = \{u\}$ is a face, where $u \in \text{Ext}(A)$. Certainly then, $F \neq \emptyset$ and F is closed as a singleton in a Hausdorff space. Consider a face G such that G is nonempty and closed, with $G \sqsubseteq F$. There is no option other than to conclude that $u \in G$, but then that means $G = F$, implying that F is a minimal face of A. $\qquad\square$

4.6.1 Supporting hyperplanes

Recalling the description of a rectangular solid from the beginning of this section, imagine that it is possible to have a plane that intersects one face of the solid such that the entire solid lies in one of the half spaces defined by the plane. For visualizing example, close this book (or some other physical book if you are reading this electronically) and stand it up vertically on a table (or desk) so that the horizontal surface of the table or desk supports the book. In this way, the rectangularly solid book lies entirely in the (upper) half space defined by the table or desk top. This little experiment motivates the following definition.

Definition 4.6.4. Let E be a Hausdorff LCS over $\mathbb{R}$ and A a convex compact subset of E. Let H be a closed hyperplane in E. Then H is a **supporting hyperplane** of A if $H \cap A \neq \emptyset$ and A is contained in one of the half spaces defined by H.

Notice that by Proposition 3.1.8, H is the null space of some $f \in E'$. The role of supporting hyperplanes is revealed in the next result.

Theorem 4.6.3. *Let E be a Hausdorff LCS over $\mathbb{R}$ and A a convex compact subset of E. Then:*
(a) $\text{Ext}(A) \neq \emptyset$.
(b) *If $f \in E'$, then there exists $u \in \text{Ext}(A)$ such that $f(u) = \sup\{f(y) : y \in A\}$.*
(c) *If H is a supporting hyperplane of A, then $H \cap \text{Ext}(A) \neq \emptyset$.*

Proof. (a) This follows directly from Theorem 4.6.2.

(b) Exercise 4.6.4.

(c) Suppose H is a supporting hyperplane of A. Then for some $c \in \mathbb{K}$, $H = \{x \in E : f(x) = c\}$. By the definition of a supporting hyperplane, A lies entirely in one half space defined by H, which we may assume without loss of generality to be $A \subset \{x \in E : f(x) \le c\}$. Part (a) alerts to the fact that there must be at least one point $u \in \mathrm{Ext}(A)$. Clearly, $H \cap A \ne \emptyset$, which implies that $c \ge \sup\{f(y) : y \in A\}$, hence, $c = \sup\{f(y) : y \in A\}$. We deduce that $f(u) = c$, which means $u \in H$ as well, and the proof is completed. $\qquad\square$

It is worth pointing out that $\mathrm{Ext}(A)$ can have properties that differ from those of A. In particular, $\mathrm{Ext}(A)$ need not be compact for a compact, convex subset of a Hausdorff LCS E, as you can verify in Exercise 4.6.6.

We are now ready for the main result of this section.

Theorem 4.6.4 (Kreĭn–Mil'man theorem). *In a Hausdorff LCS over the real numbers, every convex, compact subset is equal to the closed convex hull of the set of its extreme points; that is, if E is a Hausdorff LCS over $\mathbb{R}$ and A is a convex compact subset of E, then $A = \overline{\mathrm{conv}(\mathrm{Ext}(A))}$.*

Proof. Consider $A \subset E$, a convex compact subset of the Hausdorff LCS E over $\mathbb{R}$. Put $Y = \overline{\mathrm{conv}(\mathrm{Ext}(A))}$. Due to the previous results of this section, we immediately see that $Y \subset A$, so what remains to be shown is the opposite set containment. We proceed along a path of contradiction, namely assume that $A \not\subset Y$. This would mean there exists $x_0 \in A \setminus Y$. By the Hausdorff assumption, $\{x_0\}$ is compact, as well being obviously convex. Additionally, Y is closed and convex. Apply the Helly–Hahn–Banach theorem in the form of Corollary 4.1.1, whereby we find that there is a closed hyperplane H that strictly separates $\{x_0\}$ and Y. The closedness of H implies, via Proposition 3.1.8 that H is the null space of a continuous linear functional on E. In other words, there exist $f \in E'$ and $a \in \mathbb{R}$ such that $f(x_0) > a$ and $a > f(y)$ for all $y \in Y$. Because $f(x_0) > a$, setting $M = \sup\{f(y) : y \in A\}$ implies that $M > 0$. Put $H_M = f^{-1}(M) = \{x \in E : f(x) = M\}$. Next, restrict H_M to a set called K given by $K = \{x \in A : f(x) = M\}$. In Exercise 4.6.5, you can verify that K is a face of A. Apply Proposition 4.6.4 to deduce that there exists a minimal face F of A, then apply Theorem 4.6.2 to conclude that $F = \{u\}$, for some $u \in \mathrm{Ext}(A)$. Thus, $\{u\} = F \subset K \subset H_M$, and also $\{u\} \subset \mathrm{Ext}(A) \subset Y$. Observing that the last statement implies $\{u\} \subset H_M \cap Y$, we have established that $f(u) = M > a$, while $f(u) < a$, and a contradiction has been reached. Therefore, $A \subset Y$ and we have proven that $A = Y = \overline{\mathrm{conv}(\mathrm{Ext}(A))}$. $\qquad\square$

One more note in this section pertains to the case of $\mathbb{K} = \mathbb{C}$.

Remark 4.6.1. The results of this section are valid for $\mathbb{K} = \mathbb{C}$ by way of calculations as done in Lemma 4.1.2.

Exercises

4.6.1. Prove the following items from Proposition 4.6.1: The implication (c) $\Rightarrow$ (a), the equivalence (c) $\Leftrightarrow$ (d) (use induction), (c) $\Rightarrow$ (e) (x cannot be expressed as a convex combination of elements of A).

4.6.2. Describe a set for which the set of all extreme points is open.

4.6.3. Prove that the element u in the first part of the proof of Theorem 4.6.2 is an extreme point of the set A.

4.6.4. Prove part (b) of Theorem 4.6.3.

4.6.5. Prove that the set K of Theorem 4.6.4 is indeed a face of the set A.

4.6.6. Let A be the convex hull of the subset of $\mathbb{R}^3$ that consists of the points $\{(1,0,1),(1,0,-1),(\cos(\theta),\sin\theta,0)\}$, where $0 \le \theta \le 2\pi$.
    (a)   Prove that A is compact.
    (b)   Prove that $\mathrm{Ext}(A)$ is not compact.

4.6.7. This exercise requires information about the Banach spaces $L_p = (L_p([0,1]), \|\cdot\|_p)$, for $1 \le p < \infty$ of p-Lebesgue integrable functions on the unit interval (see A.5.1 in the Appendix). Prove that if B is the closed unit ball of L_1, then B does not contain any extreme points. Prove that for $1 < p < \infty$, the set of extreme points of the unit ball in L_p consists of the surface of the ball, namely $S = \{f \in L_p : \|f\|_p = 1\}$.

4.6.8. Let E be the space c_0 of scalar valued null sequences with the sup norm (see Definition A.5.1 in the Appendix). Determine $\mathrm{Ext}(B)$, where B is the closed unit ball, and use that result to give another proof (distinct from using Riesz's Theorem 3.2.2) that the closed unit ball of c_0 cannot be compact.

4.6.9. Let E be a LCS over $\mathbb{R}$ and suppose $A \subset E$ is convex, closed, and has nonempty interior. Prove the following:
    (a)   For every boundary point x of A (see Definition A.3.11 in the Appendix), there exists a supporting hyperplane H of A such that $x \in H$.
    (b)   Every supporting hyperplane H of A satisfies $H \cap \mathring{A} = \emptyset$, where $\mathring{A}$ is the interior of $A2$.

4.6.10. The following is adapted from [21, Ejemplo (b), p. 78]. Let E be the vector space of all 2×2 matrices over $\mathbb{R}$ such that each $A \in E$ is of the form $A = (a_{ij})$, where $a_{ij} \in \mathbb{R}$, for $i,j \in \underline{2}$. Equip E with the norm given by: For every $A = (a_{ij}) \in E$, $\|A\| = \max\{|a_{ij}| : i,j \in \underline{2}\}$. Define a set Γ by

$$\Gamma = \left\{ A = (a_{ij}) \in E : (\forall i,j \in \underline{2})\, a_{ij} \ge 0, \text{ and } \sum_{i=1}^{2} a_{ij} = 1, \sum_{j=1}^{2} a_{ij} = 1 \right\}. \tag{4.26}$$

With this information, you can verify the following:
    (a)   Prove that Γ is convex.
    (b)   Prove that Γ is compact.
    (c)   Prove that $A \in \Gamma$ if and only if

$$A = \begin{pmatrix} a & b \\ c & d \end{pmatrix} \tag{4.27}$$

       satisfies the properties that $a = d, b = c, a + b = 1, a \ge 0, b \ge 0$.
    (d)   Prove that the following are the only extreme points of Γ:

$$I = \begin{pmatrix} 1 & 0 \\ 0 & 1 \end{pmatrix}, \quad J = \begin{pmatrix} 0 & 1 \\ 1 & 0 \end{pmatrix}. \tag{4.28}$$

    (e)   Explain how the Kreĭn–Mil'man Theorem 4.6.4 applies to Γ.

5 Some special types of locally convex spaces

Now that we know something about the dual of a LCS, we can begin to look at how the interaction between a LCS and its dual can lead to new concepts, and consequently, results that involve those concepts. The first topic is about barreled spaces of Definition 5.1.1. Such spaces turn out to be useful in multiple contexts and have properties and can sometimes serve as substitutes for completeness. We also consider normed spaces that can be constructed within general LCS, specifically the spaces $(E_B, \| \cdot \|_B)$ (see Proposition 5.1.3). Between normed spaces, linear maps are continuous if and only if they map bounded sets to bounded sets, but we have already seen in Exercise 2.3.9 that in general, it is possible for a discontinuous linear map between TVS to map bounded sets to bounded sets. Hence, the topic that follows concerns determining which LCS satisfy the property that whenever a linear map sends bounded sets to bounded sets, the linear map must be continuous. Such LCS are referred to as bornological spaces of Definition 5.2.1, this term being based on the French word *borné*, which means "limited." After that, we focus on the second dual. For normed spaces (actually Banach spaces) you have probably seen the concept of reflexivity in terms of the second dual being equivalent to the original space. In the general context, there is a distinction to be made regarding the second dual, which leads to the notion of a semireflexive LCS. The penultima section is focused on when a LCS satisfies the property that every closed, bounded subset is compact. The final section is devoted to the application of the concepts of the previous sections to the context of normed and Banach spaces.

5.1 Barreled spaces

The first order of business is to define the term to which the title of this section refers.

Definition 5.1.1. An absolutely convex, closed, and absorbing set is called a **barrel**. A LCS is **barreled** if every barrel must be a zero neighborhood.

If you imagine a physical barrel (or look up an image of a barrel) and center it around the origin of $\mathbb{R}^3$, you will see that it is absolutely convex, closed (no dents or holes in it), and has a nontrivial volume (so it is absorbing). Other words for a barrel include **tonneau, keg**, and **barril**. There are other names for barreled, too, such as **tonnelé, kegly**, and **barrilado**. You may also see an alternate spelling in the form of *barrelled*, though that term is usually flagged as a misspelling (I confess that I have used that misspelling myself!). We have already seen via Theorem 2.1.1, that a LCS can always have a zero neighborhood base consisting of barrels, but what we require in Definition 5.1.1 is that *every* barrel be a zero neighborhood. Before going into more detail about barreled spaces, it behooves us to know that there exist nontrivial barreled LCS.

Proposition 5.1.1. *Every Baire LCS is barreled.*

https://doi.org/10.1515/9783111392868-005

Proof. Suppose E is a Baire LCS and let U be any barrel in E. Then $\bigcup_{n=1}^{\infty} nU = E$ because U is absorbing (and absolutely convex). As a Baire space, there must be some specific $n_0 \in \mathbb{N}$ such that $n_0 U$ has nonempty interior. In particular, this means U is a neighborhood of some point $x \in E$. The set U is balanced so $-x \in U$. Applying the convexity of U, we observe that $0 = \frac{-1}{2}x + \frac{1}{2}x \in U$. Thus, U is in fact a neighborhood of zero and Definition 5.1.1 applies. E is barreled. $\qquad\square$

Corollary 5.1.1. *Every complete, metrizable LCS is barreled; in particular, every Fréchet and every Banach space is barreled.*

So, there are plenty of barreled spaces. On the other hand, the next item represents an example of a normed space that is not barreled. Another example that does not involve Lebesgue integrals is the space c_{00}, which you can verify as not being barreled in Exercise 5.1.3.

> **Example 5.1.1.** The space of continuous real valued functions on the unit interval with the L_1 norm is not barreled.

To verify the above statement, let $E = C([0,1])$, the vector space of continuous real valued functions on $[0,1]$, equipped with the norm $(\forall f \in E)\, \|f\|_1 = \int_0^1 |f|\, d\lambda$, where λ represents the Lebesgue measure. Let $B = \{f \in E : \sup\{|f(t)| : t \in [0,1]\} \leq 1\}$. It is easy to verify that B is absolutely convex and absorbing. We will now prove that B is closed using sequences via the norm. To this end, suppose $g \in \overline{B}$ is arbitrary. Then there is a sequence (f_n) in B that converges to g in the $\|\cdot\|_1$ norm. In other words, $\int_0^1 |f_n - f|\, d\lambda \to 0$, and this means (f_n) must have a subsequence that converges to f pointwise almost everywhere. Meanwhile, each f_n is continuous and has a supremum that is at most 1, and f is also continuous. These two observations imply that f must belong to B and B is closed. Being absolutely convex, absorbing and closed, B is in fact a barrel. We finish this example by showing that B cannot be a zero neighborhood. Suppose on the contrary that B is a zero neighborhood in E. Then for some $\varepsilon > 0$, $V = \{f \in E : \int_0^1 |f|\, d\lambda \leq \varepsilon\} \subset B$. However, given any such $\varepsilon > 0$, we can find a continuous f on $[0,1]$ such that $\int_0^1 |f|\, d\lambda < \varepsilon$ but $\sup\{|f(t)| : t \in [0,1]\} > 1$, which implies that V is not contained in B; that is to say, B is not a zero neighborhood.

It can be shown that any product of barreled spaces is barreled; for details, see [103, Proposition V.27, p. 94]. On the other hand, it is possible that a closed subspace of a barreled space is not barreled, as can be seen in [130, 15-1-4, p. 249].

The next result allows us to find more barreled spaces. Feel free to check back to Definition 3.5.1 of final topologies, if necessary.

Theorem 5.1.1. *A LCS equipped with the final topology of a collection of barreled spaces is barreled. That is, suppose $\{(E_\alpha, \mathcal{T}_\alpha, f_\alpha) : \alpha \in I\}$ is a collection of barreled spaces $(E_\alpha, \mathcal{T}_\alpha)$, and for a fixed LCS E, each linear function $f_\alpha : E_\alpha \to E$ is such that E is the linear span of $\bigcup_{\alpha \in I} f_\alpha(E_\alpha)$. Then E with the final topology is barreled.*

Proof. Let D be any barrel in E. The continuity and linear properties of each f_α tell us that $(\forall \alpha \in I) f_\alpha^{-1}(D)$ is a barrel in E_α and is therefore a zero neighborhood in E_α. By the way, Proposition 3.5.1 now applies, and we conclude that D must be a zero neighborhood in E. $\qquad\square$

Given that the topologies of the spaces in the next corollary are final topologies, we discover more types of barreled spaces. In fact, we can observe that the spaces below represent some hereditary properties of being barreled.

Corollary 5.1.2. *The following types of LCS are barreled:*
(a) *Hausdorff quotients of barreled LCS.*
(b) *Inductive limits of sequences of barreled LCS.*
(c) *Direct sums of sequences of barreled LCS.*
(d) *All (LB) and (LF) spaces.*

A perk of the corollary above is that we deduce via Proposition 3.5.8 that any proper, strict (LB) spaces represents a space that is barreled but not Baire. We also deduce that proper hyperstrict (LB) spaces are barreled spaces that are complete (Theorem 3.5.5) but neither Baire nor metrizable, the last item being a consequence of Proposition 3.5.5.

There is a connection between equicontinuous sets and bounded sets that will provide more insight into barreled spaces. That connection is next.

Proposition 5.1.2. *For any LCS, equicontinuous subsets of the dual are strongly bounded.*

Proof. Suppose $\mathcal{H}$ is an equicontinuous subset of the dual E' of a LCS E. Then by Proposition 4.3.3, there is a zero neighborhood V in E such that $\mathcal{H} \subset V^\circ$. As V absorbs bounded sets, we have: If $A \subset E$ is any bounded set, then there exists $a > 0$ such that $\frac{1}{a} A \subset V$. Thus, $A^\circ \supset \frac{1}{a} V^\circ$; that is, $V^\circ \subset a A^\circ$. We then observe that $\mathcal{H} \subset V^\circ \subset a A^\circ$. Because A was an arbitrary bounded subset of E, $\mathcal{H}$ is $\beta(E', E)$-bounded (see Example 4.3.2). $\qquad\square$

Next up, we seek an equivalent condition under which a LCS is barreled. A preliminary result, which is interesting in its own right, comes first.

Theorem 5.1.2. *A subset of the dual of a Hausdorff LCS E is $\sigma(E', E)$-bounded if and only if it is contained in the polar of a barrel.*

Proof. Let $(E, \mathcal{T})$ be a Hausdorff LCS.
($\Rightarrow$): Assume $A \subset E'$ is $\sigma(E', E)$-bounded. By properties of polars (see Proposition 4.2.4), if $B = A^\circ$, then B is absolutely convex, absorbing, and $\sigma(E, E')$-closed. By Theorem 4.3.2, B is also $\mathcal{T}$-closed. Hence, B is a barrel in E, and $B^\circ = A^{\circ\circ} \supset A$.
($\Leftarrow$): Now assume $A \subset E'$ is such that $A \subset B^\circ$ for some barrel B in E. The Bipolar Theorem 4.2.1 reveals that $B = B^{\circ\circ}$. Because B is absorbing, we apply Proposition 4.2.5 and observe that B° is $\sigma(E', E)$-bounded. Hence, $A \subset B^\circ$ must also be $\sigma(E', E)$-bounded. $\qquad\square$

Corollary 5.1.3. *For a Hausdorff LCS the following are equivalent:*

(a) *E is barreled.*
(b) *Every $\sigma(E', E)$-bounded subset of E' is equicontinuous.*
(c) *The topology of E is the strong topology $\beta(E, E')$.*

Proof. Indeed, these equivalences follow by applying Theorem 5.1.2 in the case of every barrel being a zero neighborhood. $\qquad\square$

We deduce an observation from Corollary 5.1.3. The part of (a) $\Leftrightarrow$ (b) represents the **Banach–Steinhaus theorem** for dual spaces.

Corollary 5.1.4. *A Hausdorff barreled LCS carries the Mackey topology 4.3.3.*

Proof. This follows from Corollary 5.1.3 above and noting that the Mackey topology is the finest topology consistent with the duality. $\qquad\square$

Corollary 5.1.5. *In a Hausdorff barreled LCS, the closed, absolutely convex hull of a $\sigma(E', E)$-compact set remains $\sigma(E', E)$-compact.*

Proof. Exercise 5.1.9. $\qquad\square$

The result above leads to a summary of collections of subsets that are equal.

Corollary 5.1.6. *In a barreled space E, the following collections of subsets of its dual are identical:*
(a) *The equicontinuous subsets.*
(b) *The $\sigma(E', E)$-relatively compact subsets.*
(c) *The $\sigma(E', E)$-bounded subsets.*
(d) *The $\beta(E', E)$-bounded subsets.*

In particular, for a barreled LCS the weakly bounded subsets of the dual are strongly bounded, even equicontinuous.

Proof. We proceed as follows: First, (a) $\Rightarrow$ (d) by Proposition 5.1.2, (d) $\Rightarrow$ (c) by weaker topologies (the continuous linear image of a bounded set is bounded), (c) $\Rightarrow$ (a) by Corollary 5.1.3. Hence, (a), (c), and (d) are equivalent. Next, (a) $\Rightarrow$ (b) by the Alaoğlu–Bourbaki Theorem 4.4.1. To finish, we combine (b) $\Rightarrow$ (c), which holds by way of compact sets being bounded as shown in Proposition 2.5.10, with (c) $\Rightarrow$ (a). $\qquad\square$

Definition 5.1.2. A **Banach–Mackey space** is a LCS such that every weakly bounded subset is strongly bounded.

Thus, barreled spaces are Banach–Mackey spaces. Examples of nonbarreled Banach–Mackey LCS will be shown in Section 7.2.

Next is another equivalence of barreledness. The definition of lower semicontinuous functions can be found in Definition A.3.26.

Theorem 5.1.3. *A LCS is barreled if and only if every lower semicontinuous seminorm on it is continuous.*

Proof. Exercise 5.1.5. □

5.1.1 The spaces E_B

We will now construct an important kind of normed space that will be used in a variety of circumstances going forward. The construction was originally done by Grothendieck ([62]).

Definition 5.1.3. Let B be an absolutely convex subset of a LCS E. Define E_B to be the vector subspace consisting of the span of B.

Notice that the absolute convexity of B above implies that the vector space E_B can be written as $E_B = \bigcup_{n=1}^{\infty} nB$. Furthermore, on the vector space E_B we can define a seminormed topology by way of the Minkowski functional μ_B of B. When B is bounded, we obtain a norm as verified next.

Proposition 5.1.3. *Consider a LCS E and the vector subspace consisting of E_B of Definition 5.1.3 with the seminormed topology given by the Minkowski functional μ_B of B. If B is bounded, then μ_B is a norm with B as its unit ball. Moreover, the topology generated by μ_B is given by $\{\frac{1}{k}B : k \in \mathbb{N}\}$ and is finer than the relative topology on E_B.*

Proof. We already know that μ_B is a seminorm on E_B. Assume B is bounded and let $x \in E_B$ be any nonzero element. Then there exists $t > 0$ such that $x \notin tB$; that is, $\mu_B(x) > 0$, which implies that μ_B is in fact a norm. For the second statement, observe that the topology of (E_B, μ_B) is generated by $\{\frac{1}{k}B : k \in \mathbb{N}\}$. Because B is bounded, given any zero neighborhood U in E, there exists $K \in \mathbb{N}$ such that $\frac{1}{K}B \subset U$, which confirms that the topology of μ_B is finer that the relative topology from E. □

Observe that for an absolutely convex zero neighborhood U in a nonnormable LCS E, the space (E_U, μ_U) is a seminormed space that is not normed. Meanwhile, Proposition 5.1.3 prompts the following definition.

Notation 5.1.1. When B is an absolutely convex bounded subset of E, denote the norm generated by the Minkowski functional of B by $\| \cdot \|_B$. The corresponding normed space is denoted by $(E_B, \| \cdot \|_B)$.

So, we obtain a normed space $(E_B, \| \cdot \|_B)$ and this points to a variety of possibilities for properties that such a normed space could have. The three main possibilities that we will consider are defined below. Recall from Definition 2.2.4 that an absolutely convex, bounded set is called a disk.

Definition 5.1.4. Assume the set B is a disk. If the normed space $(E_B, \|\cdot\|_B)$ is complete (resp., Baire, barreled), then B is called a **Banach disk** (resp., a **Baire disk, barreled disk**).

Most of the time, it is convenient to assume that the disk B is closed. However, Valdivia proved that in every Fréchet space of infinite dimension, there exist Banach disks that are not closed. See [95, Proposition 3.2.21, p. 89] for details. Also, it is obvious that the closed unit ball of a Banach space is a Banach disk. Every Banach space is a Baire space by the Baire Category Theorem A.3.9. Proposition A.3.9 tells us that every Baire disk is a barreled disk. Going in the other direction, in [78], Kliś constructed a normed space that is Baire but not complete, and in [95, 4.7.2, p. 134] there is an example of a normed barreled space that is not Baire. Also, recall from Example 5.1.1 that there are normed spaces that are not even barreled. Considering the closed unit ball of each of these spaces, we obtain the string of relations below:

$$\text{Banach disk} \overset{\Rightarrow}{\nLeftarrow} \text{Baire disk} \overset{\Rightarrow}{\nLeftarrow} \text{barreled disk}. \tag{5.1}$$

Example 5.1.2. A Banach disk within a nonmetrizable space.

Consider a proper hyperstrict (LB) space $E = \text{ind}_n E_n$, and consider the closed unit ball B_1 of E_1. The completeness of E_1 tells us that B_1 is bounded in E. The vector space E_{B_1} and the corresponding Minkowski functional μ_{B_1} are unchanged in E. Hence, B_1 is a Banach disk in E_1. The continuity of the injection $\text{id} : E_1 \to E$ implies that B_1 is a Banach disk in E. We know from Corollary 3.5.5, that E is not metrizable.

The concept of a barreled disk is used in the following big prelude to a big result.

Theorem 5.1.4 (Banach–Mackey). *Barrels absorb closed barreled disks.*

Proof. Suppose U is a barrel in the LCS E, and let B be any closed barreled disk. Let $S = U \cap E_B$. Clearly, S is absolutely convex and absorbing in E_B. Moreover, by Proposition 5.1.3, S is closed in E_B. Thus, S is a barrel in E_B and by assumption, S is a zero neighborhood in E_B. This implies that S absorbs all bounded sets, one of which is B. We conclude that, as $S \subset U$, B is absorbed by U. $\qquad\square$

On any LCS, observe that the identity maps from the strong topology to the original topology, and from the original topology to the weak topology are both continuous. In other words, strongly bounded subsets are bounded in the original topology and sets that are bounded in the original topology are weakly bounded. Because Proposition 4.2.5 indicates that weakly bounded subsets in a LCS are connected to polars using the dual, we may wonder about when weakly bounded subsets are also originally bounded or even strongly bounded. We will address some of these questions in the rest of this section. Theorem 5.1.4 creates a pathway to observing when weakly bounded subsets are bounded with respect to the Mackey topology $\mu(E, E')$, as will be seen in the next theorem. First, however, a refinement of Theorem 5.1.4, which traditionally has been denoted as the Banach–Mackey theorem, is needed.

Proposition 5.1.4. *In a Hausdorff LCS, an absolutely convex, bounded, sequentially complete set is a Banach disk.*

Proof. Suppose B is an absolutely convex, bounded, sequentially complete subset of a Hausdorff LCS $(E, \mathcal{T})$. We first note that as a normed space, $(E_B, \|\cdot\|_B)$ is Hausdorff. Proposition 5.1.3 alerts us to the fact that $\mathcal{T}|_{E_B}$ is coarser than the topology of the norm $\|\cdot\|_B$. The base $\{\frac{1}{k}B : k \in \mathbb{N}\}$ of the topology of $\|\cdot\|_B$ consists of sets that are $\mathcal{T}$-complete. These items put us in a position to invoke Robertson's Theorem 2.5.2 to conclude that $(E_B, \|\cdot\|_B)$ is complete; that is, B is a Banach disk. $\qquad\square$

Corollary 5.1.7. *An absolutely convex, compact subset of a Hausdorff LCS is a Banach disk.*

Here comes the big result alluded to earlier.

Theorem 5.1.5 (Mackey's theorem). *The bounded sets are the same for all topologies that are consistent with a duality of Hausdorff LCS.*

Proof. Let E be such a LCS with dual E'. Let $\mathcal{T}$ be any topology that is consistent with the duality (E, E'). From the observation above, any set that is $\mathcal{T}$-bounded is $\sigma(E, E')$-bounded. What remains to be proven is the opposite implication. For this, let A be any $\sigma(E, E')$-bounded subset of E, and let $U \in \mathcal{N}_{0,\mathcal{T}}$ be arbitrary, where we assume U to be absolutely convex and closed. By Proposition 4.2.5 and properties of polars, A° is a barrel in $(E', \sigma(E', E))$. Meanwhile, U° is $\sigma(E', E)$-compact by the Alaoğlu–Bourbaki Theorem 4.4.1. Now apply Corollary 5.1.7 to observe that U° is a Banach disk. As each Banach space is barreled, Theorem 5.1.4 implies that A° absorbs U°. The Bipolar Theorem 4.2.1 tells us that $A^{\circ\circ} = \overline{\operatorname{conv}\operatorname{bal}(A)}^{\sigma(E,E')}$. Thus, $A \subset A^{\circ\circ}$, from which we deduce that U absorbs A. $\qquad\square$

As we proceed, it will become evident that the locally convex topology of many LCS that we consider important, is the Mackey topology, $\mu(E, E')$. The first corollary below represents one instance of this importance.

Corollary 5.1.8. *Metrizable LCS are Mackey spaces.*

Proof. Let $\mathcal{N}_0 = \{U_n : n \in \mathbb{N}\}$ be a base for the metrizable LCS $(E, \mathcal{T})$, where we may assume $(\forall n \in \mathbb{N})\, U_{n+1} \subset U_n$. Let V be any $\mu(E, E')$-zero neighborhood. By way of contradiction, assume $V \notin \mathcal{N}_0$. Then for every $n \in \mathbb{N}$, there is $x_n \in U_n$ for which $x_n \notin nV$. As $x_n \in U_n$ for every $n \in \mathbb{N}$, the sequence (x_n) is bounded with respect to $\mathcal{T}$. Mackey's Theorem 5.1.5 indicates that (x_n) is $\mu(E, E')$-bounded as well. This means there is $a > 0$ such that $(\forall b \geq a)(\forall n \in \mathbb{N})\, x_n \in bV$, contradicting $x_n \notin nV$. $\qquad\square$

Corollary 5.1.9. *Any incomplete metrizable LCS represents an example of an incomplete Mackey space.*

Corollary 5.1.10. *An absolutely convex, closed, bounded set in a LCS remains unchanged in all topologies consistent with the duality. In particular, a closed Banach (resp., closed Baire, closed barreled) disk in a LCS remains a closed Banach (resp., closed Baire, closed barreled) disk in all topologies consistent with the duality.*

The last corollary above is worth contemplating. It tells us that we can form Banach (normed Baire or normed barreled) subspaces in a general LCS, slide those spaces between the weak and Mackey topologies, and they will be unchanged. Some or all of the consistent topologies might not even be metrizable. Considering that closure and boundedness are topological concepts, it is surprising that we have such flexibility. We will look more closely at this phenomenon in Chapter 7.

A significant consequence of Mackey's theorem is that it reveals examples of bounded, discontinuous linear maps, thus distinguishing general LCS from normed spaces. Below is an example.

Example 5.1.3. An example of a bounded, discontinuous linear map.

Let $(E, \| \cdot \|)$ be an infinite-dimensional Banach space. Then the identity map id : $(E, \sigma(E, E')) \to (E, \| \cdot \|)$ is a bounded, discontinuous linear map. Indeed, Mackey's Theorem 5.1.5 tells us that weakly bounded sets are bounded with respect to the norm. On the other hand, Proposition 4.2.7 informs us that $\sigma(E, E')$ is strictly coarser than the normed topology on E, ensuring that the stated identity map is discontinuous.

5.1.2 Quasibarreled spaces

There is a type of LCS that is more general than a barreled space but which as comparable properties. The definition follows.

Definition 5.1.5. A subset of a TVS that absorbs all bounded sets is called **bornivorous** or a **bornivore**. If every bornivorous barrel in a LCS is a zero neighborhood, then the LCS is called **quasibarreled** (also **infrabarreled**, **evaluable**).

Certainly, if every barrel is a zero neighborhood, then so is every bornivorous barrel; that is, every barreled LCS is quasibarreled. After the next proposition, we can identify a LCS that is quasibarreled but not barreled.

Proposition 5.1.5. *Every metrizable LCS is quasibarreled; in particular, every normed space is quasibarreled.*

Proof. Suppose that in the metrizable LCS E there is a bornivorous barrel B that is not a zero neighborhood. Let $\{U_n : n \in \mathbb{N}\}$ be a decreasing base of zero neighborhoods of E. Then no U_n is contained in B; in other words, $(\forall n \in \mathbb{N})(\exists x_n \in U_n)\, x_n \notin nB$. By its construction, (x_n) is a null sequence, and consequently, $\{x_n : n \in \mathbb{N}\}$ is a bounded subset of E. By assumption, B absorbs every bounded set, so $(\exists a > 0)(\forall n \in \mathbb{N})\, x_n \in aB$. Choose $k \in \mathbb{N}$ such that $k > a$. Then the absolute convexity of B ensures that $(\forall n \in \mathbb{N})\, x_n \in kB$, a contradiction. $\qquad\square$

By the way, the case of every normed space being quasibarreled can easily be proven directly, which you can do in Exercise 5.1.15. Thus, the example below shows that being quasibarreled is a strict generalization of being barreled.

Example 5.1.4. Any nonbarreled normed space (such as Example 5.1.1) is a quasibarreled space that is not barreled.

In what follows, we see that roughly speaking, quasibarreled spaces relate to strongly bounded sets the way barreled spaces relate to weakly bounded sets.

Theorem 5.1.6. *A subset of the dual of a Hausdorff LCS is strongly bounded if and only if it is contained in the polar of a bornivorous barrel.*

Proof. ($\Rightarrow$): Assume $(E, \mathcal{T})$ is a Hausdorff LCS and that $S \subset E'$ is $\beta(E', E)$-bounded. We wish to show that S is contained in the polar of a bornivorous barrel. Of course, S is $\sigma(E', E)$-bounded. By Proposition 4.2.4(e), $S°$ is absolutely convex and $\sigma(E, E')$-closed, so by Theorem 4.3.2, $S°$ is $\mathcal{T}$-closed. Moreover, by Proposition 4.2.5, $S°$ is absorbing. In short, $S°$ is a barrel in E. As S is $\beta(E', E)$-bounded, S is absorbed by the polar of any $\sigma(E', E)$-bounded subset A of E. That is, there exists $a > 0$ such that $S \subset aA°$. Mackey's Theorem 5.1.5 signals to us that A is $\mathcal{T}$-bounded. Meanwhile, properties of polars from Proposition 4.2.4 allow us to calculate: $S \subset aA° \Rightarrow A°° \subset aS° \Rightarrow A \subset A°° \subset aS°$. The arbitrariness of the bounded set A implies that $S°$ is bornivorous. The conclusion of this part now follows from observing that $S \subset S°°$.

($\Leftarrow$): Assume $S \subset B°$ where B is a bornivorous barrel in E. Given any $\sigma(E, E')$-bounded subset A of E, Mackey's Theorem 5.1.5 applies to observe that A is $\mathcal{T}$-bounded. B is bornivorous, so there is an $a > 0$ such that $A \subset aB$, and in turn $B° \subset aA°$. As A is arbitrary, we have that $B°$ is $\beta(E', E)$-bounded. The proof is finished by noting that $S \subset B°$ implies that S is $\beta(E', E)$-bounded. $\qquad\square$

Next is an equivalence for a Hausdorff LCS to be quasibarreled, which you may compare to Corollary 5.1.3 regarding the barreled case.

Theorem 5.1.7. *A Hausdorff LCS is quasibarreled if and only if every strongly bounded subset of the dual is equicontinuous.*

Proof. ($\Rightarrow$): Suppose the Hausdorff LCS $(E, \mathcal{T})$ is quasibarreled and let $S \subset E'$ be any $\beta(E', E)$-bounded set. Theorem 5.1.6 indicates that there exists a bornivorous barrel $B \subset$

E such that $S \subset B^\circ$. By assumption, $B \in \mathcal{N}_{0,E}$. Now apply Proposition 4.3.3: S is equicontinuous.

($\Leftarrow$): Exercise 5.1.16. $\qquad\square$

The following are left as exercises.

Theorem 5.1.8. *In every quasibarreled Hausdorff LCS weakly relatively compact subsets of the dual are strongly bounded.*

Proof. Exercise 5.1.17. $\qquad\square$

Theorem 5.1.9. *Every quasibarreled Hausdorff LCS carries the Mackey topology.*

Proof. Exercise 5.1.18. $\qquad\square$

Theorem 5.1.10. *The property of being quasibarreled is preserved under the formation of final topologies. In particular, inductive limits of quasibarreled LCS are quasibarreled, locally convex direct sums of quasibarreled LCS are quasibarreled, and quotients of quasibarreled LCS are quasibarreled.*

Proof. Exercise 5.1.19. $\qquad\square$

We finish this section with the comment that the product of (quasi)barreled LCS is (quasi)barreled, the proof of which can be found in [79, 27.1 (5), p. 368]. In [79, 31.5, pp. 433–434], there is a description of a closed subspace of a barreled space that is not even quasibarreled.

Exercises

5.1.1. Prove that any LCS that is equipped with its finest locally convex topology (see Example 2.1.3) is barreled.

5.1.2. Give a direct proof that a quotient of a barreled LCS is also barreled.

5.1.3. Prove that the normed space c_{00} of finitely many nonzero scalar sequences (see Example A.5.1) is not barreled.

5.1.4. Prove that a LCS $(E, \mathcal{T})$ is barreled if and only if any locally convex topology having a base of $\mathcal{T}$-closed sets must be coarser than $\mathcal{T}$.

5.1.5. Regarding Theorem 5.1.3 for a LCS E proceed as follows:
 (a) Prove that a function $f : E \to \mathbb{R}$ is lower semicontinuous if and only if for every $a \in \mathbb{R}$ the set $\{x : f(x) \le a\}$ is closed in E.
 (b) Suppose p is a seminorm on E. Prove that the set $\{x : p(x) \le 1\}$ is a barrel if and only if p is lower semicontinuous.
 (c) Prove Theorem 5.1.3.

5.1.6. Prove that a quasicomplete Hausdorff LCS satisfies the property that weakly bounded subsets of the dual are strongly bounded; that is, every $\sigma(E', E)$-bounded subset is $\beta(E', E)$-bounded. This result will be substantially generalized in Chapter 7. *Hint:* Apply some combination of Theorem 5.1.6, Proposition 5.1.4, and Theorem 5.1.2.

5.1.7. Use Theorem 5.1.2 to give another proof of Mackey's Theorem 5.1.5. *Hint:* Use the Banach–Steinhaus theorem for normed spaces (see [67, Theorem 1.8.1, p. 62]).

5.1.8. Give a direct proof of Corollary 5.1.1 that every Fréchet space is barreled, using the technique of Corollary 5.1.8.

5.1.9. Prove Corollary 5.1.5.

5.1.10. Prove the following:
 (a) A finite codimensional subspace of a barreled space is barreled.
 (b) The closed unit ball of a dense hyperplane of an infinite-dimensional Banach space represents a barreled disk that is not a Banach disk. *Hint:* Use part (a).

5.1.11. Suppose A and B are bounded disks in a LCS E. Prove that if $A \subset B$, then $\mu_B \le \mu_A$, where μ_A and μ_B represent the Minkowski functionals of A and B, respectively.

5.1.12. Prove the following: Finite sums of Banach disks, the intersection of any family of Banach disks, and a continuous linear image of a Banach disk are again Banach disks.

5.1.13. Describe a closed bounded subset B of a LCS E such that $\bigcup_{n=1}^{\infty} nB$ is not a vector subspace of E.

5.1.14. Prove that if a barrel in a Hausdorff LCS absorbs all null sequences, then that barrel is bornivorous. *Hint:* Assume such a barrel is not bornivorous and construct a null sequence that is not absorbed by that barrel.

5.1.15. Give a direct proof the second part of Proposition 5.1.5; that is, prove directly that every normed space is quasibarreled.

5.1.16. Complete the proof of Theorem 5.1.7. *Hint:* Apply Theorem 5.1.6.

5.1.17. Prove Theorem 5.1.8.

5.1.18. Prove Theorem 5.1.9. *Hint:* Use Theorem 5.1.8.

5.1.19. Prove Theorem 5.1.10.

5.2 Bornological and ultrabornological spaces

We saw earlier (Example 5.1.3 and Exercise 2.3.9) that it is possible for a linear map between LCS to send bounded sets to bounded sets yet be discontinuous, revealing a difference between LCS and normed spaces. In this section, we look at a collection of LCS for which linear maps that send bounded sets to bounded sets must be continuous. Though it may seem that we will end up in the realm of normed spaces, we will see that this collection of LCS includes spaces that are not normable. The definition below represents the main concept of this section. In comparison with quasibarreled LCS, we now consider sets that are bornivores as possible zero neighborhoods, independently of being barrels.

Definition 5.2.1. A LCS for which every absolutely convex bornivore (see Definition 5.1.5) is a zero neighborhood is called **bornological**.

The first observation is the following.

Proposition 5.2.1. *Every bornological LCS is quasibarreled.*

Proof. Any bornivorous barrel in a bornological LCS is a zero neighborhood by Definition 5.2.1 above of a bornological LCS. $\square$

It is worth pointing out that several types of LCS are Mackey spaces, formalized below.

Remark 5.2.1. All barreled, quasibarreled, and bornological LCS are Mackey spaces: See Corollary 5.1.4, Theorem 5.1.9, and Proposition 5.2.1 above, respectively.

Of course, every normed space is bornological. More generally, metrizable LCS are bornological. The proof of this fact below will resemble that of Proposition 5.1.5.

Proposition 5.2.2. *Metrizable LCS are bornological, in particular, every normed space is bornological.*

Proof. Let $\{U_n : n \in \mathbb{N}\}$ be a decreasing base of zero neighborhoods of the metrizable LCS E. Following a path to contradiction, suppose V is an absolutely convex bornivore that is not a zero neighborhood in E. Then $(\forall n \in \mathbb{N}) \exists x_n \in U_n \setminus nV$. Then the sequence (x_n) converges to 0 which means the set $A = \{x_n : n \in \mathbb{N}\}$ is bounded in E. This means A must be contained in some scalar multiple of V. We conclude that for some $k \in \mathbb{N}$, $A \subset kV$ and we have reached the desired contradiction. Hence, E is bornological. $\square$

Some observations: First, the existence of normed spaces that are not barreled (see Example 5.1.1 and Exercise 5.1.3) indicates that there are LCS that are bornological but not barreled. On the other hand, there exist barreled spaces that are not bornological but such constructions are more involved. See [91, p. 449] for more details and references regarding the existence of such spaces. Meanwhile, though we saw in Proposition 5.2.1

that every bornological LCS is quasibarreled, there exist quasibarreled LCS that are not bornological by way of the previous sentence.

Our next task is to verify that bornological spaces are precisely those LCS for which linear maps that send bounded sets to bounded sets must be continuous. To prepare for that result, some calculations with images and preimages of linear maps are helpful. Such calculations are done in the next proposition.

Proposition 5.2.3. *Suppose T is a linear map from a vector space E to a vector space F. If $A \subset E$ and $C \subset F$ are both balanced then C absorbs $T(A)$ in F if and only if $T^{-1}(C)$ absorbs A in E.*

Proof. ($\Rightarrow$): With the notation of the statement, suppose C absorbs $T(A)$. Then for some $a > 0$, $T(A) \subset aC$. Properties of images and inverse images by functions tell us that $A \subset T^{-1}(T(A)) \subset aT^{-1}(C)$, by the linearity of T.

($\Leftarrow$): Now suppose $T^{-1}(C)$ absorbs the set A. Then $A \subset aT^{-1}(C)$, for some $a > 0$. Consequently, $T(A) \subset aT(T^{-1}(C)) \subset aC$. $\qquad\qquad\square$

What follows is the characterization of bornological spaces with regard to linear maps.

Theorem 5.2.1. *A LCS is bornological if and only if every linear map from that LCS to another LCS that sends bounded sets to bounded sets must be continuous.*

Proof. ($\Rightarrow$): Assume $(E, \mathcal{T})$ is a bornological LCS. Let F be any other LCS and $T : E \to F$ a linear map. We intend to prove that if T sends bounded subsets of E to bounded subsets of F, then T must be continuous. To this end, let $V \in \mathcal{N}_{0,F}$ be arbitrary, where we additionally assume that V is absolutely convex. As a zero neighborhood, V is bornivorous. Also, $T^{-1}(V)$ is absolutely convex by Theorem 1.1.2. Let $B \subset E$ be any bounded set. By assumption, $T(B) \subset F$ is bounded, so V absorbs $T(B)$. Applying Proposition 5.2.3 above to V, we see that $T^{-1}(V)$ absorbs B. We have shown that $T^{-1}(V)$ is a bornivorous, absolutely convex subset of E. Because E is bornological, $T^{-1}(V) \in \mathcal{N}_{0,E}$, and T is indeed continuous.

($\Leftarrow$): Now assume that for any LCS F every linear map $T : E \to F$ that sends bounded sets to bounded sets is continuous. We will show that E must be bornological. Let $U \subset E$ be absolutely convex and bornivorous. Consider the Minkowski functional μ_U of U. By Proposition 2.1.3, μ_U is a seminorm on E. Therefore, we may consider (E, μ_U) as a LCS. Let us examine the linear map $\mathrm{id}_{\mu_U} : (E, \mathcal{T}) \to (E, \mu_U)$. We will show that id_{μ_U} sends bounded sets to bounded sets. Let $B \subset E$ be any $\mathcal{T}$-bounded set. As U is bornivorous, there is some $a > 0$ for which $B \subset aU = a\{x : \mu_U(x) \leq 1\}$. Hence, $B = \mathrm{id}_{\mu_U}(B)$ is μ_U-bounded. The assumption is that such functions are continuous, and this means $U = \mathrm{id}_{\mu_U}^{-1}(U) \in \mathcal{N}_{0,E}$. We have established that E is bornological. $\qquad\qquad\square$

A seminorm version of Theorem 5.2.1 is indicated next.

Proposition 5.2.4. *LCS is bornological if and only if every seminorm that sends bounded sets to bounded sets is continuous.*

Proof. Exercise 5.2.1. □

The next result provides verification of some hereditary properties of bornological LCS.

Theorem 5.2.2. *The property of being bornological is preserved under the formation of final topologies.*

Proof. Let $\{(E_\alpha, \mathcal{T}_\alpha, f_\alpha) : \alpha \in I\}$ be a collection of bornological LCS and let E be a vector space for which we equip E with the final locally convex topology with respect to the linear maps $f_\alpha : E_\alpha \to E$ (see Definition 3.5.1). The goal is to prove that E is bornological with respect to the final topology. Let $U \subset E$ be an absolutely convex bornivore. Then $(\forall \alpha \in I) f_\alpha^{-1}(U)$ is absolutely convex in E_α. We will see that $f_\alpha^{-1}(U)$ is also bornivorous in E_α. If $B_\alpha \subset E_\alpha$ is any bounded set, then the continuity of f_α indicates that $f_\alpha(B_\alpha)$ is bounded in E. As such, $f_\alpha(B_\alpha)$ is absorbed by U. Thus, $(\exists a > 0) f_\alpha(B_\alpha) \subset aU$. Hence, $f_\alpha^{-1}(U)$ is a bornivore in E_α, which means $f_\alpha^{-1}(U)$ is a zero neighborhood in E_α, because E_α is bornological. By Proposition 3.5.1, U is a zero neighborhood in E. We conclude that E is bornological. □

Corollary 5.2.1. *The following types of LCS are bornological:*
(a) *Quotients (see Definition 3.1.1) of bornological LCS.*
(b) *Inductive limits (see Definition 3.5.3) of sequences of bornological LCS.*
(c) *Direct sums (see Definition 3.5.2) of collections of bornological LCS.*

The situation for initial topologies, including products of bornological LCS, is more complicated. There exist closed subspaces of bornological LCS that are not bornological. Such examples may be found for instance, in [79, 28.4, p. 387]. Countable products of bornological LCS are bornological, as is proven in [79, 28.4 (4), p. 387]. For uncountable products, whether or not such a product of bornological LCS is bornological depends on some deep results of set theory. A detailed discussion of this situation can be found in [70, Section 13.5, pp. 281–283]. For subspaces of finite codimension, we have the following.

Proposition 5.2.5. *A finite codimensional subspace of a bornological LCS is bornological.*

Proof. Exercise 5.2.6. □

The next definition and upcoming results will be useful for establishing some equivalences for a LCS to be bornological.

Definition 5.2.2. Denote by $\mathcal{T}_b$ the locally convex topology generated by the collection of all absolutely convex bornivores in a LCS $(E, \mathcal{T})$.

Note that this collection does indeed form a base for a locally convex topology by virtue of Theorem 2.1.2. The following reveals a relation between $\mathcal{T}_b$ and $\mathcal{T}$.

Proposition 5.2.6. *Consider a LCS $(E, \mathcal{T})$ and the locally convex topology $\mathcal{T}_b$ generated by the collection of all absolutely convex bornivores in E. Then:*
(a) *The topology $\mathcal{T}_b$ is finer than $\mathcal{T}$.*
(b) *A subset of E is $\mathcal{T}$-bounded if and only if it is $\mathcal{T}_b$-bounded.*

Proof. (a) If $U \in \mathcal{N}_{0,\mathcal{T}}$ is any absolutely convex set, then U is also a bornivore and as such, $U \in \mathcal{T}_b$.

(b) By part (a), any $\mathcal{T}_b$-bounded set is $\mathcal{T}$-bounded by way of the continuity of the linear map id : $(E, \mathcal{T}_b) \to (E, \mathcal{T})$. Conversely, every $\mathcal{T}$-bounded set is $\mathcal{T}_b$-bounded by the definition of $\mathcal{T}_b$. $\qquad\qquad\square$

Theorem 5.2.3. *For a Hausdorff LCS $(E, \mathcal{T})$, the following are equivalent:*
(a) *$(E, \mathcal{T})$ is bornological.*
(b) *The topology $\mathcal{T}$ is equivalent to the topology $\mathcal{T}_b$.*
(c) *$\mathcal{T}$ is the final locally convex topology with respect to a collection of normed spaces.*

Part (c) above is often stated in terms of a general (meaning possibly uncountable) *inductive limit of normed spaces.*

Proof. First, we prove (a) $\Rightarrow$ (b), noting that (b) $\Rightarrow$ (a) follows immediately from the definition of $\mathcal{T}_b$. Assume $(E, \mathcal{T})$ is bornological. If V is any absolutely convex zero neighborhood with respect to $\mathcal{T}_b$, then V is an absolutely convex bornivore. The assumption that $(E, \mathcal{T})$ is bornological implies that V is a $\mathcal{T}$-zero neighborhood as well.

(a) $\Rightarrow$ (c): Assume $(E, \mathcal{T})$ is bornological. Let $\mathcal{B}$ denote the collection of all $\mathcal{T}$-bounded disks in E. Because singleton sets are bounded and because the absolutely convex hull of a bounded set is bounded by Propositions 2.2.3 and 2.2.4, we observe that $E = \bigcup \{B : B \in \mathcal{B}\}$. Now consider the final topology $\mathcal{T}_{\text{final}}$ with respect to the collection $\{(E_B, \|\cdot\|_B, \text{id}_B) : B \in \mathcal{B}\}$, as defined in Notation 5.1.1. Proposition 5.1.3 indicates that the topology generated by $\|\cdot\|_B$ is finer than the relative topology on E_B for each $B \in \mathcal{B}$. Therefore, it follows that $\mathcal{T}_{\text{final}}$ is finer than $\mathcal{T}$. Conversely, we examine the identity map id : $(E, \mathcal{T}) \to (E, \mathcal{T}_{\text{final}})$. If $A \subset E$ is $\mathcal{T}$-bounded, then there exists $B \in \mathcal{B}$ (e. g., the absolutely convex hull of A) such that $A \subset B$. This means A is $\|\cdot\|_B$-bounded and, therefore, A is $\mathcal{T}_{\text{final}}$-bounded. We have shown that the linear map id : $(E, \mathcal{T}) \to (E, \mathcal{T}_{\text{final}})$ sends bounded sets to bounded sets. Applying Theorem 5.2.1 to the bornological LCS $(E, \mathcal{T})$, we conclude that id : $(E, \mathcal{T}) \to (E, \mathcal{T}_{\text{final}})$ is continuous. That is to say, $\mathcal{T}$ is finer than $\mathcal{T}_{\text{final}}$.

(c) $\Rightarrow$ (a): This follows instantly from Theorem 5.2.2. $\qquad\qquad\square$

Corollary 5.2.2. *A Hausdorff LCS is bornological if and only if it has the final topology with respect to the collection of its closed bounded disks. That is, a Hausdorff LCS E is bornological if and only if E has the final topology with respect to the collection $\{(E_B, \|\cdot\|_B, \text{id}_B) : B$ is a bounded disk and $\text{id}_B : E_B \to E$ is the canonical injection$\}$.*

Proof. The same argument of Theorem 5.2.3 applies to the collection $\mathcal{B}_c$ of closed, bounded disks in a LCS E. The assumption that E is Hausdorff ensures that the spaces $(E_B, \|\cdot\|_B)$ are normed as opposed to being seminormed. $\qquad\square$

Further examination of the topology $\mathcal{T}_b$ reveals the following property.

Proposition 5.2.7. *The topology $\mathcal{T}_b$ on a LCS $(E, \mathcal{T})$ is the finest locally convex topology having the same bounded sets as $\mathcal{T}$.*

Proof. By Definition 5.2.2, $(E, \mathcal{T}_b)$ is bornological. Suppose $\mathcal{T}^\sharp$ is a locally convex topology on E such that the $\mathcal{T}^\sharp$-bounded sets are the same as the $\mathcal{T}$-bounded sets. Consider the identity map $\mathrm{id} : (E, \mathcal{T}_b) \to (E, \mathcal{T}^\sharp)$. By Proposition 5.2.6, the identity id maps bounded sets to bounded sets. Applying Theorem 5.2.1 reveals that id is continuous, and thus, $\mathcal{T}_b$ is finer than $\mathcal{T}^\sharp$. $\qquad\square$

Theorem 5.2.3 motivates the following definition.

Definition 5.2.3. The **associated bornological topology** for a LCS is the finest locally convex topology having the same collection of bounded sets as the original topology of the LCS.

Hence, we obtain another corollary of Theorem 5.2.3, as stated next.

Corollary 5.2.3. *A Hausdorff LCS is bornological if and only if its topology is equivalent to the associated bornological topology.*

The next result is a completeness result related to bornological LCS.

Theorem 5.2.4. *The strong dual of a bornological LCS is complete.*

Proof. Let $\mathfrak{F}$ be a Cauchy filter on E' with respect to $\beta(E', E)$ for a bornological LCS E. For a fixed $x \in E$, define $\mathfrak{F}(x)$ to be $\{f(x) : f \in S, S \in \mathfrak{F}\}$. In Exercise 5.2.11, you can prove that $\mathfrak{F}(x)$ is a filter on the scalar field $\mathbb{K}$. In fact, $\mathfrak{F}(x)$ is a Cauchy filter on $\mathbb{K}$, as follows: By the definition of the strong topology (see Example 4.3.2), for any $\varepsilon > 0$ and any bounded subset B of E that contains x, there exists $S \in \mathfrak{F}$ such that $S - S \subset \varepsilon B^\circ$. For the fixed $x \in B$, the last statement amounts to $(\forall f, g \in S)\, |f(x) - g(x)| < \varepsilon$. Because $\mathbb{K}$ is complete, we obtain a pointwise limit; that is, for every $x \in E$, $\mathfrak{F}(x)$ converges to some $f_0(x) \in \mathbb{K}$. It is easy to check that f_0 given by $\langle f_0, x \rangle = f_0(x)$ defines a linear functional on E. Furthermore, we have the following.

Claim 5.2.1. *The linear functional f_0 is continuous.*

Proof of Claim 5.2.1. Let B be any bounded subset of E. Then there is $S \in \mathfrak{F}$ such that $S - S \subset B^\circ$, which in particular, means $(\forall f, g \in S)(\forall x \in B)\, |f(x) - g(x)| \leq 1$. In other words, for any $f \in S$, f is bounded on B. Hence, $|f(x)| \leq \frac{a}{2}$, for some $a > 0$. Meanwhile, given $x \in B$, there exists $g_x \in S$ for which $|f_0(x) - g_x(x)| \leq \frac{a}{2}$, for $x \in B$. We observe that

$$|f_0(x)| \le |f_0(x) - g_x(x)| + |g_x(x)| \le \frac{a}{2} + \frac{a}{2} = a, \tag{5.2}$$

and this holds for all $x \in B$. We have shown that f_0 sends bounded sets in E to bounded sets in $\mathbb{K}$. Applying the hypothesis that E is bornological and Theorem 5.2.1, f_0 is continuous, proving the claim.

What remains to prove is that $\mathfrak{F} \to f_0$ with respect to $\beta(E', E)$. First, $\mathfrak{F}$ being $\beta(E', E)$-Cauchy implies that for every bounded $B \subset E$ and every $\varepsilon > 0$, there is $S \in \mathfrak{F}$ for which $(\forall x \in B)(\forall f, g \in S)\, |f(x) - g(x)| < \frac{\varepsilon}{2}$. In addition, for every $x \in B$ there is $g \in S$ such that $|f_0(x) - g(x)| < \frac{\varepsilon}{2}$. Combining, we obtain $(\forall x \in B)(\forall g \in S)$,

$$|f_0(x) - g(x)| \le |f_0(x) - f(x)| + |f(x) - g(x)| < \frac{\varepsilon}{2} + \frac{\varepsilon}{2} = \varepsilon. \tag{5.3}$$

That is, $S \subset f_0 + \varepsilon B^\circ$. In general, given a finite collection $\{B_j : j \in \underline{n}\}$ of bounded subsets of E, and any $\varepsilon > 0$, there is an $S \in \mathfrak{F}$ such that $S \subset f_0 + \varepsilon \bigcap_{j=1}^{n} B_j^\circ$, which tells us that $\mathcal{F} \to f_0$ with respect to $\beta(E', E)$. $\qquad\square$

5.2.1 Ultrabornological spaces

In Theorem 5.2.3, we may consider the case where the bounded disks are Banach disks. That is, we give a name to any LCS whose topology is the finest locally convex topology with respect to a collection of Banach spaces, as stated next.

Definition 5.2.4. A LCS that has the final topology with respect to a collection of Banach spaces is called **ultrabornological**.

Clearly, every (LB) space is ultrabornological. More generally, we have the following observation.

Proposition 5.2.8. *A quasicomplete bornological LCS is ultrabornological.*

Proof. We saw in Proposition 5.1.4 that a bounded disk that is complete in a LCS is in fact a Banach disk. Thus, quasicompleteness of a LCS means we may consider the LCS as having the final topology with respect to the closed bounded disks, which comprise Banach spaces per Corollary 5.2.2. $\qquad\square$

Corollary 5.2.4. *Fréchet and strict (LF) spaces are ultrabornological.*

Proof. Exercise 5.2.10. $\qquad\square$

As you may already suspect, augmenting some more general completeness assumption to a bornological LCS might result in the LCS being ultrabornological. That is indeed the case, as indicated next.

Proposition 5.2.9. *A sequentially complete bornological LCS is ultrabornological.*

Proof. The proof is a straightforward application of Proposition 5.1.4 and is left as Exercise 5.2.12. $\qquad\square$

Proposition 5.2.9 will be generalized even further in Chapter 7. The concept of an associated ultrabornological topology on a LCS is described as follows.

Definition 5.2.5. A subset of a LCS that absorbs all Banach disks is called **infrabornivorous** or an **infrabornivore**.

Thus, bounded sets are replaced by Banach disks (which are also bounded). Proceeding as in Theorem 5.2.3 and Proposition 5.2.7 using Banach disks and infrabornivores, we obtain the following.

Definition 5.2.6. The **associated ultrabornological topology** $\mathcal{T}_{ub}$ for a LCS is the finest locally convex topology having the same collection of Banach disks as the original topology on the LCS.

Correspondingly, we obtain an ultrabornological version of Theorem 5.2.3, as stated next.

Theorem 5.2.5. *For a Hausdorff LCS $(E, \mathcal{T})$, the following are equivalent:*
(a) *$(E, \mathcal{T})$ is ultrabornological.*
(b) *$\mathcal{T}$ is equivalent to $\mathcal{T}_{ub}$.*
(c) *$\mathcal{T}$ is the final locally convex topology with respect to a collection of Banach spaces.*

The concept of a barreled LCS is important enough that variations of this concept have been developed and studied. Such developments include more restrictive definitions called **strong barreledness conditions**, which you can read about in [95, Chapter 9]. There are also more general concepts of barreledness, called **weak barreledness conditions**, and you can read about them in the summary work of Saxon and Sánchez Ruiz [110].

Exercises

5.2.1. Prove Proposition 5.2.4.

5.2.2. Suppose that for a LCS E, every linear map from E to any normed space that maps null sequences to bounded sets must be continuous. Prove that E is bornological. In particular, if every seminorm on E that sends null sequences to bounded sets is continuous, then E is bornological.

5.2.3. Consider LCS $(E, \mathcal{T})$ and (F, η). Prove that a linear map $T : (E, \mathcal{T}) \to (F, \eta)$ sends bounded sets to bounded sets if and only if $T : (E, \mathcal{T}_b) \to (F, \eta)$ is continuous, where $\mathcal{T}_b$ is the associated bornological topology of Definition 5.2.3.

5.2.4. Suppose a linear map between two LCS sends null sequences to bounded sequences. Prove that the linear map sends bounded sets to bounded sets. *Hint:* Use the ideas of the proof of Proposition 5.2.2.

5.2.5. Prove that for E a bornological LCS and F is any LCS, if a linear map $T : E \to F$ sends null sequences to bounded sequences, then T must be continuous. *Hint:* Use the Exercise 5.2.4 above and Theorem 5.2.1.

5.2.6. Prove Proposition 5.2.5.

5.2.7. Prove that a LCS is bornological if it is a Mackey space and every linear functional that sends bounded sets to bounded sets is continuous. *Hint:* Compare the associated bornological topology $\mathcal{T}_b$ of Definition 5.2.3 to the Mackey topology.

5.2.8. Prove directly that a finite product of bornological (resp., ultrabornological) LCS is again bornological (resp., ultrabornological).

5.2.9. Give a direct proof of the ultrabornological version of Corollary 5.2.2; in other words, show that a Hausdorff LCS E is ultrabornological if and only if E carries the final topology with respect to the family $\{(E_B, \| \cdot \|_B, \mathrm{id}_B) : B$ is a Banach disk and $\mathrm{id}_B : E_B \to E$ is the canonical injection$\}$.

5.2.10. Prove Corollary 5.2.4.

5.2.11. Prove that $\mathfrak{F}(x)$ defined in the proof of Theorem 5.2.4 is a filter on the scalar field $\mathbb{K}$.

5.2.12. Prove Proposition 5.2.9.

5.2.13. Prove the following equivalences for a Hausdorff LCS:

(a) $(E, \mathcal{T})$ is ultrabornological.

(b) Every seminorm on E that is bounded on each Banach disk must be continuous.

(c) For every LCS F and every linear map $T : E \to F$, if T sends Banach disks to bounded sets, then T is continuous.

5.3 Reflexivity and Montel spaces

Consider the dual E' of a LCS E and what information we might obtain by determining its dual, that is, $(E')'$, the dual of the dual, where we consider some locally convex topology on such a space. In the context of normed spaces (see, e. g., [89, Section 1.11, pp. 97–109] for more details), recall that if $1 < p < \infty$, then $l_p'' = l_p$, whereas $c_0'' = l_\infty$ (where "=" means equivalent as normed spaces). This tells us that sometimes but not always, the second dual of a normed space is equal to the original. We will study this situation in the context of LCS, which we assume to be Hausdorff in this section. In addition, we assume E' is equipped with the strong topology, $\beta(E', E)$. The details are in the next definition.

Definition 5.3.1. The **bidual** (also **second dual**, **second adjoint**) is the dual of the strong dual of a Hausdorff LCS. That is, if E is a Hausdorff LCS then the bidual is $E'' = (E', \beta(E', E))'$.

In order to describe elements of E'', observe that for $x \in E$ we may define $\tilde{x}$ on E' by way of duality, as $\tilde{x}(y) = \langle x, y \rangle$, for all $y \in E'$. By examining the proof of Proposition 4.2.2, we conclude that $\tilde{x}$ is $\sigma(E', E)$-continuous, hence also continuous with respect to the finer topology $\beta(E', E)$ on E'. In other words, $\tilde{x} \in E''$. Now consider the map from E to E'' that sends x to $\tilde{x}$. This prompts the following definition.

Definition 5.3.2. The map $J : E \to E''$ given by $J(x) = \tilde{x}$ is called the **canonical embedding** (also: **canonical inclusion**).

This map is clearly linear. In Exercise 5.3.6, you can verify that this map is also injective. Definition 5.3.2 above is algebraic and indicates that we may identify E with a subspace of E''; that is, $E \subset E'' \subset (E')^*$. In particular, we can consider the possibility that J is surjective, which would imply that E and E'' are algebraically equivalent. This also leads us to the topological consideration of when J might be an isomorphism (see Definition 2.3.3) when we equip E'' with the topology $\beta(E'', E')$. This discussion motivates the following definition.

Definition 5.3.3. A Hausdorff LCS is **semireflexive** if the canonical embedding of Definition 5.3.2 is surjective. If the canonical embedding is additionally an isomorphism when E'' is equipped with the topology $\beta(E'', E')$, then the LCS is called **reflexive**.

In particular, *every reflexive LCS is semireflexive.* An example of a semireflexive LCS that is not reflexive will be given in Corollary 5.4.2. In order to describe some properties of semireflexive and reflexive LCS, we can take a look at boundedness properties of some specific topologies of a dual pair, as seen next.

Proposition 5.3.1. *On a dual pair (E, E'), the bounded sets are the same for the following topologies: $\sigma(E', E'')$, $\mu(E', E'')$, and $\beta(E', E)$.*

Proof. By Definition 5.3.1 of E'', $\sigma(E',E'')$, $\mu(E',E'')$, and $\beta(E',E)$ are topologies of the dual pair (E',E''). Thus, Mackey's Theorem 5.1.5 applies, and we conclude that the bounded sets are same for all three of these topologies. $\qquad\square$

Corollary 5.3.1. *If E is semireflexive, then $\beta(E',E)$ is a topology of the dual pair (E',E).*

Proof. This follows from Proposition 5.3.1 and Definition 5.3.3 of a semireflexive LCS. $\quad\square$

We know by way of Corollary 3.2.6 that the Heine–Borel property never holds in infinite-dimensional normed spaces. The next result shows that semireflexive LCS are precisely those for which the weak topology does satisfy the Heine–Borel property. Notice that the result below generalizes Corollary 4.5.1.

Theorem 5.3.1. *Let $(E,\mathcal{T})$ be a Hausdorff LCS. Then the following are equivalent:*
(a) *E is semireflexive.*
(b) *The strong topology on E' is a topology of the dual pair; that is, $\beta(E',E) = \mu(E',E)$.*
(c) *The Heine–Borel property holds for the weak topology; that is, every $\sigma(E,E')$-bounded subset of E is $\sigma(E,E')$-relatively compact (See Definition A.3.20).*

Proof. We prove each equivalence separately.

(a) $\Rightarrow$ (b): Assume E is semireflexive. The statement that the strong topology on E' being a topology of the dual pair then follows immediately by Corollary 5.3.1. Because $\beta(E',E)$ is stronger than $\mu(E',E)$ and $\mu(E',E)$ is the finest topology consistent with the duality by Equation (4.14), these topologies are equivalent.

(b) $\Rightarrow$ (a): Now assume $\beta(E',E) = \mu(E',E)$. Then $E'' = (E',\beta(E',E))' = E$ by the Mackey–Arens Theorem 4.4.2, and we have shown that E is semireflexive.

(a) $\Rightarrow$ (c): Assume E is semireflexive. By (a), $\beta(E',E) = \mu(E',E)$. Suppose $B \subset E$ is $\sigma(E,E')$-closed and bounded. Then its polar, $B^\circ \in \beta(E',E)$ by the definition of the strong topology (see Example 4.3.2). Hence, $B^\circ \in \mu(E',E)$. The last statement implies that there is a $\sigma(E,E')$-compact and convex set $C \subset E$, which we may assume to be balanced by Theorem 4.3.1(d), such that $C^\circ \subset B^\circ$. We then obtain $B \subset B^{\circ\circ} \subset C^{\circ\circ} = C$. We have established that B is $\sigma(E,E')$-compact. To finish this part, we observe that by Mackey's Theorem 5.1.5, the $\sigma(E,E')$-bounded subsets of E are the same as the $\mathcal{T}$-bounded subsets, which means B can be any $\mathcal{T}$-bounded subset in E. That is, every weakly closed and bounded subset of E is weakly compact.

(c) $\Rightarrow$ (a): Assume every weakly closed and bounded subset of E is weakly compact. By the Mackey–Arens Theorem 4.4.2 and Theorem 4.3.1, $\beta(E',E)$ is the topology of uniform convergence on all $\sigma(E,E')$-absolutely convex closed, bounded subsets of E. By assumption, these sets are $\sigma(E,E')$-compact. Therefore, the strong topology $\beta(E',E)$ is equivalent to the Mackey topology $\mu(E',E)$. In other words, $\beta(E',E)$ is consistent with the duality (E,E'), and a glance at part (a) shows that the proof is completed. $\qquad\square$

Corollary 5.3.2. *Every semireflexive Hausdorff LCS is weakly quasicomplete.*

Proof. If E is a Hausdorff LCS that is semireflexive, then by Theorem 5.3.1(b) above, the closure of any $(\sigma(E, E'))$-bounded set is $\sigma(E, E')$-compact, hence complete. □

In Theorem 5.2.4, we saw that the strong dual of a bornological LCS is complete. There is a barreled version of this for semireflexive LCS, as seen below.

Proposition 5.3.2. *The strong dual of a semireflexive Hausdorff LCS is barreled.*

Proof. Exercise 5.3.8. □

Notice that the converse of Proposition 5.3.2 above is false. An example is given by any nonreflexive Banach space E (such as c_0). In this case, the strong dual of such a space is a Banach space that by Corollary 5.1.1 is barreled, hence quasibarreled.

It is worth noting that semireflexivity is preserved by certain other constructions as shown below.

Proposition 5.3.3. *Products and direct sums of semireflexive LCS are again semireflexive.*

Proof. Exercise 5.3.10. □

Inductive limits of semireflexive LCS are not so easily determined. Some results in this context can be found in [95, Chapter 8] and references therein.

5.3.1 Reflexive spaces

We now turn our attention to the more restrictive collection of reflexive LCS. For convenience of notation, we define the following.

Definition 5.3.4. Let $\varepsilon(E'', E')$ denote the topology of uniform convergence on the equicontinuous subsets of E'.

Remark 5.3.1. Observe that by Theorem 4.3.3, on a LCS $(E, \mathcal{T})$, $\mathcal{T}$ is the $\mathcal{S}$-topology of equicontinuous subsets of E'. In particular, $\varepsilon(E'', E')$ induces on E its original topology $\mathcal{T}$.

Notice that by its definition, $\beta(E'', E')$ is finer than $\varepsilon(E'', E')$. These two topologies agree precisely when the LCS is quasibarreled, as established next. This result will be useful in understanding reflexive LCS.

Proposition 5.3.4. *On a Hausdorff LCS, $\beta(E'', E')$ agrees with $\varepsilon(E'', E')$ if and only if E is quasibarreled.*

Proof. The topologies $\beta(E'', E')$ and $\varepsilon(E'', E')$ agree if and only if every $\beta(E', E)$-bounded set is equicontinuous, and by Theorem 5.1.7 this happens if and only if E is quasibarreled. □

Theorem 5.3.2. *A Hausdorff LCS is reflexive if and only if it is semireflexive and quasibarreled.*

Proof. ($\Rightarrow$): If $(E, \mathcal{T})$ if reflexive then it is already semireflexive. Moreover, E being semireflexive implies that $\beta(E'', E')$ induces the original topology $\mathcal{T}$ on E. By Proposition 5.3.4 above, E is quasibarreled.

($\Leftarrow$): Suppose E is semireflexive and quasibarreled. Then $\text{id} : E \to E''$ is surjective. Because E is quasibarreled, Proposition 5.3.4 implies that $\text{id} : E \to E''$ is even injective and is a strict morphism (see Definition 2.3.3(c)). Hence, E is reflexive. $\qquad\square$

Corollary 5.3.3. *A semireflexive LCS is reflexive if and only if it is barreled.*

Proof. Exercise 5.3.9. $\qquad\square$

A result comparable to Proposition 5.3.2 holds for reflexive LCS, as indicated below.

Proposition 5.3.5. *The strong dual of a reflexive LCS is reflexive.*

Proof. Assume E is reflexive. By Proposition 5.3.2, the strong dual of E is barreled. Let $B \subset E'$ be $\sigma(E', E'')$-closed and bounded. By virtue of E being reflexive, B is in fact $\sigma(E', E)$-closed and bounded. The barreledness of E allows us to make use of Corollary 5.1.6 to conclude that B is in fact $\sigma(E', E)$-compact. Apply Theorem 5.3.1 to conclude that $(E', \beta(E', E))$ is semireflexive. Theorem 5.3.2 tells us that $(E', \beta(E', E))$ is reflexive, and the proof is completed. $\qquad\square$

There are some deep results regarding the reflexivity of normed (hence, Banach) spaces, which will be touched on in Section 5.4.

5.3.2 Semi-Montel and Montel spaces

In this subsection, we continue our interest in the Heine–Borel property; that is, LCS for which bounded sets are relatively compact (Definition A.3.20). The collection of such spaces excludes infinite-dimensional normed spaces because of Corollary 3.2.6. On the other hand, there are nontrivial LCS for which the property that bounded sets are relatively compact, and such LCS even have a name, which we state next.

Definition 5.3.5. A Hausdorff LCS is **semi-Montel** if all of its bounded subsets are relatively compact. A **Montel** LCS is a quasibarreled semi-Montel space.

Horváth [67, Section 3.9, pp. 231–243] provides an in-depth discussion of (semi-)Montel spaces, including proofs of the following nontrivial examples.

Theorem 5.3.3. *The spaces $\mathcal{E}(\Omega)$ of Example 2.1.8 and the space of test functions, $\mathcal{D}$ of Definition 3.5.4, are Montel LCS. Additionally, the strong dual of a Montel LCS is Montel.*

As compact subsets of Hausdorff LCS are complete, we can state the following.

Proposition 5.3.6. *Every (semi-)Montel LCS is quasicomplete.*

Another quick observation is next.

Proposition 5.3.7. *Every semi-Montel LCS is semireflexive and every Montel LCS is reflexive.*

Proof. As each bounded subset of a semi-Montel space is compact, the fact that the weak topology is coarser than the original topology tells us that such sets are also weakly compact. Moreover, Mackey's Theorem 5.1.5 reminds us that weakly bounded subsets are the same as those bounded in the original topology. The result now follows from Theorem 5.3.1. □

Corollary 5.3.4. *Every Montel LCS is barreled.*

Proof. This follows from Corollary 5.3.3. □

In Exercise 5.3.1, you can describe spaces that are barreled but not Montel as well as reflexive but not Montel. Our next objective is to show that every (semi-)Montel space satisfies Schur's property of Definition 4.5.2. A preliminary result leads the way.

Proposition 5.3.8. *In a semi-Montel LCS, the weak and original topologies agree on bounded subsets.*

Proof. Let $(E, \mathcal{T})$ be a Hausdorff semi-Montel LCS. It suffices to prove the conclusion for $\mathcal{T}$-closed sets because if $\mathcal{T}$ agrees with $\sigma(E, E')$ on the $\mathcal{T}$-closures of bounded sets, then these topologies agree on the original bounded sets. Let $B \subset E$ be any $\mathcal{T}$-closed, bounded subset. Then the topology of $\sigma(E, E')$ induced on B is Hausdorff and coarser than the topology induced by topology $\mathcal{T}$. At the same time, the assumption that E is semi-Montel implies that B is compact with respect to the finer topology $\mathcal{T}$. Therefore, we are in a situation in which we have two comparable Hausdorff topologies such that B is compact with respect to the finer topology. This is exactly the situation in which we may apply Theorem A.3.3 from which we deduce that $\mathcal{T}$ and $\sigma(E, E')$ are equivalent on B. □

The following corollary is immediate.

Corollary 5.3.5. *Semi-Montel LCS satisfy Schur's property 4.5.2.*

Below are a few hereditary properties of semi-Montel spaces that can be obtained via exercises. For a more in depth discussion of properties of (semi-)Montel spaces, see [67, Section 3.9, pp. 231–243].

Proposition 5.3.9. *The property of being semi-Montel is preserved under the following constructions:*
(a) *Closed subspaces.*
(b) *Arbitrary products.*
(c) *Strict inductive limits.*

Proof. Exercise 5.3.12. □

Exercises

5.3.1. Describe examples of the following:
  (a) A barreled space that is not Montel.
  (b) A reflexive space that is not Montel.

5.3.2. Prove that a subset of the dual of a LCS is strongly bounded if and only if its polar absorbs all weakly bounded absolutely convex subsets of the LCS; that is, for a LCS E, $D \subset E$ is $\beta(E'E)$-bounded if and only if D° absorbs every $\sigma(E, E')$-bounded absolutely convex subsets of E.

5.3.3. Prove that for a LCS $(E, \mathcal{T})$, $\mathcal{T} \subset \beta(E'', E) \cap E$.

5.3.4. Prove that for a LCS $(E, \mathcal{T})$ the canonical embedding of Definition 5.3.2 is a homeomorphism if and only if E is quasibarreled. Here, we assume E'' is equipped with the topology $\beta(E'', E')$.

5.3.5. Prove that every finite-dimensional TVS is reflexive, and is even a Montel space.

5.3.6. Prove that the map $J : E \to E''$ given by $J(x) = \tilde{x}$, is injective.

5.3.7. Prove that for the map $J : E \to E''$ given by $J(x) = \tilde{x}$, the null space of J is the closure of $\{0\}$ in E.

5.3.8. Prove Proposition 5.3.2. *Hint:* Use polars.

5.3.9. Prove Corollary 5.3.3.

5.3.10. Prove Proposition 5.3.3.

5.3.11. Prove that every reflexive LCS is quasicomplete.

5.3.12. Prove Proposition 5.3.9. *Hint:* For part (b), use Tychonov's Theorem A.3.5.

5.4 Barreled, bornological, and reflexive properties of normed spaces

The first result gives another indication that infinite-dimensional Banach spaces that are equipped with their weak topologies are marvelously complicated. You will see that the proof of each part is short, but keep in mind that the shortness involves citing deep results.

Theorem 5.4.1. *Any infinite-dimensional Banach space E, equipped with its weak topology $\sigma(E, E')$ satisfies each of the following:*
(a) *$(E, \sigma(E, E'))$ is not a Mackey space so in fact, is not even quasibarreled.*
(b) *$(E, \sigma(E, E'))$ is not bornological.*
(c) *$(E, \sigma(E, E'))$ is not Montel.*

Proof. In fact, each of these results can be quickly verified using material from previous sections, and is left as Exercise 5.4.1. □

Corollary 5.4.1. *For $1 \leq p \leq \infty$, the Banach spaces l_p satisfy the items of Theorem 5.4.1 above.*

Another significant result that will be useful from time to time is next.

Theorem 5.4.2. *On every infinite-dimensional Banach space there exists a finer norm that is not barreled.*

Proof. See [95, Proposition 4.6.7(iv), pp. 130–131]. □

We also have the following property of infinite-dimensional normed spaces.

Proposition 5.4.1. *An infinite-dimensional normed space, equipped with its weak topology, is incomplete.*

Proof. See Megginson [89, Proposition 2.5.15, p. 215]. □

In Exercise 5.4.2, you can verify that for an infinite-dimensional normed space the weak topology is strictly coarser than the Mackey topology. This provides another proof that such spaces are not barreled and are not even quasibarreled. A particular case of the Banach space l_1 with its weak topology produces another counterexample, as seen next.

Proposition 5.4.2. *The LCS given by l_1 equipped with its weak topology represents a space that satisfies Schur's property of Definition 4.5.2 but is not a Montel space.*

Proof. Corollary 3.2.6. □

As alluded to earlier, some of the deepest results regarding reflexive LCS appear in the context of normed spaces. Below is a short discussion of reflexivity for normed spaces. Proofs of some of the results are beyond the scope of this book, however, accessible discussions (and references) of reflexive normed spaces can be found in [89] and Chapter 15 of [91]. The first result below indicates that there is no such thing as an incomplete reflexive normed space.

Theorem 5.4.3. *Every semireflexive normed space is a reflexive Banach space.*

Proof. By Theorem 5.2.4 and the obvious fact that normed spaces are bornological, the strong dual of any normed space is complete. Hence, a reflexive normed space must be complete by Proposition 5.3.5. □

The next result provides a way to determine when a normed space is reflexive.

Theorem 5.4.4 (Banach–Bourbaki). *A normed space is reflexive if and only if its closed unit ball is weakly compact.*

Proof. We prove the case for the space $(E, \|\cdot\|)$ being a Banach space, the proof of the general normed space version being left as Exercise 5.4.6. To this end, let B denote the closed unit ball of E.

($\Rightarrow$): Assume E is reflexive. Then by Theorem 5.3.1, $B \subset K$, where K is $\sigma(E, E')$-compact. B is closed with respect to the norm and of course is convex, so by Theorem 4.3.2, B is also $\sigma(E, E')$-closed and, therefore, $\sigma(E, E')$-compact.

($\Leftarrow$): Now assume that B is $\sigma(E, E')$-compact. Then $(\forall a > 0)\, aB$ is also $\sigma(E, E')$-compact. As E is normed, every bounded subset of E is contained in aB, for some $a > 0$. Hence, E is semireflexive by Theorem 5.3.1. The completeness of E as a Banach space now implies that E is barreled by way of Corollary 5.1.1, and then by Corollary 5.3.3, E is in fact reflexive. □

Corollary 5.4.2. *Any infinite-dimensional reflexive Banach space, when equipped with its weak topology, is a semireflexive LCS that is not reflexive.*

Proof. By the Banach–Bourbaki Theorem 5.4.4 above, such a space is semi-reflexive by Theorem 5.3.1, but is not barreled by Theorem 5.4.1(a) above. The result follows by applying Corollary 5.3.3. □

We also have (first proved by Pettis in 1938; see [89, Theorem 1.11.16, p. 104]) the following.

Proposition 5.4.3. *A closed subspace of a reflexive Banach space is reflexive.*

Proof. Exercise 5.4.7. □

Yet another deep result regarding reflexivity of Banach spaces is James' theorem, stated below. Details about the proof of this result can be found in Megginson [89, Section 1.13, pp. 115–136, Section 4.5, pp. 411–423] and the references therein. A general discussion of James' theorem can be found in Narici and Beckenstein [91, Chapter 15]. Before we state the theorem, a definition is needed.

Definition 5.4.1. Suppose $(E, \| \cdot \|)$ is a Banach space. A functional $f \in E'$ is called **norm-attaining** if there exists an element $x \in E$ of norm one (a unit vector) such that $f(x) = \|f\|$. In this case, the element x is called a **maximal element** of f.

Theorem 5.4.5 (James' theorem). *A Banach space is reflexive if and only if every continuous linear functional on the space is norm-attaining.*

Exercises

5.4.1. Prove Theorem 5.4.1.

5.4.2. Prove that on an infinite-dimensional normed space the weak topology is strictly coarser than the Mackey topology.

5.4.3. Contemplating Theorem 5.4.2 leads to the items below. For notational purposes, on the vector space E denote the topology of the original norm by $\|\cdot\|$ and the topology of the strictly finer norm by $\|\cdot\|_{sf}$. As a hint for both parts, recall that every normed space is bornological.

 (a) Prove that the topology of $\|\cdot\|_{sf}$ is not consistent with the duality (E, E') of the original norm $\|\cdot\|$.

 (b) Prove that every $\|\cdot\|_{sf}$-bounded subset of E is $\|\cdot\|$-bounded, and that there exists a subset A of E that is $\|\cdot\|$-bounded but not $\|\cdot\|_{sf}$-bounded.

5.4.4. Prove that a normed space E is reflexive if and only if its closed unit ball is $\sigma(E, E')$-complete.

5.4.5. Prove that a Banach space is reflexive if and only if its (strong) dual is reflexive.

5.4.6. Prove the case of general normed spaces of Theorem 5.4.4.

5.4.7. Prove Proposition 5.4.3.

5.4.8. Consider the space $C(K)$ of continuous functions from a compact Hausdorff topological space K to the scalar field $\mathbb{K}$, equipped with the sup norm $\|f\|_\infty = \sup\{|f(x)| : x \in K\}$. Prove that if K is finite, then $C(K)$ is reflexive. The converse, that if $C(K)$ is reflexive then K is finite, also holds. See [91, Example 15.5.2, p. 502].

5.4.9. Regarding James' Theorem 5.4.5 Prove or describe the following:

 (a) On every Banach space there is a continuous linear functional for which every unit vector is a maximal element. What is that linear functional?

 (b) Verify James' Theorem for finite-dimensional LCS.

 (c) Prove that if E is reflexive, then every continuous linear functional on E is norm-attaining. *Hint:* Use the Banach–Bourbaki Theorem 5.4.4.

 (d) Use the previous part to prove that $L_1([0, 1])$ (the space of Lebesgue integrable functions on $[0, 1]$) is not reflexive.

6 Sequences and series in locally convex spaces

Starting from calculus, we know that sequences and series are important in the general world of analysis and topology. Certainly in metrizable spaces this is the case. For LCS the situation is different. It is different because sequences are insufficient for proving properties in many exciting spaces that are not even metrizable as we saw in Example 3.3.4. Types of LCS that are nonmetrizable include proper (LB) spaces of Proposition 3.5.10, uncountable products of normed spaces from Theorem 3.3.3, and the (LF) space $\mathcal{D}$ of test functions from Corollary 3.5.6. Thus, sequences and series in general LCS seem not to be useful. However, we will discover in this and later chapters that in fact sequences and series do serve a useful purpose, even in nonmetrizable spaces.

As for this chapter, the first section consists of a description of a property in which every null sequence has a subsequence whose series converges (Definition 6.1.2). This type of so-called subseries convergence turns out to be quite useful for obtaining deep results such as the Principle of Uniform Boundedness, as we will see in Section 7.2. Next, we study LCS for which every null sequence converges in some normed space of the form $(E_B, \|\cdot\|_B)$ of Proposition 5.1.3. This property is called the Mackey convergence condition, and appears in a variety of contexts, such as differential calculus on TVS, as you can see, for example, in [18, Chapter 4] as well as [82, Chapter 1], which we will consider in Section 10.2. After that, we consider the structure of a web, which consists of countably many sets such that certain sequences formed from elements of those sets define null sequences and convergent series. Webs were initially defined by De Wilde in [35] as a tool for proving versions of the closed graph theorem, which we will examine in Chapter 8. We finish this chapter with some examples and counterexamples of these concepts, several of which are of the form of infinite-dimensional Banach spaces equipped with their weak topologies, like we saw in Section 5.4. Several concepts in this chapter are valid in the general context of TVS, too. Without further ado, let us begin.

6.1 Property $\mathcal{K}$

In this section, we will examine LCS in which every null sequence has a subsequence whose series converges. The first task is to define the convergence of a series in a TVS.

Definition 6.1.1. For a series $\sum_{n=1}^{\infty} x_n$ in a TVS the Nth-**partial sum** is given by $S_N = \sum_{n=1}^{N} x_n$. The series $\sum_{n=1}^{\infty} x_n$ **converges** (or **is convergent**) if the limit of S_N as $N \to \infty$ exists.

The definition above now permits us to define the following.

Definition 6.1.2. A null sequence in a TVS is $\mathcal{K}$-**convergent** (also **subseries convergent**) if every one of its subsequences contains a further subsequence whose series converges.

https://doi.org/10.1515/9783111392868-006

If every null sequence in a TVS is subseries convergent, we say the TVS satisfies **property** $\mathcal{K}$ or that the TVS is a $\mathcal{K}$-**space**.

Thus, a TVS satisfies property $\mathcal{K}$ is and only if every null sequence has a subsequence whose series converges. The use of the notation $\mathcal{K}$ comes from the fact that this property was studied systematically by mathematicians in the Katowice Branch of the Mathematics Institute of the Polish Academy Sciences and has been influenced by people such as Antosik, J. Mikusiński, Swartz, and Burzyk, as well as others earlier on. Many of the ideas of property $\mathcal{K}$ described here can be found in [7] as well as in parts of [116] and [117]. Introductory sections of the first three chapters of [117] include details of the historical development of this topic. In Exercise 6.1.1, you can verify that $\mathbb{R}$ satisfies property $\mathcal{K}$. Even more is true, as seen in the first (nontrivial) result of this section below.

Proposition 6.1.1. *Any complete, metrizable TVS satisfies property $\mathcal{K}$.*

Proof. Suppose (E, d) is a metrizable, complete TVS. Any null sequence (x_n) in E contains a subsequence (x_{n_k}) such that for every $k \in \mathbb{N}$, $d(x_{n_k}, 0) < 2^{-k}$. By way of translation invariance of the metric d, the partial sums of this subsequence form a Cauchy sequence in E, which converges by the assumption of completeness. $\square$

The converse of Proposition 6.1.1 is false. Kliś' example [78], cited in connection with Definition 5.1.4, is in fact an incomplete normed space that satisfies property $\mathcal{K}$. Furthermore, an example of a nonmetrizable LCS that satisfies property $\mathcal{K}$ follows from the next result, and another example will be given in Example 6.4.1.

Proposition 6.1.2. *Property $\mathcal{K}$ is preserved under the formation of sequentially retractive inductive limits (see Definition 3.5.7(c)) and finite products.*

Proof. Exercise 6.1.6. $\square$

Definition 6.1.2 reveals that a sequence that is $\mathcal{K}$-convergent is necessarily a null sequence of the relevant TVS. The converse is false, as indicated next.

Example 6.1.1. A null sequence that is not $\mathcal{K}$-convergent.

Consider the space c_{00} of finitely nonzero sequences with the sup norm $\|\cdot\|_\infty$. We make use of the sequence $(e^{(n)})$ of unit vectors of c_{00} having a 1 in the nth entry and zeros elsewhere. Let $(x^{(n)})$ be the sequence given by $\frac{1}{n}e^{(n)}$. Then $(x^{(n)})$ converges to 0 in c_{00}, however, any subseries of $(x^{(n)})$ would have infinitely many nonzero terms, and thus, could not converge in c_{00}.

The following version of boundedness, based on Proposition 2.2.5, will be useful.

Definition 6.1.3. A subset B of a TVS is $\mathcal{K}$-**bounded** if for any sequence (x_n) in B and any null sequence (a_n) in $\mathbb{K}$, the sequence $(a_n x_n)$ is $\mathcal{K}$-convergent.

Certainly, every $\mathcal{K}$-bounded set is bounded by Proposition 2.2.5 (see also Exercise 6.1.3), however, the example below distinguishes these two concepts.

Example 6.1.2. A bounded set that is not $\mathcal{K}$-bounded.

Consider again the space c_{00} of finitely nonzero sequences with the sup norm $\|\cdot\|_\infty$ as in Example 6.1.1. Let B denote the collection of unit vectors of c_{00} having a 1 in the nth entry and zeros elsewhere. Clearly, B is bounded with respect to $\|\cdot\|_\infty$. On the other hand, if (a_n) is a null sequence in $\mathbb{K}$ for which $a_n \neq 0$ for each $n \in \mathbb{N}$, then $S_N = \sum_{n=1}^{N} a_n e^{(n)}$ is a Cauchy sequence that does not converge to an element of c_{00}. Thus, B is not $\mathcal{K}$-bounded.

By the way, an unexpected phenomenon of $\mathcal{K}$ convergence is that a $\mathcal{K}$-convergent sequence need not be $\mathcal{K}$-bounded. An example of such a sequence appears in [116, Example 4.9, p. 41]. The following indicates that there are nontrivial LCS for which every bounded set is $\mathcal{K}$-bounded.

Proposition 6.1.3. *In any Fréchet space, bounded sets are $\mathcal{K}$-bounded.*

Proof. Exercise 6.1.4. □

The next result indicates that Banach disks of Definition 5.1.4 are examples of bounded sets that are $\mathcal{K}$-bounded.

Proposition 6.1.4. *Banach disks are $\mathcal{K}$-bounded.*

Proof. Let B be a Banach disk in a LCS E, and let (x_n) be any sequence of elements of B. Let (a_n) be any null sequence in $\mathbb{K}$. We seek a subsequence $(a_{n_k} x_{n_k})$ of $(a_n x_n)$ such that the series $\sum_{k=1}^{\infty} a_{n_k} x_{n_k}$ converges in E. In fact, a stronger result will be obtained, that being which the convergence is in $(E_B, \|\cdot\|_B)$. First, the completeness of $\mathbb{K}$ allows us to find a subsequence (a_{n_k}) of (a_n) such that $\sum_{k=1}^{\infty} |a_{n_k}| < \infty$. We will use this preliminary calculation to show that the sequence (S_N) of partial sums of $\sum_{k=1}^{\infty} a_{n_k} x_{n_k}$ is Cauchy in $(E_B, \|\cdot\|_B)$. To this end, let $r > 0$ be arbitrary. Then there exists $N \in \mathbb{N}$ such that $\sum_{k=N}^{\infty} |a_{n_k}| < r$. This implies that for all $m > j \geq N$,

$$S_m - S_j = r \cdot \sum_{k=j+1}^{m} \left(\frac{a_{n_k}}{r} \right) x_{n_k} \in rB \tag{6.1}$$

by the absolute convexity of B. Hence, (S_N) is indeed Cauchy and, therefore, converges in the Banach space $(E_B, \|\cdot\|_B)$. □

Corollary 6.1.1. *Every absolutely convex, compact subset of a Hausdorff LCS is $\mathcal{K}$-bounded.*

Spaces for which every bounded set is $\mathcal{K}$-bounded can be used in proofs of the principle of uniform boundedness, which we will see in Section 7.2. This encourages us to give a name to such spaces as stated next.

Definition 6.1.4. A LCS for which every bounded set is $\mathcal{K}$-bounded is called an $\mathcal{A}$-**space**.

Thus, every Fréchet space is an $\mathcal{A}$-space.

One of the primary ways that property $\mathcal{K}$ can be applied is by way of a $\mathcal{K}$-matrix, the definition of which is next.

Definition 6.1.5. Consider a TVS E and a collection $\mathcal{M} = \{x_{ij} : i, j \in \mathbb{N}\}$ of elements of E. $\mathcal{M}$ is called a $\mathcal{K}$-**matrix**, denoted by $\mathcal{M} = [x_{ij}]$ if the following hold:
(a) $\lim_{i \to \infty} x_{ij} = x_j$ exists for each $j \in \mathbb{N}$.
(b) The rows of $\mathcal{M}$ are $\mathcal{K}$-convergent; that is, in each row i, for each increasing sequence (m_j) of positive integers there is a subsequence (n_j) of (m_j) for which $\sum_{j=1}^{\infty} x_{in_j}$ is convergent.

With this definition in mind, we have the following fundamental result.

Theorem 6.1.1 (Basic matrix theorem). *For a $\mathcal{K}$-matrix of Definition* 6.1.5 *the following conclusions hold:*
(a) *For each $j \in \mathbb{N}$, the convergence $\lim_{i \to \infty} x_{ij} = x_j$ is uniform in i.*
(b) *In particular,*

$$\lim_{i \to \infty} \lim_{j \to \infty} x_{ij} = \lim_{j \to \infty} \lim_{i \to \infty} x_{ij} = 0, \quad and \quad \lim_{i \to \infty} x_{ii} = 0. \tag{6.2}$$

Proof. Details of the proof may be seen in [7, Theorem 2.2, pp. 7–9], or [116, Theorem 9.2, pp. 92–94]. □

Incidently, the basic matrix theorem is also referred to as the **Antosik–Mikusiński theorem**. There is no reason to wait before we apply the basic matrix theorem, and said application is next. The application is for the case $E = \mathbb{K}$, where the notion of $\mathcal{K}$ convergence is the same as being convergent to 0. Its proof indirectly depends on the Alaoğlu–Bourbaki Theorem 4.4.1.

Theorem 6.1.2. *In a normed space every weakly $\mathcal{K}$-convergent sequence is norm convergent.*

Proof. Assume that in the normed space $(E, \|\cdot\|)$, the sequence (x_n) is weakly $\mathcal{K}$-convergent. We intend to conclude that $x_n \to 0$ in norm. Replacing E by $\overline{\text{span}\{x_n : n \in \mathbb{N}\}}$, we may assume E is separable. For each $n \in \mathbb{N}$, we find $y_n \in E'$ such that $\|y_n\|' = 1$, where the norm is taken in the dual of E, and furthermore, satisfies $\langle y_n, x_n \rangle = \|x_n\|$, which is possible by Corollary 4.1.5 of the Helly–Hahn–Banach theorem. By separability, Corollary 4.5.2 tells us that (y_n) has a subsequence (y_{n_k}) that is $\sigma(E', E)$-convergent to some element $y \in E'$. We form the matrix $\mathcal{M} = [\langle y_{n_k}, x_{n_l} \rangle]$, for $k, l \in \mathbb{N}$. By way of the $\sigma(E', E)$ convergence of (y_{n_k}) and the assumption of $\sigma(E', E)$ $\mathcal{K}$ convergence of (x_n), $\mathcal{M}$ is a $\mathcal{K}$-matrix. Apply the Basic Matrix Theorem 6.1.1 to conclude that $\lim_{k \to \infty} \langle y_{n_k}, x_{n_k} \rangle = \lim_{k \to \infty} \|x_{n_k}\| = 0$. This argument shows that any subsequence of (x_n) contains a further subsequence that converges in norm to 0. We conclude that $x_n \to 0$ with respect to the norm on E. $\qquad\square$

We note below that in general $\mathcal{K}$ convergence does not imply convergence with respect to the strong topology, as seen next.

Example 6.1.3. For $E = (l_\infty, \sigma(l_\infty, l_1))$, $\mathcal{K}$ convergence does not imply convergence in the strong topology.

Observe that in this case we are considering the dual pairing of l_∞ with l_1. In this context, the strong topology $\beta(E, E')$ is the norm topology on l_∞. Meanwhile, the sequence of unit vectors, $(e^{(n)})$ is $\mathcal{K}$-convergent but not strongly convergent, as you can explain in Exercise 6.1.5.

Another application of the basic matrix theorem shows that despite Example 6.1.3, a $\mathcal{K}$-convergent sequence is at least strongly bounded. In other words, we have the following.

Proposition 6.1.5. *A $\mathcal{K}$-convergent sequence in a LCS E is strongly bounded; that is, if (x_n) is $\mathcal{K}$-convergent, then $\{x_n : n \in \mathbb{N}\}$ is bounded with respect to $\beta(E, E')$.*

Proof. Exercise 6.1.7. $\qquad\square$

The following indicates that property $\mathcal{K}$ can be used to obtain significant results. Its proof can be understood from the material we have seen up to now but in the interest of brevity, that proof is left for you to read in a reference.

Theorem 6.1.3 (Antosik–Burzyk). *A LCS is both barreled and bornological if the two conditions below are satisfied:*
(a) *Every sequentially continuous seminorm on the LCS is continuous.*
(b) *The LCS satisfies property $\mathcal{K}$.*

Proof. See [6]. $\qquad\square$

An important feature of property $\mathcal{K}$ is given below. In Chapter 7, it will allow us to clarify where $\mathcal{K}$-spaces are located with respect to some other types of LCS.

Theorem 6.1.4. *Every metrizable LCS that satisfies property $\mathcal{K}$ is a Baire space.*

Proof. The proof here is for the case of normed spaces and follows that of the original in [30]. See also [117, Theorem 3.2.9, pp. 23–24]. For general metrizable spaces, one may use the concept of an F-norm and the proof otherwise differs little from the normed space version. Details of the proof for metrizable TVS can be seen in [95, Theorem 1.2.17, p. 20]. The strategy is to prove that any countable intersection of open dense subsets is dense, per Theorem A.3.10. Let $\{U_n : n \in \mathbb{N}\}$ be a sequence of open, dense subsets of a normed $\mathcal{K}$-space $(E, \|\cdot\|)$. Without loss of generality, we assume the sequence $\{U_n : n \in \mathbb{N}\}$ is decreasing. To prove that $\bigcap_{n=1}^\infty U_n$ is dense, it suffices to prove that 0 belongs to the closure of $\bigcap_{n=1}^{\infty} U_n$. This, of course, can be established via sequences. To this end, we first define some temporary notation, namely for any $A \subset E$, let $A^0 = E$ and $A^1 = A$. We proceed as follows.

Claim 6.1.1. *Given any $a > 0$, there exist a sequence (x_n) in E and a sequence $\{V_n : n \in \mathbb{N}\}$ of open subsets of E such that the following hold for each $n \in \mathbb{N}$:*
(i) $\|x_n\| < a2^{-n}$.
(ii) $\overline{V_n} \subset U_n$.
(iii) $\sum_{j=1}^{n} e_j x_j \in \bigcap_{j=1}^{n} V_j^{e_j}$, *for all $e_j = 0, 1$.*

Observe that once we prove the claim, (x_n) will be a null sequence by (i), which by the assumption of property $\mathcal{K}$, has a subsequence (x_{n_k}) for which the series $\sum_{k=1}^{\infty} x_{n_k}$ converges to some $x \in E$. Hence, $\|x\| \leq a$ (by (i) again). Applying (iii) will reveal that $\sum_{k=1}^{N} x_{n_k} \in V_{n_j}$ for all $j \in \underline{N}$. This leads to the observation that $x \in \bigcap_{k=1}^{\infty} \overline{V_{n_k}} \subset \bigcap_{k=1}^{\infty} V_{n_k}$, and by (ii) we obtain $x \in \bigcap_{k=1}^{\infty} U_{n_k} = \bigcap_{n=1}^{\infty} U_n$. The arbitrariness of $a > 0$ completes the proof. Thus, we now prove the claim.

Proof of Claim 6.1.1. We employ proof by induction. For $n = 1$, choose $x_1 \in U_1$ such that $\|x_1\| < a/2$. There exists an open set V_1 such that $x_1 \in V_1$ and $\overline{V_1} \subset U_1$. Next, assume $x_1, x_2, \ldots, x_n$ have been chosen and the open sets $V_1, V_2, \ldots, V_n$ have been constructed with the desired properties of (ii) and (iii). Define V by

$$V = \bigcap \left\{ \bigcap_{j=1}^{n} V_j^{e_j} - \sum_{j=1}^{n} e_j x_j : e_j = 0, 1 \right\}. \tag{6.3}$$

By the assumption of (iii), V is a zero neighborhood in E. In addition,

$$U = \bigcap \left\{ U_{n+1} - \sum_{j=1}^{n} e_j x_j : e_j = 0, 1 \right\}, \tag{6.4}$$

is a dense, open subset of E. Next, there exists $x_{n+1} \in U \cap V$ such that $\|x_{n+1}\| < a2^{-(n+1)}$. Let $U^{\sharp}$ be an open subset of E such that $x_{n+1} \in U^{\sharp}$ and $\overline{U^{\sharp}} \subset U \cap V$. Now put

$$V_{n+1} = \bigcup \left\{ U^{\sharp} + \sum_{j=1}^{n} e_j x_j : e_j = 0, 1 \right\}. \tag{6.5}$$

Because $\overline{U^{\sharp}} \subset U$, we have

$$\overline{V_{n+1}} = \overline{U^{\sharp}} + \sum_{j=1}^{n} e_j x_j \subset U_{n+1}, \tag{6.6}$$

for $e_j = 0, 1$. That is, item (ii) holds for $n + 1$. To verify item (iii), the definition of V and the fact that $x_{n+1} \in U^{\sharp} \subset V$ imply that

$$\sum_{j=1}^{n} e_j x_j + x_{n+1} \in \left[\bigcap_{j=1}^{n} V_j^{e_j} \right] \bigcap V_{n+1}, \tag{6.7}$$

for $e_j = 0, 1$. With this, we have verified (iii) for the case $n + 1$, the claim has been established, and the proof is completed. $\qquad\square$

We will make use of property $\mathcal{K}$ going forward. For example, Theorem 7.2.8 in Chapter 7 is a version of the Banach–Steinhaus theorem that does not require completeness of the relevant spaces. Its proof relies on property $\mathcal{K}$.

Exercises

6.1.1. In $\mathbb{R}$, please do the following:
  (a) For $x_n = \frac{1}{\sqrt[3]{n}}$, describe a subsequence x_{n_k} that is subseries convergent.
  (b) Prove that $\mathbb{R}$ satisfies property $\mathcal{K}$.

6.1.2. Verify that property $\mathcal{K}$ is not a property of a dual pair by exhibiting a LCS such that property $\mathcal{K}$ holds for one topology consistent with the duality but does not hold for another topology consistent with the duality. *Hint:* Take a look at Theorem 6.1.2.

6.1.3. Prove that, like in calculus, if a series $\sum_{n=1}^{\infty} x_n$ converges in a LCS E, then the sequence (x_n) is a null sequence in E.

6.1.4. Prove Proposition 6.1.3. *Hint:* Proposition 6.1.1.

6.1.5. Provide details of Example 6.1.3.

6.1.6. Prove Proposition 6.1.2. *Hint:* Take a look at Theorem 3.5.5.

6.1.7. Prove Proposition 6.1.5. *Hint:* Use the Basic Matrix Theorem 6.1.1.

6.1.8. Suppose E and F are TVS and $T : E \to F$ is linear and sequentially continuous (see Definition A.3.22). Prove that T maps $\mathcal{K}$-convergent sequences to $\mathcal{K}$-convergent sequences, and maps $\mathcal{K}$-bounded sets to $\mathcal{K}$-bounded sets.

6.1.9. Consider E, the space of all real-valued functions on c_0, equipped with the topology of pointwise convergence on c_0. Prove that E is a complete LCS that does not satisfy property $\mathcal{K}$. *Hint:* To prove that E does not satisfy property $\mathcal{K}$, define a sequence $(f^{(m)})$ of elements of E by: For any $x = (x_n) \in c_0$, $f^{(m)}(x) = x_m$. Prove that $(f^{(m)})$ is a null sequence in E and that no subsequence of $(f^{(m)})$ has a convergent series.

6.2 Mackey convergence

In this section, we consider another type of convergence, namely when a null sequence in a LCS E converges with respect to a specific finer, normed topology. Recall from Definition 2.2.4 that an absolutely convex bounded set is called a disk, and the definition (see 5.1.1) of the normed space $(E_B, \|\cdot\|_B)$. Note that by the translation invariance of locally convex topologies, it suffices to define convergence to zero in the definition below, as you can verify in Exercise 6.2.2. Mackey convergence turns out to be a useful property in a variety of contexts. For example, we will encounter such usefulness in Section 10.2.

Definition 6.2.1. Let $(E, \mathcal{T})$ be a LCS. A null sequence (x_n) is **Mackey convergent to** 0 (also **Mackey null, locally null, locally convergent**) if there exists a closed disk B in E such that each x_n belongs to E_B and $x_n \to 0$ in the normed topology $\|\cdot\|_B$. If every null sequence is Mackey convergent, then we say the space satisfies the **Mackey convergence condition**, abbreviated as **MCC**.

Note that Mackey's Theorem 5.1.5 implies that Mackey convergence of a sequence in Definition 6.2.1 depends only on the duality (E, E'). However, this statement does not apply to the concept of the Mackey convergence condition (why?). Moreover, the continuity of the injection from $(E_B, \|\cdot\|_B)$ to E implies that the normed topology of E_B is finer than the relative topology on the vector space E_B, so Mackey convergence of a sequence involves convergence with respect to a finer topology, which of course is not always guaranteed. Our first task below regarding this concept is to obtain a way to calculate Mackey convergence, which also indicates a way to define Mackey convergence in TVS that are not necessarily locally convex.

Proposition 6.2.1. *In a LCS E, a sequence (x_n) is Mackey null if and only if there exists a sequence (a_n) of scalars such that $a_n \to \infty$ and $a_n x_n \to 0$ in E.*

Proof. ($\Rightarrow$): Assume the sequence (x_n) is Mackey null in a LCS E. Then there exists a closed disk B such that (x_n) is a null sequence in $(E_B, \|\cdot\|_B)$. Hence, for every $n \in \mathbb{N}$, there is some $\varepsilon_n > 0$ such that $\varepsilon_n \to 0$ and $x_n \in \varepsilon_n B$. Choosing $a_n = \frac{1}{\sqrt{\varepsilon_n}}$, we conclude that $a_n \to \infty$ as $n \to \infty$ and $a_n x_n \to 0$ in E_B. Because the topology generated by $\|\cdot\|_B$ is finer than the relative topology on the vector space E_B, $a_n x_n \to 0$ in E as well.

($\Leftarrow$): Let (x_n) be any null sequence in E and assume we can find a sequence (a_n) of scalars that diverges to $+\infty$, with $a_n x_n \to 0$ as $n \to \infty$ in E. As a convergent sequence, $\{a_n x_n : n \in \mathbb{N}\}$ is a bounded set, so we can form the closed disk $B = \overline{\operatorname{conv bal}\{a_n x_n : n \in \mathbb{N}\}}$. Recalling that $\{\frac{1}{m} B : m \in \mathbb{N}\}$ is a base of zero neighborhoods for the normed space E_B, it is clear that $x_n \to 0$ in E_B. $\qquad\square$

A general TVS that satisfies the sequential condition above is referred to as being a **braked** space, for example, in [117, Remark 3.5.7, p. 34] and [118, p. 75]. Thus, in the context of LCS the term braked refers equivalently to the Mackey convergence condition.

Of course, every normed space satisfies the MCC. There are more spaces that satisfy the MCC, the first type of which is established next.

Proposition 6.2.2. *Every metrizable TVS is braked; in particular, every metrizable LCS satisfies the Mackey convergence condition.*

Proof. We verify the sequential condition of Proposition 6.2.1 above. Suppose (x_n) is a null sequence in the metrizable TVS E. Let $\{U_n : n \in \mathbb{N}\}$ be a decreasing sequence of zero neighborhoods that forms a base for the topology of E. Given any $k \in \mathbb{N}$, there exists $N_k \in \mathbb{N}$ for which $x_n \in \frac{1}{k}U_k$ whenever $n \geq N_k$. Define (a_n) in $\mathbb{R}$ by: For each $n \in \mathbb{N}$, $a_n = k$, for $N_k \leq n < N_{k+1}$. Then $a_n \to \infty$ as $n \to \infty$, and $a_n x_n \to 0$ in E. $\qquad\square$

Next, there are several connections between Mackey convergence and bornological LCS. Three of those connections are next.

Theorem 6.2.1. *A LCS E is bornological if and only if every absolutely convex set that absorbs Mackey null sequences is a zero neighborhood.*

Proof. ($\Rightarrow$): Assume E is bornological and let V be any absolutely convex subset of E that absorbs all Mackey null sequences. Taking a path through contradiction, assume V is not a zero neighborhood. This could only mean that there is a bounded subset B of E that V cannot absorb. Without loss of generality, for every $n \in \mathbb{N}$, there is $x_n \in B$ such that $x_n \notin n^3 V$. In other words, $\frac{1}{n^2}x_n \notin nV$. Put $y_n = \frac{1}{n^2}x_n$ and let $a_n = n$, for each $n \in \mathbb{N}$. Undoubtedly, $a_n \to \infty$ as $n \to \infty$. By Proposition 2.2.5, $a_n y_n \to 0$ as $n \to \infty$. Oh, but this means (y_n) is a Mackey null sequence that is not absorbed by V, a contradiction.

($\Leftarrow$): Assume that in the LCS E any absolutely convex set that absorbs all Mackey null sequences is a zero neighborhood. Let V be an absolutely convex bornivore; that is, if (x_n) is any Mackey null sequence, then (x_n) is a null sequence in E, and as such, $\{x_n : n \in \mathbb{N}\}$ is a bounded subset of E. This means V absorbs the sequence (x_n) and by assumption, V is indeed a zero neighborhood. $\qquad\square$

It is worth noting that the proof above also shows that being bornological is equivalent to the property that every absolutely convex set, which absorbs null sequences must be a zero neighborhood.

Proposition 6.2.3. *Suppose a LCS satisfies the property that every sequentially continuous seminorm is continuous. If the space satisfies the MCC, then it is bornological.*

Proof. By Exercise 5.2.2, it suffices to show that every seminorm that sends null sequences to bounded sets must be continuous. To this end, suppose the seminorm p is such that for any null sequence (x_n) in the LCS E, $\{p(x_n) : n \in \mathbb{N}\}$ is bounded in $\mathbb{K}$. Given such a sequence (x_n), there is a sequence (a_n) of scalars such that $a_n \to \infty$ as $n \to \infty$ and $a_n x_n \to 0$ as $n \to \infty$. In particular, this means there exists $M > 0$ such that for each $n \in \mathbb{N}$, $p(a_n x_n) \leq M$. This last statement implies $p(x_n)$ converges to 0 in $\mathbb{K}$; in other words, p is sequentially continuous. By assumption, p is continuous. Hence, E is bornological. $\qquad\square$

A converse of Proposition 6.2.3, is next.

Proposition 6.2.4. *If E is bornological, then every sequentially continuous seminorm is continuous.*

Proof. Exercise 6.2.7.   □

There is a completeness aspect to Mackey convergence as defined below.

Definition 6.2.2. A null sequence (x_n) in a LCS is **fast convergent** if there exists a Banach disk B such that (x_n) converges to zero in $(E_B, \| \cdot \|_B)$. If every null sequence in a LCS is fast convergent, then the LCS is said to satisfy the **fast convergence condition**, denoted by **FCC**.

Definition 6.2.2 is taken from [95, Definition 6.1.20, p. 171]. Obviously, every LCS that satisfies the FCC also satisfies the MCC. Meanwhile, in [36] a sequence is defined as being fast convergent if it converges in $(E_K, \| \cdot \|_K)$, where K is absolutely convex and compact. There is no need to fret about this difference of definitions because, in fact, they are equivalent as we now verify.

Proposition 6.2.5. *For a sequence in a Hausdorff LCS E, the following are equivalent:*
(a) *The sequence converges in $(E_B, \| \cdot \|_B)$, where B is a Banach disk.*
(b) *The sequence converges in $(E_K, \| \cdot \|_K)$, where K is a compact disk.*

Proof. Without loss of generality, we may assume convergence to zero.

(a) $\Rightarrow$ (b): If $x_n \to 0$ in $(E_B, \| \cdot \|_B)$ for some Banach disk B, then there exists a sequence (a_n) of scalars such that $a_n \to \infty$ and $a_n x_n \to 0$ in E. Put $K = \overline{\text{convbal}}\{a_n x_n : n \in \mathbb{N}\}$. Then K is compact in E_B. By the continuity of id : $E_K \to E$, K is compact in E. Moreover, $x_n \to 0$ in $(E_K, \| \cdot \|_K)$.

(b) $\Rightarrow$ (a): Compact sets are bounded by Corollary 2.5.1 and complete by Proposition 2.5.9.   □

Proposition 6.2.5 reveals a difference between the FCC and MCC, namely that the FCC always allows for convergence in a compact disk whereas for the MCC the relevant disk need not be compact. By Proposition 5.1.4, Mackey convergent sequences in any sequentially complete LCS are fast convergent. We will examine some additional properties of fast convergence in Chapter 7.

It is probably not surprising, but the concepts of $\mathcal{K}$ convergence and Mackey convergence are distinct. In Exercise 6.2.5, you can describe a LCS that satisfies the MCC but not property $\mathcal{K}$. On the other hand, in the LCS $(l_\infty, \sigma(l_\infty, l_1))$, the sequence $e^{(n)}$ of Example 6.1.1 is $\mathcal{K}$-convergent but not Mackey convergent, as you can verify in Exercise 6.2.6. Nevertheless, one property common between $\mathcal{K}$ convergence and Mackey convergence occurs regarding weak convergence. In Theorem 6.1.2, we saw that for a normed space any weak $\mathcal{K}$-convergent sequence is convergent in the original normed topology. This property holds for Mackey convergence and, in fact, is true for any LCS, as we will now see.

Theorem 6.2.2. *In any LCS, a sequence that is Mackey convergent with respect to the weak topology converges with respect to the original topology.*

Proof. Suppose (x_n) is Mackey null with respect to the weak topology. Then by Proposition 6.2.1, there exists a sequence (a_n) of real numbers that diverges to $+\infty$ such that $a_n x_n \to 0$ with respect to the weak topology. The set $\{a_n x_n : n \in \mathbb{N}\}$ is weakly bounded. Mackey's Theorem 5.1.5 then implies that $\{a_n x_n : n \in \mathbb{N}\}$ is bounded with respect to the original topology of the given LCS. The conclusion follows upon noting that as $(a_n x_n)$ is bounded and $a_n^{-1} \to 0$ as $n \to \infty$, we have that $x_n = (1/a_n)(a_n x_n) \to 0$. $\qquad\square$

We also have the following hereditary properties of the MCC.

Proposition 6.2.6. *The MCC is preserved under the formation of:*
(a) *Subspaces*
(b) *Sequentially retractive inductive limits (Definition 3.5.7(c))*
(c) *Countable products and countable direct sums.*

Proof. In Exercise 6.2.9, you can prove parts (a) and (b), as well as give a direct proof of a finite product of spaces that satisfy the MCC. For countably infinite products and countable direct sums of part (c), the proofs are more involved, and you can find a proof of part (c) in [95, Proposition 5.1.31, pp. 158–159]. $\qquad\square$

Exercises

6.2.1. In $\mathbb{R}$, please do the following:
   (a) For the sequence $x_n = \frac{1}{\sqrt[3]{n}}$, describe a sequence (a_n) of positive numbers such that $a_n \to \infty$ as $n \to \infty$ and $a_n x_n \to 0$.
   (b) Prove that $\mathbb{R}$ satisfies the MCC.

6.2.2. Mackey and fast convergence can be defined for convergence to points other than zero as follows. Define a sequence (x_n) to be Mackey convergent to x_0 if $(x_n - x_0)$ is Mackey convergent in the sense of Definition 6.2.1. Prove that these two versions of convergence are equivalent. State and prove the corresponding statement for fast convergence.

6.2.3. Prove that $K = \overline{\mathrm{convbal}\{a_n x_n : n \in \mathbb{N}\}}$ of the proof of Proposition 6.2.5(a) is compact. *Hint:* Take a peek at Proposition 2.5.12 and Proposition 2.5.13.

6.2.4. Let $T : E \to F$ be a linear map between Hausdorff LCS E and F. Prove that the following are equivalent:
   (a) T maps bounded sets to bounded sets.
   (b) T maps Mackey null sequences to Mackey null sequences.
   (c) T maps Mackey null sequences to null sequences.
   (d) T maps Mackey null sequences to bounded sequences.
   Hint: To prove (d) $\Rightarrow$ (a) use the ideas of the first part of the proof of Theorem 6.2.1.

6.2.5. Describe a LCS that satisfies the MCC but not property $\mathcal{K}$.

6.2.6. Prove that in $(l_\infty, \sigma(l_\infty, l_1))$, the sequence $e^{(n)}$ of unit vectors as seen in Example 6.1.1 is $\mathcal{K}$-convergent but not Mackey convergent.

6.2.7. Prove Proposition 6.2.4. *Hint:* Proceed by contradiction.

6.2.8. Prove Proposition 6.2.5 for the case where $x_n \to x_0$, where $x_0 \neq 0$.

6.2.9. Regarding Proposition 6.2.6, prove parts (a) and (b), and give a direct proof that a finite product of spaces satisfying the MCC satisfies the MCC.

6.3 Spaces with webs

In this section, we examine an important structure that is traditionally a tool for proving versions of the closed graph theorem (as we will see in Chapter 8), called a web. For detailed reading about webs, see [36, 105, 103], as well as [70, Chapter 5], and [80, Section 35]. Much of what we will discuss regarding webs is valid in general TVS, which you can check out in the references listed in the previous sentence. Prior to the formal definition, an indication of the general idea might be helpful. The goal is to construct a countable collection of subsets of a LCS that have properties similar to those of a finer topology than the original, even though the sets in the collection need not be neighborhoods. The definition follows next.

Definition 6.3.1. A **(Robertson) web** $\mathcal{W}$ on a vector space E is a countable collection of absolutely convex sets, indexed by finite sequences of positive integers and arranged in **layers** as follows:

(a) The **first layer** of $\mathcal{W}$ consists of a sequence $(W_{n_1} : n_1 \in \mathbb{N})$ such that $\bigcup\{W_{n_1} : n_1 \in \mathbb{N}\}$ is absorbing in E.

(b) For each $n_1 \in \mathbb{N}$, there is a sequence $(W_{n_1 n_2} : n_2 \in \mathbb{N})$ of sets in $\mathcal{W}$ that is **determined by** W_{n_1} in the sense that each $W_{n_1 n_2} \subset \frac{1}{2} W_{n_1}$. In this case, we assume that $\bigcup\{W_{n_1 n_2} : n_2 \in \mathbb{N}\}$ is absorbing in W_{n_1}. The **second layer** of the web then consists of $\{W_{n_1 n_2} : n_1, n_2 \in \mathbb{N}\}$.

(c) In general, for the kth **layer** of $\mathcal{W}$, there is a sequence $(W_{n_1 n_2 \cdots n_k} : n_k \in \mathbb{N})$ of sets in $\mathcal{W}$ such that for each $n_k \in \mathbb{N}$,

$$W_{n_1 n_2 \cdots n_k} \subset \frac{1}{2} W_{n_1 n_2 \cdots n_{k-1}}, \tag{6.8}$$

and the following absorption property holds:

$$\bigcup_{n_k=1}^{\infty} W_{n_1 n_2 \cdots n_k} \text{ is absorbing in } W_{n_1 n_2 \cdots n_{k-1}}. \tag{6.9}$$

The kth layer then consists of $\{W_{n_1 n_2 \cdots n_k} : n_1, n_2, \ldots, n_k \in \mathbb{N}\}$.

For example, starting with the second layer, for each set $W_{n_1 n_2}$ a sequence $(W_{n_1 n_2 n_3} : n_3 \in \mathbb{N})$ determined by $W_{n_1 n_2}$ satisfies that for each $n_3 \in \mathbb{N}$, $W_{n_1 n_2 n_3} \subset \frac{1}{2} W_{n_1 n_2}$, and $\bigcup\{W_{n_1 n_2 n_3} : n_3 \in \mathbb{N}\}$ is absorbing in $W_{n_1 n_2}$. The third layer then consists of $\{W_{n_1 n_2 n_3} : n_1, n_2, n_3 \in \mathbb{N}\}$.

More generally, the sets of a web need not be absolutely convex, but are always assumed to be balanced. In this context, Equation (6.8) is modified to

$$W_{n_1 n_2 \cdots n_k} + W_{n_1 n_2 \cdots n_k} \subset W_{n_1 n_2 \cdots n_{k-1}}. \tag{6.10}$$

Most of what is done with a web involves looking at sequences of its sets for which there is an element in each layer of the web. This idea prompts the next definition.

Definition 6.3.2. Given a sequence $(n_1, n_2, \ldots) = (n_k : k \in \mathbb{N})$ of positive integers, the corresponding sequence $(W_{n_1}, W_{n_1 n_2}, \ldots)$ of sets of a web $\mathcal{W}$ is called a **strand** if for each $k \in \mathbb{N}$, $W_{n_1 n_2 \cdots n_k}$ is in the kth layer and $W_{n_1 n_2 \cdots n_{k+1}}$ is determined by $W_{n_1 n_2 \cdots n_k}$.

For a given strand, the following notation is useful.

Notation 6.3.1. Whenever confusion is unlikely, the notation used for a strand of a web is (W_k).

In particular, for any strand (W_k) of a web $\mathcal{W}$, we have that for every $k \in \mathbb{N}$,

$$W_{k+1} \subset \frac{1}{2} W_k. \tag{6.11}$$

Observe that a *web always has infinitely many layers but need not have infinitely many strands*.

Figure 6.1 shows a general diagram of a web for which there are infinitely many strands. The top row of sets, denoted by $W_1, W_2, \ldots$, represents the sets of the first layer. In the second row, sets denoted by $W_{11}, W_{12}, \ldots$ under W_1 are determined by W_1, whereas $W_{21}, W_{22}, \ldots$ are determined by W_2, and so on. These sets represents the elements of the second layer. Sets marked with three indices are sets in the third layer, and this pattern continues for all layers. For each $j \in \mathbb{N}$, the set W_{1j} is a subset of $\frac{1}{2} W_1$; for each $m \in \mathbb{N}$, the set W_{21m} is a subset of $\frac{1}{2} W_{21}$, and so on. In general, the indices are arranged so as to indicate the sets that are determined by a particular set of the previous layer. For instance, W_{21} is not a subset of W_1, so these two sets cannot form part of a strand. An example of a strand in the figure would be $\{W_2, W_{21}, W_{212}, \ldots\}$. Another strand would be $\{W_1, W_{11}, W_{111}, W_{1112}, \ldots\}$. The sets in Figure 6.1 are made to look "smaller" in each layer to represent the property $W_{k+1} \subset \frac{1}{2} W_k$, for each $k \in \mathbb{N}$, where k represents the kth layer of the web. It turns out that to describe a web it is often convenient to define each of its layers, or each of its strands.

Definition 6.3.1 is due to W. Robertson [105]. Detailed discussions of this version of a web can be found in [105] and [103]. The concept of a web was originally defined by De Wilde in [35], a detailed exposition of which can be seen in [36]. A comparison between the definitions by De Wilde and W. Robertson can be found in [105, pp. 716–726]. Except where noted, properties and results in this book are valid with respect to both definitions. A layer of a web is also referred to as a **slice** and a strand is also referred to as a **thread**, for example, in [18, p. 196]. Despite what appears to be a complicated definition, in practice webs are often constructed intuitively by making use of the structure of the given LCS. Speaking of examples, up next is a simple one.

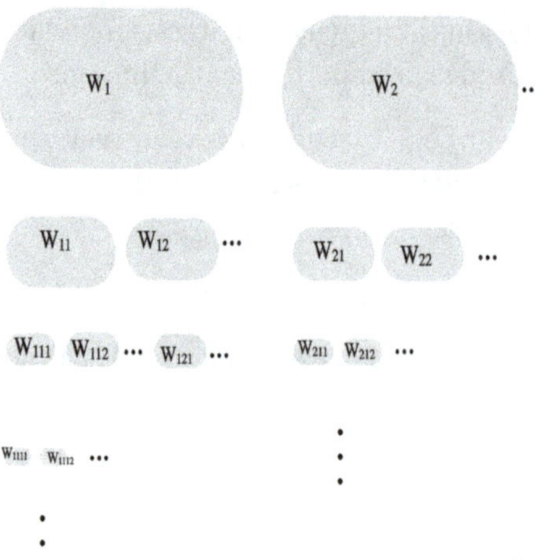

Figure 6.1: A web in a locally convex space.

Example 6.3.1. If $\{U_n : n \in \mathbb{N}\}$ is a base of absolutely convex zero neighborhoods of a metrizable LCS such that for each $n \in \mathbb{N}$, $U_{n+1} \subset \frac{1}{2}U_n$, then $\mathcal{W} = \{U_n : n \in \mathbb{N}\}$ is a web on that metrizable space. In particular, if B is the unit ball of a normed space, then $\mathcal{W} = \{2^{-n}B : n \in \mathbb{N}\}$ is a web on the normed space.

Observe that the web for a metrizable LCS consists of only one strand. For a metrizable LCS, there is but one set in the kth layer ($k \geq 2$), namely $W_k = 2^{1-k}U_{k-1}$. In the case of a normed space the set of the kth layer ($k \geq 2$) is $W_k = 2^{1-k}B$, where B is the unit ball of the space. More examples of webs in LCS will be constructed in Theorem 6.3.1 shortly.

The next goal is to exhibit a connection between web structures and the topologies of the given LCS. That is, if we intend to make use of webs in a LCS, then there needs to be a connection between the web structure and the topology of a LCS. Such a connection is defined next.

Definition 6.3.3. A web $\mathcal{W}$ on a LCS E is **compatible** (with the topology) if, given any zero neighborhood U in E, and any strand (W_k) of $\mathcal{W}$, there exists $K \in \mathbb{N}$ such that $W_K \subset U$. Hence, in this case, $(\forall k \geq K)\, W_k \subset U$.

Clearly, the webs defined in Example 6.3.1 above are compatible. For more general cases, the following describes concrete ways to determine if a web is compatible.

Proposition 6.3.1. *Let $\mathcal{W}$ be a web on a Hausdorff LCS E. Then the following are equivalent:*
(a) *$\mathcal{W}$ is compatible.*
(b) *Given any strand (W_k) of $\mathcal{W}$, and any sequence (x_k) in E such that for each $k \in \mathbb{N}$, $x_k \in W_k$, the sequence $S_N = \sum_{k=1}^{N} x_k$ of partial sums of (x_k) is a Cauchy sequence.*
(c) *Given (W_k) and (x_k) as in (b), the sequence (x_k) is a null sequence in E.*

Proof. (a) $\Rightarrow$ (b): Assume the web $\mathcal{W}$ is compatible. Given any strand (W_k) of $\mathcal{W}$ and any $U \in \mathcal{N}_0$, there exists $N \in \mathbb{N}$ such that for all $k \geq N$, $W_k \subset U$. For the partial sum $S_N = \sum_{k=1}^{N} x_k$ of a sequence (x_k) such that $(\forall k \in \mathbb{N})x_k \in W_k$, we have, for $m > n \geq N$,

$$S_m - S_n = x_{n+1} + x_{n+2} + \cdots + x_m \in W_{n+1} + W_{n+2} + \cdots + W_m$$
$$\subset \frac{1}{2}W_n + \frac{1}{2^2}W_n + \cdots + \frac{1}{2^{m-n}}W_n$$
$$\subset W_n \subset W_N \subset U. \tag{6.12}$$

(b) $\Rightarrow$ (c): Obvious.

(c) $\Rightarrow$ (a): Assume (c) and suppose $\mathcal{W}$ is not compatible. Then there exists a zero neighborhood U of E and a strand (W_k) of $\mathcal{W}$ such that for each $k \in \mathbb{N}$ there exists $x_k \in W_k \setminus U$. Then certainly, (x_k) is not a null sequence in E, and with this contradiction, the proof is completed. $\square$

Notice that for any strand (W_k) of a compatible web, we have $\bigcap_{k=1}^{\infty} W_k = \{0\}$, which you can verify in Exercise 6.3.4.

The next item shows that it is possible to construct compatible webs on a variety of LCS, including some types (such as inductive limits), that need not be metrizable.

Theorem 6.3.1. *For Hausdorff LCS, possessing a compatible web is preserved under the following constructions:*
(a) *Linear subspaces*
(b) *Continuous linear images*
(c) *Hausdorff quotients (Definition 3.1.1)*
(d) *Countable inductive limits (Definition 3.5.3). In particular, a countable direct sum (Definition 3.5.2) of LCS having compatible webs has a compatible web.*
(e) *Countable products (Definition 3.3.1)*
(f) *Countable projective limits (Definition 3.4.3)*
(g) *The strong dual of a metrizable LCS (Definition 4.3.4).*

Proof. (a) If $\mathcal{W}$ is a compatible web on a LCS E, and L is linear subspace of E, then a compatible web on L is given by $\{W \cap L : W \in \mathcal{W}\}$, which is easily verified in Exercise 6.3.5(a).

(b) For a continuous linear map $T : E \to F$, which we might as well assume is surjective, and assuming $\mathcal{W}_E$ is a compatible web on E, define a web on F as the images under T of the sets of $\mathcal{W}_E$. That is, let $\mathcal{W}_F = \{T(W) : W \in \mathcal{W}_E\}$. That each set of $\mathcal{W}_F$ is absolutely convex is a consequence of the linearlity of T (see Theorem 1.1.2). Thus, (Y_k) is a strand in $\mathcal{W}_F$ if and only if $Y_k = T(W_k)$ for all $k \in \mathbb{N}$, where (W_k) is a strand in $\mathcal{W}_E$. Property (a) of Definition 6.3.1 follows from the surjectivity of T. Properties (b) and (c) are valid because of the definition of $\mathcal{W}_F$. Next, we establish that $\mathcal{W}_F$ is compatible with the topology of F. Let V be an arbitrary zero neighborhood in F, and let (Y_k) be any strand of $\mathcal{W}_F$. The continuity of T implies that there exists a zero neighborhood U in E such that $T(U) \subset V$. For the strand (W_k) such that $T(W_k) = Y_k$, there exists

$K \in \mathbb{N}$ such that $W_K \subset U$, because $\mathcal{W}_E$ is compatible with the topology of E. Hence, $Y_K = T(W_K) \subset T(U) \subset V$.

(c) If F is a closed subspace of E, then a web on the quotient E/F is given by $\{\varphi(W) : W \in \mathcal{W}\}$, where $\varphi : E \to E/F$ is the canonical quotient map and $\mathcal{W}$ is a compatible web on E. That this defines a compatible web on E/F follows from part (b), in light of the continuity and surjectivity of the canonical quotient map on a subspace of a LCS.

(d) Let $E = \operatorname{ind}_n E_n$ be an inductive limit of LCS $(E_n, \mathcal{T}_n)$ each of which has a compatible web $\mathcal{W}^{(n)}$, for $n \in \mathbb{N}$. Form a web on E such that the kth layer of $\mathcal{W}$ is composed of all the kth layers of the webs $\mathcal{W}^{(n)}$ for $n \in \mathbb{N}$ and each strand of $\mathcal{W}$ is a strand of some $\mathcal{W}^{(m)}$. Of course, we must verify the properties of Definition 6.3.1 of a web. Clearly, the sets are all absolutely convex. Consider

$$\bigcup \{ W_{n_1}^{(m)} : n_1, m \in \mathbb{N} \}, \tag{6.13}$$

where each $W_{n_1}^{(m)}$ belongs to the first layer of the web $\mathcal{W}^{(m)}$ on E_m. By the construction of $\mathcal{W}$, given any $x \in E$, there exists $m \in N$ such that $x \in E_m$, which means x is absorbed by $\bigcup \{ W_{n_1}^{(m)} : n_1 \in \mathbb{N} \}$. Hence, x is absorbed by the union over all $m \in \mathbb{N}$ of such sets; in other words, x is absorbed by $\bigcup \{ W_{n_1}^{(m)} : n_1, m \in \mathbb{N} \}$, and property (a) of Definition 6.3.1 holds. For property (b), given any $A \in \mathcal{W}$, there is some $n \in \mathbb{N}$ such that $A = W_k^{(n)}$, for some $k \in \mathbb{N}$. Because $\mathcal{W}^{(n)}$ is a web on E_n, we have $W_{k+1}^{(n)} \subset \frac{1}{2} W_k^{(n)}$, and this holds for any such $A \in \mathcal{W}$. The absorbing property follows from the fact that $\mathcal{W}^{(n)}$ is a web on E_n. The argument for property (c) of Definition 6.3.1 is similar, and is left for you as Exercise 6.3.8(c). To prove compatibility, let U be any zero neighborhood in E and let (W_k) be any strand in $\mathcal{W}$. Then there is $m \in \mathbb{N}$ such that (W_k) is a strand of $\mathcal{W}^{(m)}$ on E_m. By properties of the inductive limit topology, (see Proposition 3.5.6), $U_m = \operatorname{id}^{-1}(U)$ is a zero neighborhood in E_m. $\mathcal{W}^{(m)}$ is compatible with the topology $\mathcal{T}_m$ of E_m, so there exists $K_m \in \mathbb{N}$ such that for all $k \geq K_m$, $W_k \subset U_m$, which in turn implies that $W_k \subset U$. Note that because a direct sum can be considered as an inductive limit, the result on direct sums follows as well.

(e) You can construct a compatible web on a finite product of LCS that have compatible webs in Exercise 6.3.8(b), by taking the products of the sets of the webs of each layer of the spaces. For countably infinite products $E = \prod_{n=1}^{\infty} E_n$ of LCS E_n that have compatible webs $\mathcal{W}^{(n)}$, we need to construct a web in concert with the product topology. The idea is that for each layer k, we introduce a new strand of one of the spaces E_n, while keeping the sets of the strands that have already been introduced. Describing the desired web by layers is done as follows: Let the first layer be defined as shown in Equation (6.14) below:

$$W_{n_1}^{(1)} \times E_2 \times E_3 \times \cdots, \tag{6.14}$$

where $\{ W_{n_1}^{(1)} : n_1 \in \mathbb{N} \}$ are sets in the first layer of the web $\mathcal{W}^{(1)}$ on E_1. For the second layer, use

$$W^{(1)}_{n_1 n_2} \times W^{(2)}_{m_1} \times E_3 \times E_4 \times \cdots, \tag{6.15}$$

where $\{W^{(1)}_{n_1 n_2} : n_1, n_2 \in \mathbb{N}\}$ and $\{W^{(2)}_{m_1} : m_1 \in \mathbb{N}\}$ are sets in the second and first layer of the webs $\mathcal{W}^{(1)}$ and $\mathcal{W}^{(2)}$, respectively. The kth layer is constructed as shown in Equation (6.16).

$$W^{(1)}_{n_1 n_2 \cdots n_k} \times W^{(2)}_{m_1 m_2 \cdots m_{k-1}} \times \cdots \times W^{(k)}_{r_1} \times E_{k+1} \times E_{k+2} \times \cdots, \tag{6.16}$$

for sequences of positive integers $(n_1, n_2, \ldots)$, $(m_1, m_2, \ldots)$, $\cdots$, $(r_1, r_2, \ldots)$, $\cdots$. In Exercise 6.3.8(c) you can write the routine verification that this construction satisfies parts (a)–(c) of Definition 6.3.1 of a web on the product space E. Meanwhile, we now verify that the web $\mathcal{W}$ formed by this collection of sets is compatible with the product topology on E. Let U be any zero neighborhood in E. By definition of the product topology (Definition A.3.17), U can be written in the form

$$U = \prod_{j=1}^{N} U^{(j)} \times \prod_{j=N+1}^{\infty} E_j, \tag{6.17}$$

for some $N \in \mathbb{N}$ and $U^{(j)} \in \mathcal{N}_{0,E_j}$, for $j = 1, 2, \ldots, N$.

To describe a general strand (W_k) of $\mathcal{W}$, assume $(W^{(n)}_k : k \in \mathbb{N})$ is a strand of the web $\mathcal{W}^{(n)}$ on E_n, for each $n \in \mathbb{N}$. Then a strand in the product E is given by Equation (6.18):

$$W_k = W^{(1)}_k \times W^{(2)}_{k-1} \times W^{(3)}_{k-2} \times \cdots \times W^{(k)}_1 \times E_{k+1} \times E_{k+2} \times \cdots. \tag{6.18}$$

The web $\mathcal{W}^{(j)}$ is compatible in E_j for each $j \in \mathbb{N}$. Therefore, given a strand (W_k) of the web $\mathcal{W}$ as shown in Equation (6.18), for each $j = 1, 2, \ldots, N$ there exists $K_j \in \mathbb{N}$ such that $W^{(j)}_{K_j} \subset U^{(j)}$. Put $K = \max\{K_j : j = 1, 2, \ldots, N\}$. Then for all $k \geq K$, $W^{(n)}_k \subset U$; that is, $W^{(n)}_k \subset W^{(n)}_K \subset W^{(n)}_{K_j} \subset U^{(j)}$. Hence,

$$\begin{aligned}
W^{(1)}_{N+K-1} \times W^{(2)}_{N+K-2} &\times \cdots \times W^{(N+K-1)}_1 \times E_{N+K} \times E_{N+K+1} \times \cdots \\
&\subset U^{(1)} \times U^{(2)} \times \cdots \times U^{(N)} \times E_{N+1} \times E_{N+2} \times \cdots \\
&\subset U,
\end{aligned} \tag{6.19}$$

and we have established that $\mathcal{W}$ is compatible with the topology of the product topology.

(f) This follows from the fact that a countable projective limit is a subspace of a countable product, and parts (a) and (e) above.

(g) Let $\{U_n : n \in \mathbb{N}\}$ be a basis of absolutely convex zero neighborhoods in a metrizable LCS E. We will construct a compatible web on the strong dual of E that consists of polars of the zero neighborhoods U_n. Define $\mathcal{W}$ on $(E', \beta(E', E))$ such that the kth layer consists of

$$\left\{ \frac{1}{2^k} U_n^{\circ} : n \in \mathbb{N} \right\}, \tag{6.20}$$

where U_n° is the polar of U_n. Properties (a)–(c) of Definition 6.3.1 follow from the definition of the web $\mathcal{W}$ and the fact that $\bigcap_{k=1}^{\infty} U_k = \{0\}$, so that $\bigcup_{k=1}^{\infty} U_k^{\circ} = E'$ by the continuity of the elements of E'. What remains for this part is to prove compatibility. Given any bounded subset B of E, its polar B° is a zero neighborhood in $(E', \beta(E', E))$ by the definition of the strong topology. Thus, for any $U_n \in \mathcal{N}_{0,E}$ there exists $a_n > 0$ such that $B \subset a_n U_n$, and then Proposition 4.2.4 implies that $U_n^{\circ} \subset a_n B^{\circ}$. In other words, each U_n° is $\beta(E', E)$-bounded, hence absorbed by any zero neighborhood of the strong topology on E'. $\square$

We are now ready to investigate some connections between webs and Mackey convergence. The gist of part (c) of Proposition 6.3.1 is that if terms of a sequence are eventually elements of all but finitely many sets of a strand of a compatible web, then the sequence converges to zero. The idea of the definition below is that every null sequence arises this way.

Definition 6.3.4. A web $\mathcal{W}$ on a LCS is **sequential** (or the LCS has a **sequential web**) if $\mathcal{W}$ is compatible, and given any null sequence (x_n) in E there is a strand (W_k) of $\mathcal{W}$ for which, given any $k \in \mathbb{N}$, there is $N_k \in \mathbb{N}$ such that $x_n \in \overline{W_k}$, for all $n \geq N_k$.

Certainly every normed space has a sequential web given by the zero neighborhood base $\{2^{-k} B : k \in \mathbb{N}\}$, where B is the closed unit ball. Moreover, every metrizable LCS has a sequential web given by the neighborhood base $\{U_k : k \in \mathbb{N}\}$, of absolutely convex, closed sets U_k such that for each $k \in \mathbb{N}$, $U_{k+1} \subset 2^{-1} U_k$. More examples of LCS that have sequential webs will be given in Subsection 6.3.1. In the meantime, the main result of this section is next.

Theorem 6.3.2. *Every LCS that has a sequential web satisfies the MCC.*

Proof. Assume E has a sequential web $\mathcal{W}$ and let (x_n) be any null sequence in E. Per Proposition 6.2.1, it suffices to find a sequence (a_n) of scalars such that $a_n \to \infty$ as $n \to \infty$ and $a_n x_n \to 0$ in E. The assumption that $\mathcal{W}$ is sequential implies that there exists a strand (W_k) of $\mathcal{W}$ such that $(\forall k \in \mathbb{N})(\exists N_k \in \mathbb{N})$ satisfying $(\forall n \geq N_k) x_n \in \overline{W_k}$. Equation (6.11) of Definition 6.3.1 allows for the following calculations: $\overline{W_{k+1}} \subset \frac{1}{2}\overline{W_k}$, $\overline{W_{k+2}} \subset \frac{1}{4}\overline{W_k}, \ldots$, and in general we have

$$(\forall l \in \mathbb{N}) \quad \overline{W_{k+l}} \subset 2^{-l}\overline{W_k}. \tag{6.21}$$

Given any $k \in \mathbb{N}$, we can find $l_k \in \mathbb{N}$ such that $2^{-l_k} < \frac{1}{k}$. Thus,

$$\overline{W_{k+l_k}} \subset 2^{-l_k}\overline{W_k} \subset \frac{1}{k}\overline{W_k}. \tag{6.22}$$

Choose $N_{k+l_k} \in \mathbb{N}$ such that for all $n \geq N_{k+l_k}$, $x_n \in \frac{1}{k}\overline{W_k}$. That is, $kx_n \in \overline{W_k}$, and this holds for all $k \in \mathbb{N}$. Define $a_n \in (0, \infty)$, $n \in \mathbb{N}$, by

$$a_n = k, \quad N_{k+l_k} \leq n < N_{k+l_{k+1}}. \tag{6.23}$$

Obviously, $\lim_{n\to\infty} a_n = \infty$ as $n \to \infty$, and it only remains to show that $a_n x_n \to 0$ in E. To verify this, let $U \in \mathcal{N}_0$ be absolutely convex and closed. The compatibility of the web $\mathcal{W}$ ensures that there exists $K \in \mathbb{N}$ such that $\overline{W_K} \subset U$. By the construction of the sequence (a_n), there exists $N_K \in \mathbb{N}$ such that for all $n \geq N_K$, we have $a_n x_n \in \overline{W_K} \subset U$, and we have completed the proof. $\qquad\square$

In Chapter 8, we will require a concept of completeness regarding webs. We introduce that completeness concept in the Definition 6.3.5 below. Such completeness is also useful for determining Mackey or $\mathcal{K}$ convergence, as we shall see shortly.

Definition 6.3.5. A web $\mathcal{W}$ on a LCS E is **completing** (or that E has a **completing web**, or is a **webbed space**), if given any strand (W_k) of $\mathcal{W}$ and any sequence (x_k) with $x_k \in W_k$ for each $k \in \mathbb{N}$, the series $\sum_{k=1}^{\infty} x_k$ converges in E. If in addition each such convergent series satisfies $\sum_{l=k+1}^{\infty} x_l \in W_{k-1}$ for each k, then $\mathcal{W}$ is called **strict** (or E has a **strict web**, or is **strictly webbed**).

Other terms for a completing web include that the web is of **type (c)**, and another term for a strict web is to say that the web is **tight**. A simple way to conclude that a completing web is strict will be given shortly, and an example of a completing web that is not strict can be found in [105].

The property of a web being completing is connected to the compatibility of the web, as seen next.

Proposition 6.3.2. *Every completing web is compatible and every compatible web in a sequentially complete LCS is completing.*

Proof. Let $\mathcal{W}$ be a web on a LCS E and consider any strand (W_k) of $\mathcal{W}$. Assume (x_k) is any sequence such that for each $k \in \mathbb{N}$, $x_k \in W_k$. To prove the first statement, the assumption that $\mathcal{W}$ is completing implies that the series $\sum_{k=1}^{\infty} x_k$ converges in E. Therefore, the partial sums of the series form a Cauchy sequence. Proposition 6.3.1(b) then implies that $\mathcal{W}$ is compatible. For the second statement, if E is sequentially complete and $\mathcal{W}$ is compatible, then Proposition 6.3.1 applies again to conclude that the partial sums of the series $\sum_{k=1}^{\infty} x_k$ form a Cauchy sequence, which of course converges by the sequential completeness assumption. $\qquad\square$

The following represents examples of LCS having strict webs.

Proposition 6.3.3. *Every Fréchet space is strictly webbed. In particular, every Banach space is strictly webbed.*

Proof. The appropriate web is given by a zero neighborhood base of absolutely convex closed sets described in Example 6.3.1. If in both cases we choose the sets to be closed, then it is easy to see that the webs are strict. $\qquad\square$

The statement in the proof of Proposition 6.3.3 about assuming sets of a web are closed reveals what turns out to be a simple way to observe when a completing web is strict, as seen next. See Definition A.3.11 of sequentially closed sets for the proposition below.

Proposition 6.3.4. *A completing web that consists of (sequentially) closed sets, is strict.*

Proof. Assume a LCS E has a completing web $\mathcal{W}$ that consists of (sequentially) closed sets. Let (W_k) be any strand of $\mathcal{W}$ and let (x_k) be any sequence that satisfies $x_k \in W_k$, for each $k \in \mathbb{N}$. By the completing assumption of $\mathcal{W}$, $\sum_{k=1}^{\infty} x_k$ converges in E. That each W_k is (sequentially) closed indicates that the limit of the series $\sum_{l=k+1}^{\infty} x_l$ belongs to W_{k-1} for each $k \in \mathbb{N}$, and conclusion follows. $\qquad\square$

When we augment the concept of a compatible web with the concept of a completing web, we again have a significant list of hereditary properties, which will now be detailed.

6.3.1 Hereditary properties of webbed spaces

A salient feature of webs that are either completing or strict is that they enjoy several hereditary properties, which we now proceed to discuss. The concept of a (strict) completing web is crucial to proofs of results that we will see in Chapter 8. In the result below, we assume a few properties that may seem extraneous for this section but will make for shorter proofs in Chapter 8.

Theorem 6.3.3. *For Hausdorff LCS, the completing (resp., strict) property of webs is preserved under the following constructions:*
(a) *Closed subspaces*
(b) *Continuous linear images*
(c) *Hausdorff quotients (Definition 3.1.1)*
(d) *Countable inductive limits (Definition 3.5.3) and countable direct sums (Definition 3.5.2)*
(e) *Countable products (Definition 3.3.1)*
(f) *Countable projective limits (Definition 3.4.3)*
(g) *The strong dual of a metrizable LCS (Definition 4.3.4).*

Proof. We obtain the results for completing webs. The case where completing webs are strict involves mostly routine calculations and is left as Exercise 6.3.10. Assume the constructions and properties of compatibility of the webs from Theorem 6.3.1.

(a) Suppose $\mathcal{W}$ is a completing web on a LCS E, and L is closed subspace of E. For the compatible web on L of Theorem 6.3.1(a) given by $\{W \cap L : W \in \mathcal{W}\}$, this web on L is completing by the assumption that the subspace is closed.

(b) For a continuous surjective linear map $T : E \to F$ and for which we assume E is webbed, define a web on F as in Theorem 6.3.1(b), namely a strand (Y_k) in $\mathcal{W}_F$ is given by $Y_k = T(W_k)$, for some strand (W_k) in $\mathcal{W}_E$. We prove that $\mathcal{W}_F$ is completing. To this end, let (Y_k) be any strand of $\mathcal{W}_F$ and suppose the sequence (y_k) satisfies that for all $k \in \mathbb{N}, y_k \in Y_k$. The surjectivity of T implies that there is a sequence (x_k) in E such that $T(x_k) = y_k$, and additionally, $x_k \in W_k$ for each $k \in \mathbb{N}$. As $\mathcal{W}_E$ is completing in E, $\sum_{k=1}^{\infty} x_k$ converges in E, and the continuity of T implies that $\sum_{k=1}^{\infty} y_k = \sum_{k=1}^{\infty} T(x_k)$ converges in F.

(c) This follows from part (b) and properties of the canonical quotient map on a closed subspace of a LCS.

(d) Let $E = \operatorname{ind}_n E_n$ be an inductive limit of LCS E_n with the compatible web $\mathcal{W}$ on E as described in the proof of Theorem 6.3.1(d). If for any strand (W_k) of $\mathcal{W}$ one has for each $k \in \mathbb{N}, x_k \in W_k$, then for some $m \in \mathbb{N}, x_k \in W_k$ for all $k \in \mathbb{N}$, where (W_k) is a strand in the completing web $\mathcal{W}^{(m)}$ and, therefore, $\sum_{k=1}^{\infty} x_k$ converges in E_m. It then follows that $\sum_{k=1}^{\infty} x_k$ converges in E by the continuity of $\mathrm{id} : E_m \to E$. The case of countable direct sums follows by the observation made in Theorem 6.3.1(d).

(e) The finite case is left as Exercise 6.3.8(b). For countably infinite products $E = \Pi_{n=1}^{\infty} E_n$ of LCS E_n that have completing webs $\mathcal{W}^{(n)}$, use the web constructed in the proof of Theorem 6.3.1(e). To verify the completing property, let π_n denote the projection from E onto each space $E_n, n \in \mathbb{N}$. Let (W_k) be any strand of $\mathcal{W}$ and suppose (x_k) satisfies $x_k \in W_k$, for each $k \in \mathbb{N}$. Let U be an arbitrary zero neighborhood of the form given in Equation (6.17). Certainly, $\pi_n(x_k) \in E_n$, and once $k \geq n, \pi_n(x_k) \in W_k^{(n)}$. The web $\mathcal{W}^{(n)}$ is completing in E_n, so the series $\sum_{k=1}^{\infty} \pi_n(x_k)$ converges to some element $y^{(n)} \in E_n$. This holds for each $n \in \mathbb{N}$. Hence, $\sum_{k=1}^{\infty} x_k$ converges in the product space E.

(f) Exercise 6.3.8(c).

(g) Using the compatible web on the strong dual of a metrizable LCS E from Theorem 6.3.1(g), what needs to be proven is that $\mathcal{W}$ is completing. Every metrizable LCS is bornological, so by Theorem 5.2.4, the strong dual of E is complete. Now apply Proposition 6.3.2 to conclude that the web $\mathcal{W}$ is completing. $\qquad\square$

Using a combination of the methods of parts (d) and (g) of Theorem 6.3.3, we also have the following.

Proposition 6.3.5. *The strong dual of a strict inductive limit of metrizable LCS has a strict web.*

Proof. Exercise 6.3.9. $\qquad\square$

In what follows, we discover new types of LCS that satisfy property $\mathcal{K}$ and the MCC.

Proposition 6.3.6. *If a LCS has a web that is both sequential and completing, then the LCS satisfies property $\mathcal{K}$.*

Proof. We need to prove that the space in question satisfies property $\mathcal{K}$. Suppose E has a web $\mathcal{W}$ that satisfies the hypotheses. The sequential assumption of $\mathcal{W}$ allows us to cite Definition 6.3.4: Given a null sequence (x_n) in E, there is a strand (W_k) of $\mathcal{W}$ such that for all $k \in \mathbb{N}$, there is $N_k \in \mathbb{N}$ such that $x_n \in \overline{W_k}$, for all $n \geq N_k$. This condition implies that there is a subsequence (x_{n_k}) of (x_n) such that for all $k \in \mathbb{N}$, $x_{n_k} \in \overline{W_k}$. Because $\mathcal{W}$ was assumed to be completing, the series $\sum_{k=1}^{\infty} x_{n_k}$ converges in E. $\qquad\square$

Corollary 6.3.1. *Any sequentially retractive inductive limit of LCS, each of which has a web that is sequential and completing, satisfies property $\mathcal{K}$ and the MCC. In particular, every strict (LF) space satisfies property $\mathcal{K}$ and the MCC.*

Proof. Exercise 6.3.11. $\qquad\square$

More generally, we have the following two hereditary properties of LCS that have sequential webs.

Corollary 6.3.2. *Having a sequential web is preserved under the formation of:*
(a) *Closed subspaces*
(b) *Sequentially retractive inductive limits (Definition 3.5.7(d)) $E = \mathrm{ind}_n E_n$ such that for each $n \in \mathbb{N}$, E_n is closed in E. In particular, this holds for strict inductive limits of LCS, each of which has a sequential web.*

Proof. Exercise 6.3.12. $\qquad\square$

In [45], it is shown that a regular inductive limit of Fréchet spaces is sequentially retractive if and only if it satisfies the MCC. Hence, we can state via Theorem 6.3.2, the result below.

Corollary 6.3.3. *A regular inductive limit of Fréchet spaces is sequentially retractive if and only if it has a sequential web.*

6.3.2 Strictly barreled LCS

Webs can also be used to define a type of barreled LCS, as we shall now see.

Definition 6.3.6. Given two sequences of positive integers $\alpha = (a_n)$ and $\beta = (b_n)$, define an ordering as $\alpha \leq \beta$ if for each $n \in \mathbb{N}$, $a_n \leq b_n$.

Definition 6.3.7. A web $\mathcal{W}$ having strands (W_k) is **ordered** if given sequences of natural numbers $\alpha = (r_n)$, and $\beta = (s_n)$ such that $\alpha \leq \beta$, then for all $k \in \mathbb{N}$,

$$W_{r_1 r_2 \cdots r_k} \subset W_{s_1 s_2 \cdots s_k}. \tag{6.24}$$

Definition 6.3.7 above was formally defined in [124, p. 150]. Every metrizable LCS has an ordered web given by a single strand consisting of a decreasing zero neighborhood base of the form $\{U_n : n \in \mathbb{N}\}$. In the case of a normed space, we may use $\{2^{-n}B : n \in \mathbb{N}\}$, where B is the unit ball of the space.

> **Example 6.3.2.** An ordered web on an inductive limit of normed spaces.

An ordered a web can be constructed on an inductive limit $E = \mathrm{ind}_n E_n$ of normed spaces $(E_n, \|\cdot\|_n)$, as follows: By way of adjustments by positive scalars, we may assume that the unit ball B_n of E_n, satisfies $B_n \subset B_{n+1}$, for each $n \in \mathbb{N}$. On each E_n, construct a strand of the form $\{2^{-k}B_n : k \in \mathbb{N}\}$, where $2^{-k}B_n$ is the set in the kth layer of the web. The kth layer of the web then consists of the collection $\{2^{-k}B_n : n \in \mathbb{N}\}$. This construction is quickly seen to form an ordered web on the inductive limit.

With the concept of ordered webs in mind, we define the following, which also comes from [124, p. 160].

Definition 6.3.8. A LCS is **strictly barreled** every ordered web on it has at least one strand whose closures are all zero neighborhoods.

We observe the following properties of strictly barreled LCS next.

Proposition 6.3.7. *Every Baire space is strictly barreled. Every strictly barreled LCS is barreled and there exist barreled spaces that are not strictly barreled.*

Proof. You can prove the first two statements in Exercise 6.3.13. For the third statement, consider any proper strict (LB) space E such as that of Example 3.5.4. As an inductive limit of barreled LCS, E is barreled by Corollary 5.1.2. An ordered web can then be constructed on E as seen in Example 6.3.2. Because the inductive limit is proper, the closure of every set of the constructed web has empty interior. Hence, E is not strictly barreled. $\qquad\qquad\square$

Corollary 6.3.4. *Every Fréchet space, in particular every Banach space, is strictly barreled.*

Valdivia [124, Proposition 17, p. 160] proved that there exist strictly barreled LCS that are not Baire. Putting these items together, we obtain the relations shown below:

$$\text{Baire} \;\overset{\Rightarrow}{\not\Leftarrow}\; \text{strictly barreled} \;\overset{\Rightarrow}{\not\Leftarrow}\; \text{barreled.} \tag{6.25}$$

Next, we can establish some hereditary properties of strictly barreled LCS.

Theorem 6.3.4. *The property of being strictly barreled is preserved under the following formations:*
(a) *If a dense subspace of a LCS is strictly barreled, then the LCS is strictly barreled.*
(b) *Finite products*
(c) *Quotients by closed subspaces*

Proof. (a) Suppose F is a dense subspace of a LCS E such that F is strictly barreled. Let $\mathcal{W}$ be any ordered web on E. Then $\mathcal{W}_F = \{A \cap F : A \in \mathcal{W}\}$ is an ordered web on F. By assumption, there is a strand $(W_{F,k})$ of $\mathcal{W}_F$ such that for every $k \in \mathbb{N}$, the closure in F of $W_{F,k}$ is a zero neighborhood in F. Observe that the strand $(W_{F,k})$ is of the form $W_{F,k} = W_k \cap F$, where (W_k) is a strand of $\mathcal{W}$ on E. The density of F now implies that the closure in E of W_k is a zero neighborhood in E, for each $k \in \mathbb{N}$.

(b) Suppose E and F are strictly barreled. Let $\mathcal{W}$ be an ordered web on $E \times F$. Every set W of $\mathcal{W}$ is of the form $W = W_E \times W_F \subset E \times F$. Any set W_k of the kth layer of $\mathcal{W}$ is of the form $W_k = W_{E,k} \times W_{F,k}$. This information allows us to conclude that if $\mathcal{W}_E$ is the collection of the sets W_E, then $\mathcal{W}_E$ is an ordered web on E. Likewise, letting $\mathcal{W}_F$ be the collection of the sets W_F, then $\mathcal{W}_F$ is an ordered web on F. The verification of items (a)–(c) of Definition 6.3.1 are straightforward and are left as Exercise 6.3.13. By assumption, there is a sequence $\alpha = (n_k : k \in \mathbb{N})$ of positive integers for which the corresponding strand (W_{E,n_k}) of $\mathcal{W}_E$ satisfies the condition that the closure of each W_{E,n_k} is a zero neighborhood in E. Likewise, there is a sequence $\beta = (m_k : k \in \mathbb{N})$ of positive integers for which the corresponding strand (W_{F,m_k}) of $\mathcal{W}_F$ satisfies the closure of each W_{F,m_k} being a zero neighborhood in F. Define $\gamma = (r_k : k \in \mathbb{N})$ such that for each $k \in \mathbb{N}$, $r_k = \max\{n_k, m_k\}$. Then the strand (W_{r_k}) of $\mathcal{W}$ satisfies the condition that the closure in $E \times F$ of each W_{r_k} is a zero neighborhood in $E \times F$. That is,

$$\overline{W_{E,r_k}} \times \overline{W_{F,r_k}} = \overline{W_{E,r_k} \times W_{F,r_k}} = \overline{W_{r_k}} \tag{6.26}$$

by the closure property of a product topology. It follows that for each $k \in \mathbb{N}$, $\overline{W_{r_k}}$ is a zero neighborhood in $E \times F$. Hence, (W_{r_k}) is the desired strand of $\mathcal{W}$. The case of general finite products follows by induction.

(c) This result follows from properties of the canonical quotient map (see Definition 1.1.11), and is left to you as Exercise 6.3.13. $\qquad\square$

Below is a connection between strictly barreled LCS and the MCC.

Proposition 6.3.8. *A strictly barreled LCS for which there is a compatible ordered web satisfies the MCC.*

Proof. Assume E is a strictly barreled LCS and let $\mathcal{W}$ be a compatible, ordered web on E. By Definition 6.3.8 of a strictly barreled LCS, there is a strand of $\mathcal{W}$ for which the closure of every set of that strand is a zero neighborhood. Theorem 6.3.2 now applies, and we conclude that such a space satisfies the MCC. $\qquad\square$

Proposition 6.3.9. *Every Baire LCS that has a compatible web satisfies the MCC.*

Proof. This follows from Propositions 6.3.7 and 6.3.8. $\qquad\square$

6.3.3 Strict Mackey convergence

As is typical, there is a strict version of Mackey convergence, which we now explore.

Definition 6.3.9. A LCS E satisfies the **strict Mackey convergence condition** (abbreviated: **SMCC**) if for every bounded subset A of E there exists a closed disk B containing A such that the topologies of E and of $\| \cdot \|_B$ agree on A.

Of course, every normed space satisfies the SMCC. Also, it is clear that every LCS satisfying the SMCC satisfies the MCC. We can make use of webs as a tool for discovering nonnormable LCS that satisfy the SMCC. The details are in the definition below.

Definition 6.3.10. A compatible web $\mathcal{W}$ on a LCS E is **boundedly compatible** if for any bounded subset A of E there exists a strand (W_k) of $\mathcal{W}$ such that the following properties hold:

(a) For every $k \in \mathbb{N}$, there is a scalar a_k such that $A \subset a_k W_k$.

(b) For every $k \in \mathbb{N}$, there is a closed absolutely convex zero neighborhood U_k such that $A \cap U_k \subset W_k$.

Theorem 6.3.5. *A LCS having a boundedly compatible web satisfies the SMCC.*

Proof. Let $\mathcal{W}$ be a boundedly compatible web on the LCS $(E, \mathcal{T})$. Assume $A \subset E$ is any bounded set. Let (W_k) and (a_k) be the respective strand and sequence of scalars of Definition 6.3.10. We find $b_k \in \mathbb{N}$ such that $(\forall k \in \mathbb{N})\ b_k > 0$, $b_k^{-1}|a_k| < 1$, and $b_k^{-1}|a_k| \to 0$, as $k \to \infty$. Define $B = \overline{\bigcap_{k=1}^{\infty} b_k W_k}$.

Observe that B is closed and $A \subset B$. B is absolutely convex by virtue of the assumption of absolute convexity of the sets of $\mathcal{W}$. Moreover, B is bounded because if $U \in \mathcal{N}_0$ is arbitrary, then the compatibility of $\mathcal{W}$ implies that there exists $K \in \mathbb{N}$ such that $W_K \subset U$, so $b_K W_K \subset b_K U$. What we will show now is that on A, $\mathcal{T}$ is equivalent to $\| \cdot \|_B$. Because $\| \cdot \|_B$ is finer than the relative topology on E_B inherited by $\mathcal{T}$, it suffices to prove the reverse; that is, that the relative topology on E_B is finer than $\| \cdot \|_B$ on the set A. To show this, recall that a base for the normed space $(E_B, \| \cdot \|_B)$ is given by $\{\frac{1}{n}B : n \in \mathbb{N}\}$ (See Proposition 5.1.3). Thus, let $r > 0$ be arbitrary. The goal is to find $U \in \mathcal{N}_{0,\mathcal{T}}$ such that $A \cap U \subset rB$. We find $K \in \mathbb{N}$ such that for all $k \geq K$, $b_k^{-1}|a_k| < r$. Then for all $k \geq K$,

$$A \subset a_k W_k \subset r b_k W_k. \tag{6.27}$$

Next, we check what happens for $k = 1, 2, \ldots, K - 1$. There exists $M \in \mathbb{N}$ such that $2^{-M} < rb$, where $b = \min\{b_j : j = 1, 2, \ldots, K - 1\}$. In addition, there is a $P \in \mathbb{N}$ for which $W_P \subset 2^{-M} W_K$ by Definition 6.3.1. We calculate as follows: For $k = P$, there exists a closed $U_P \in \mathcal{N}_{0,\mathcal{T}}$ that satisfies (b) of Definition 6.3.10, and this implies

$$A \cap U_P \subset W_P \subset 2^{-M} W_K \subset rbW_K$$
$$\subset 2^{-1} rbW_{K-1} \subset \cdots \subset 2^{-K+1} rbW_1. \tag{6.28}$$

Thus,

$$A \cap U_P \subset 2^{-1} rbW_{K-1} \subset rbW_{K-1} \subset r[b_{K-1} W_{K-1}]. \tag{6.29}$$

Applying Equation (6.29) again, we get

$$A \cap U_P \subset 2^{-2} rbW_{K-2} \subset rbW_{K-2} \subset r[b_{K-2} W_{K-2}], \tag{6.30}$$

and so forth for $W_{K-2}, W_{K-3}, \ldots, W_k$. Combining Equations (6.27) and (6.30), we conclude that for *every* $k \in \mathbb{N}$, $A \cap U_P \subset rb_k W_k$, which implies that $A \cap U_P \subset rB$. With $U = U_P$, E satisfies the SMCC. $\qquad\square$

We could replace items (a) and (b) of Definition 6.3.10 with the assumption that those items are valid for the closures, $\overline{W_k}$ of the set W_k of a strand in the web. In Exercise 6.3.15, you can prove that Theorem 6.3.5 remains valid with this generalization. Theorem 6.3.5 allows us to indicate nonnormable examples of LCS that satisfy the SMCC.

Corollary 6.3.5. *Every metrizable LCS has a boundedly compatible web, hence satisfies the SMCC.*

Proof. A boundedly compatible web is given by the zero neighborhood base: $\{U_n : n \in \mathbb{N}\}$ for which each U_n is closed, absolutely convex, and such that for every $n \in \mathbb{N}$, $U_{n+1} \subset \frac{1}{2} U_n$. $\qquad\square$

Corollary 6.3.6. *A compatible ordered web on a strictly barreled LCS is boundedly compatible, thus any such LCS having a compatible web satisfies the SMCC. In particular, this holds for any Baire space that has a compatible web.*

Proof. The proof of this corollary is similar to that of Proposition 6.3.8, and you can write the details in Exercise 6.3.16. $\qquad\square$

Bierstedt [15] studied connections between the SMCC and various properties of regularity of inductive limits. Several such results involve spaces that can be easily seen to have boundedly compatible webs. For the next corollary, see [16] for the definition of the dual density condition.

Corollary 6.3.7. *Any LCS having a boundedly compatible web satisfies the dual density condition.*

Proof. In [16], it is shown that if a LCS satisfies the SMCC, then it satisfies the dual density condition. $\qquad\square$

The next item provides an equivalence for an inductive limit to be boundedly retractive (Definition 3.5.7(c)).

Proposition 6.3.10. *Suppose $E = \mathrm{ind}_n E_n$ is an inductive limit of LCS such that each E_n has a boundedly compatible web $\mathcal{W}^{(n)}$. Let $\mathcal{W}$ be the web formed such that each strand of $\mathcal{W}$ is a strand of $\mathcal{W}^{(m)}$ for some $m \in \mathbb{N}$. Then E with $\mathcal{W}$ is boundedly retractive if and only if $\mathcal{W}$ is boundedly compatible.*

Proof. ($\Rightarrow$): If E is boundedly retractive (see Definition 3.5.7(c)) and A is any bounded subset of E, then A is contained in and bounded in some E_m. Moreover, the topologies of E and E_m agree on A. One then compares relevant topologies and properties of the boundedly compatible web $\mathcal{W}^{(m)}$ on E_m to conclude that the web $\mathcal{W}$ is boundedly compatible on E.

($\Leftarrow$): Now assume $\mathcal{W}$ is boundedly compatible and let A be any bounded subset of E. Then there is a strand (W_k) of $\mathcal{W}$ for which property (a) of Definition 6.3.10 holds for the strand (W_k). That is, for each $k \in \mathbb{N}$ there is a positive scalar a_k such that $A \subset a_k W_k$. By the construction of $\mathcal{W}$, the strand (W_k) is a strand of $\mathcal{W}^{(m)}$ for some $m \in \mathbb{N}$. Thus, $A \subset E_m$. We show that A is bounded in E_m. If V is any closed absolutely convex zero neighborhood in E_m, then by the compatibility of $\mathcal{W}^{(m)}$, there exists $K \in \mathbb{N}$ for which $W_K \subset V$. For the corresponding scalar a_K, we have $A \subset a_K W_K \subset a_K V$, and A is indeed bounded in E_m. To finish, with the same W_K and V apply Definition 6.3.10(b) to find a closed absolutely convex zero neighborhood U_K in E such that $A \cap U_K \subset W_K$. In fact, $A \cap U_K \subset W_K \subset V$. This shows that the induced topology of E on A is finer than the induced topology of E_m on A. Meanwhile, by Definition 3.5.3, the topology of E is always coarser than the topology on E_m, so these two topologies agree on A, and E is boundedly retractive. $\square$

The next item in this section is to verify what you might have already suspected.

Proposition 6.3.11. *A boundedly compatible web is sequential.*

Proof. Suppose $\mathcal{W}$ is a boundedly compatible web on a LCS E. Consider any null sequence (x_n) in E. If $A = \{x_n : n \in \mathbb{N}\}$, then of course A is bounded and there is a strand (W_k) of $\mathcal{W}$ and corresponding zero neighborhoods U_k such that property (b) of Definition 6.3.10 holds. In particular, we have that for each $k \in \mathbb{N}$ there is $N_k \in \mathbb{N}$ satisfying $x_n \in A \cap U_k \subset W_k$ whenever $n \geq N_k$, and we conclude that $\mathcal{W}$ is sequential. $\square$

It turns out that for a LCS to have a sequential web is strictly more general than to have a boundedly compatible web, as the next example shows.

Example 6.3.3. A sequential web that is not boundedly compatible.

Let E be a nonseparable reflexive Banach space, equipped with the topology $\mathcal{T}$ of uniform convergence on the separable bounded subsets of the dual E'. Let B denote the closed unit ball of E. Define a web $\mathcal{W}$ on E as $\mathcal{W} = \{2^{-k}B : k \in \mathbb{N}\}$. In Exercise 6.3.14, you can verify that $\mathcal{W}$ is a compatible web on $(E, \mathcal{T})$. By [95, 8.3.38, p. 259], any null sequence in $(E, \mathcal{T})$ is a null sequence in $(E_B, \|\cdot\|_B)$. Hence, $\mathcal{W}$ is a sequential web. On the other hand, if $\mathcal{W}$ were boundedly compatible, it would imply that $\mathcal{T}$ is equivalent to the normed topology of E, which is impossible (see [95, 8.3.38, p. 259] for precise details).

More results about (strict) Mackey convergence and webs can be found in [49, 50, 54, 56, 57], and references therein. We will apply some properties of LCS that satisfy the (S)MCC in Section 10.2 and in Section 10.3.

Exercises

6.3.1. Prove that $\mathbb{R}$ satisfies the SMCC.

6.3.2. Give a direct proof that on a LCS E a web $\mathcal{W}$ is compatible if and only if, given any strand (W_k) of $\mathcal{W}$ and any sequence (x_k) in E such that for every $k \in \mathbb{N}$ $x_k \in W_k$, then (x_k) converges to 0 in E.

6.3.3. Let $(E, \|\cdot\|)$ be an infinite-dimensional Banach space. Let D be an absolutely convex zero neighborhood for $\sigma(E, E')$. Prove that $\mathcal{W} = \{\frac{1}{2^k} D : k \in \mathbb{N}\}$ is a web on E but is not compatible in $(E, \|\cdot\|)$.

6.3.4. Prove that if $\mathcal{W}$ is a compatible web on a Hausdorff LCS E, then for any strand (W_k) from $\mathcal{W}$, we have $\bigcap_{k=1}^{\infty} W_k = \{0\}$.

6.3.5. From the proof of Theorem 6.3.1, prove the following:
   (a) That the web defined on a linear subspace of a LCS that has a compatible web, is compatible.
   (b) A direct proof that a countable direct sum of LCS, each of which has a compatible web, has a compatible web.

6.3.6. Suppose $\mathcal{W}$ is a compatible web on E, and that (a_k) is any sequence of nonzero scalars. Prove that the set $B = \bigcap_{k=1}^{\infty} a_k W_k$ is bounded for any strand (W_k) of $\mathcal{W}$.

6.3.7. Assume that $x_n \to 0$ such that there is a strand (W_k) of a completing web E, and there is a subsequence (x_{n_k}) of (x_n) such that $(\forall k \in \mathbb{N})$ $x_{n_k} \in W_k$. Prove that (x_n) is a $\mathcal{K}$-sequence.

6.3.8. From the proof of Theorem 6.3.3, verify the following:
   (a) The details of the webs defined for the types of LCS in parts (a) and (b).
   (b) The proof of a completing web on a finite product of LCS, each of which has a completing web.
   (c) Give a direct proof that a countable projective limit of LCS having completing webs has a completing web. *Hint:* To prove compatibility, refer to Definition 3.4.3 and proceed with essentially the same process as for part (e) of Theorem 6.3.1.
   (d) That the construction in Equation (6.16) satisfies the relevant parts of Definition 6.3.1 of a web.

6.3.9. Describe a strict web on the following types of LCS:
   (a) A LCS with a completing web that has exactly three strands.
   (b) A countable product $E = \prod_{j=1}^{\infty} E^{(j)}$, where each $E^{(j)}$ is an (LB) space; that is, for each $j \in \mathbb{N}$, $E^{(j)} = \mathrm{ind}_n E_n^{(j)}$ such that each $E_n^{(j)}$ is a Banach space.
   (c) An inductive limit $E = \mathrm{ind}_n E_n$, where each E_n is countable product given by $E_n = \prod_{j=1}^{\infty} E_n^{(j)}$ of spaces, each of which has a completing web.
   (d) The strong dual of a strict inductive limit of metrizable LCS, as stated in Proposition 6.3.5.

6.3.10. Prove that the completing webs described in Theorem 6.3.3 are strict.

6.3.11. Prove Corollary 6.3.1.

6.3.12. Prove Corollary 6.3.2. *Hint:* For part (b), use the construction of a web on the inductive limit from the proof of Theorem 6.3.1(b). Apply the assumption that each E_n is closed in E to deduce that the closures in E_n of the sets of the web on E_n are also closed with respect to the inductive limit topology on E.

6.3.13. Some properties of strictly barreled LCS.
   (a) Prove that every Baire LCS that has a compatible web is strictly barreled. *Hint:* Theorem A.3.10 is useful for this.
   (b) Prove that every strictly barreled LCS is barreled from Proposition 6.3.7. *Hint:* Let U be any barrel in such a space and form a web using scalar multiples of U.
   (c) From part (b) of the proof of Theorem 6.3.4, prove that the webs $\mathcal{W}_E$ and $\mathcal{W}_F$ are ordered webs on E and F, respectively. *Hint:* For the absorbing properties of each layer, consider points of the form $(x, 0) \in E \times \{0\}$ for $\mathcal{W}_E$ and points of the form $(0, y) \in \{0\} \times F$ for $\mathcal{W}_F$.
   (d) Prove part (c) of Theorem 6.3.4 that a quotient by a closed subspace of a strictly barreled space is strictly barreled.

6.3.14. Verify that $\mathcal{W}$ of Example 6.3.3 is indeed a compatible web on the LCS E.

6.3.15. Prove that the conclusion of Theorem 6.3.5 holds if we use the closures $\overline{W_k}$ in E of the elements of the compatible web on E.

6.3.16. Prove that every strictly barreled LCS that has a compatible ordered web has a boundedly compatible web, and hence, satisfies the SMCC.

6.3.17. Write a proof of Theorem 6.3.5 using the assumption of the closures of the sets of the strands in items (a) and (b) of Definition 6.3.10.

6.4 Examples of property $\mathcal{K}$, the MCC, and webs

In this section, we consider additional examples and counterexamples of the types of spaces considered in this chapter. You will see that infinite-dimensional Banach spaces provide a variety of such examples and counterexamples (for a more advanced treatment of sequences and series in Banach spaces, check out Diestel's classic [37]). In Chapter 8, we will discover LCS that serve as additional examples and counterexamples of the concepts of this chapter.

Example 6.4.1. The space $(l_1, \sigma(l_1, l_\infty))$ is a nonmetrizable LCS that satisfies property $\mathcal{K}$ and the MCC.

To verify the statement above, we observe that by Schur's Theorem 4.5.4, a null sequence (x_n) with respect to the weak topology is a null sequence with respect to the norm on l_1. As $(l_1, \| \cdot \|_1)$ is a Banach space, (x_n) is subseries convergent with respect to the topology of the norm. Thus, (x_n) is also weakly subseries convergent. As for the MCC, if $x_n \to 0$ with respect to $\sigma(l_1, l_\infty)$, then applying Schur's Theorem 4.5.4 again, we have that $x_n \to 0$ with respect to the l_1 norm.

Proposition 6.4.1. *For* $1 < p, q < \infty$ *such that* $\frac{1}{p} + \frac{1}{q} = 1$, *the LCS* $(l_p, \sigma(l_p, l_q))$ *is a strictly webbed LCS that satisfies exactly **none** of the following:*
(a) *Property* $\mathcal{K}$
(b) *The MCC*
(c) *The property of having a sequential web.*

Proof. Exercise 6.4.1. $\qquad\qquad\qquad\qquad\qquad\qquad\qquad\qquad\qquad\qquad\qquad$ $\square$

The above example and proposition, as well as earlier results involving l_1 with its weak topology $\sigma(l_1, l_\infty)$ (See Exercise 4.5.3 and Proposition 5.4.2), indicate that among the l_p spaces for $1 \le p < \infty$, l_1 is something of a platypus in the sense that, with its weak topology, it shares some properties of the other l_p spaces such as those discussed in Section 5.4, yet satisfies additional properties that are satisfied by no other l_p space, $1 < p < \infty$.

Let us now look at some distinguishing examples between Mackey and $\mathcal{K}$ convergence.

Example 6.4.2. A LCS that satisfies the MCC but not property $\mathcal{K}$.

Recall that in [95, 4.7.2, p. 134] there is an example of a normed (barreled) space that is not Baire. Any normed space clearly satisfies the MCC. On the other hand, a normed space that is not Baire cannot satisfy property $\mathcal{K}$, as this would contradict Theorem 6.1.4. An example of a specific sequence that is Mackey convergent but not $\mathcal{K}$-convergent was displayed in Example 6.1.1.

Example 6.4.3. A sequence in a LCS that is $\mathcal{K}$-convergent but not Mackey convergent.

For this example, consider $E = (l_\infty, \sigma(l_\infty, l_1))$. Observe that by the construction of this weak$*$ topology, linear functionals with respect to $\sigma(l_\infty, l_1)$ are of the form $f_y : l_\infty \to \mathbb{K}$ such that for $y = (y^{(n)}) \in l_1$, $x = (x^{(n)}) \in l_\infty$, $f_y(x) = \sum_{k=1}^\infty y^{(k)} x^{(k)}$. As you can verify in Exercise 6.4.2(a), a sequence $(x^{(n)})$ in E is convergent to $x = (x_k)$ if and only if the following two items hold:

(i) $x_k^{(n)} \to x_k$ as $n \to \infty$ for each $k \in \mathbb{N}$; that is, convergence is coordinatewise.

(ii) $x^{(n)}$ is bounded in both indices; that is, there exists $M > 0$ such that $|x_k^{(n)}| \le M$, for each $k, n \in \mathbb{N}$.

Define $(x^{(n)}) \in E$ by $(x_k^{(n)}) = 1$ whenever $k = n$ and is zero otherwise. This sequence clearly converges to 0 and its series converges; that is, $(x^{(n)})$ is $\mathcal{K}$-convergent. On the other hand, you can check in Exercise 6.4.2(b) that by way of items (i) and (ii) above, there can be no sequence (a_n) of positive numbers such that $a_n \to \infty$ as $n \to \infty$ and $a_n x^{(n)} \to 0$ in E. Hence, $(x^{(n)})$ is not Mackey convergent.

It is worth noting that this space does not exhibit an example of a LCS that has property $\mathcal{K}$ but not the MCC because, as you can show in Exercise 6.4.2(c), there exists a null sequence in E that is not $\mathcal{K}$-convergent. It turns out that there does exist a LCS having property $\mathcal{K}$ but not the MCC. The details of such a space are beyond the scope of what we wish to do here, however, an outline of how to construct such a space is given in Exercise 6.4.3. On the other hand, if an $\mathcal{A}$-space (see Definition 6.1.4) satisfies the MCC, then it is a $\mathcal{K}$-space, as we see next.

Proposition 6.4.2. *An $\mathcal{A}$-space that satisfies the MCC is a $\mathcal{K}$-space.*

Proof. Suppose E is an $\mathcal{A}$-space and satisfies the MCC. Given any null sequence (x_n) in E, there exists a sequence (a_n) in $\mathbb{R}$ such that $a_n \to \infty$ as $n \to \infty$, and $a_n x_n \to 0$ in E. In Exercise 6.4.4, you can use this fact to prove that E satisfies property $\mathcal{K}$. $\qquad\square$

Example 6.4.4. An example of an $\mathcal{A}$-space that is not a $\mathcal{K}$-space.

You can verify in Exercise 6.4.5 that for $1 < p, q < \infty$ and $\frac{1}{p} + \frac{1}{q} = 1$, $(l_p, \sigma(l_p, l_q))$ is such a space. Thus, the collection of $\mathcal{A}$-spaces is strictly larger than that of $\mathcal{K}$-spaces. Meanwhile, there exist $\mathcal{A}$-spaces that are not complete, formally stated below.

Example 6.4.5. The example of Kliś [78] represents an $\mathcal{A}$-space that is not complete.

We finish this section with some examples involving webs. The first item is to exhibit an infinite collection of nonmetrizable LCS that are strictly webbed, as seen next.

Proposition 6.4.3. *Any Banach space, equipped with a topology that is consistent with the duality, is strictly webbed. In particular, every infinite-dimensional Banach space equipped with its weak topology is nonmetrizable and strictly webbed.*

Proof. The case where the space is of finite dimension is clear. Consider any infinite-dimensional Banach space $(E, \|\cdot\|)$. We already know that $\mathcal{W} = \{2^{-k}B : k \in \mathbb{N}\}$, where B is the closed unit ball, is a strict web on E. Now let $\mathcal{T}$ be any topology that is consistent with the duality (E, E'). As a Banach space, E is barreled, so Corollary 5.1.4 tells us that $(E, \|\cdot\|)$ carries its Mackey topology. This means all other consistent topologies are coarser than the normed topology. In particular, the identity map id $: (E, \|\cdot\|) \to (E, \mathcal{T})$ is continuous and surjective. The result now follows by Theorem 6.3.3(b). $\quad\square$

Proposition 6.4.3 allows us to find a LCS that is strictly webbed and not even sequentially complete, indicated below.

Example 6.4.6. The LCS given by c_0, equipped with its weak topology, is a strictly webbed LCS that is not sequentially complete.

See Exercise 4.5.4. In Chapter 8, we will discover strictly webbed LCS that surprisingly satisfy no completeness whatsoever.

Exercises

6.4.1. Prove Proposition 6.4.1.

6.4.2. Verify the following items regarding the LCS $E = (l_\infty, \sigma(l_\infty, l_1))$ of Example 6.4.3:
   - (a) Properties (i) and (ii).
   - (b) There is no sequence (a_n) of positive numbers such that $a_n \to \infty$ as $n \to \infty$ and $a_n x^{(n)} \to 0$ in E.
   - (c) Consider the sequence $(x^{(n)})$ in E that consists of zeros for the first n entries and then 1 for all other entries (so of the form $x^{(n)} = (0, 0, \ldots, 0, 1, 1, \ldots)$, where the zeros appear n times). Prove that $(x^{(n)})$ is not $\mathcal{K}$-convergent.

6.4.3. The goal of this exercise is to outline the construction of an example of a LCS that satisfies property $\mathcal{K}$ but not the MCC. Let E be a Fréchet space that is not quasinormable (see [70, Section 10.7, pp. 214–216] for the definition of a quasinormable LCS), and consider $(E', \beta(E', E))$, the strong dual of E.
   - (a) One can then prove that if E satisfies a property called Heinrich's density condition (see [17]), then the bounded subsets of $(E', \beta(E', E))$ are metrizable with respect to the relative topology.
   - (b) Deduce from part (a) that $(E', \beta(E', E))$ satisfies property $\mathcal{K}$. That $(E', \beta(E', E))$ does not satisfy the MCC follows from the fact that E is not quasinormable, as can be seen in [19].

6.4.4. Complete the proof of Proposition 6.4.2. *Hint:* Express the sequence (x_n) using the sequence $(a_n x_n)$.

6.4.5. Prove the claim of Example 6.4.4.

6.4.6. In the items below, describe an example of a strictly webbed LCS that satisfies the stated condition.
   - (a) Is nonmetrizable and is not a Banach space equipped with its weak topology. *Hint:* Take a look at Theorem 6.3.3.
   - (b) Is neither a Baire space nor a Banach space that is equipped with its weak topology.

6.4.7. Let E be a Banach space with weakly null sequences that are not convergent in norm. Describe an example of such a sequence, which contains a $\mathcal{K}$-convergent subsequence.

6.4.8. Determine which, if any, of the LCS of Examples 6.4.1 and 6.4.2 satisfy the SMCC.

7 Local completeness

In Chapter 3, we saw a variety of types of LCS and addressed the completeness or lack thereof of most of those types. Completeness is, of course, an important property due to the fact that it allows us to conclude that a filter or sequence has a limit without having to know what that limit is. Later on, in Chapter 5, it turned out that LCS that are barreled can replace completeness in some contexts. In this chapter, we consider completeness and barreledness in terms of disks (absolutely convex, bounded sets). More precisely, we consider completeness and barreledness in normed spaces $(E_B, \| \cdot \|_B)$ formed by Banach, Baire, or barreled disks, respectively (see Definition 5.1.4 for more details). This approach presents some advantages when compared to checking such properties with respect to a given LCS topology. One such advantage is that the normed space structure is, at least symbolically, easier to deal with compared to, say, a nonmetrizable topology. Second, such normed spaces $(E_B, \| \cdot \|_B)$ can be formed in LCS that need not be metrizable, allowing us to make use of properties of normed spaces in a context that otherwise would not be possible. Third, this approach will lead to obtaining quite general versions of significant results, such as the Principle of Uniform Boundedness (Theorem 7.2.1) and the Banach Steinhaus theorem (Theorem 7.2.6). Along the way, several properties of such "locally-something" LCS are obtained, such as hereditary properties for when the disk B is a Banach disk, as well as connections with other properties we have seen such as property $\mathcal{K}$. You will also notice that Robertson's Theorem 2.5.2 gets applied in several places. There is no need to wait longer, so let us begin.

7.1 Locally complete, locally Baire, locally barreled spaces

Recall from Section 5.1.1 that a Banach disk is an absolutely convex, bounded subset B of a LCS E such that the normed space $(E_B, \| \cdot \|_B)$ is complete; that is, a Banach space, where $\| \cdot \|_B$ represents the norm given by the Minkowski functional of B. Correspondingly, we saw the definition of Baire and barreled disks. All three of these types of disks are announced in Definition 5.1.4. These items lead to the following definition.

Definition 7.1.1. A LCS is **locally complete** (resp., **locally Baire**, **locally barreled**) if every one of its bounded subsets is contained in a disk that is a Banach disk (resp., Baire disk, barreled disk).

It is sometimes useful to assume the relevant disk is also closed, as we will see shortly in Proposition 7.1.1. Below is the first result of this section. Its importance stems from the fact that the properties of disks of Definition 5.1.4 allow us to consider the concepts of Definition 7.1.1 independently of topologies that are consistent with the duality.

Theorem 7.1.1. *Locally complete, locally Baire, and locally barreled LCS of Definition 7.1.1 are invariant with respect to topologies consistent with the duality.*

https://doi.org/10.1515/9783111392868-007

Proof. This follows directly from the fact that the closed, absolutely convex, bounded subsets are the same for all topologies consistent with the duality of a LCS. □

The concept of local completeness was first studied in detail by De Wilde [36]. Alternate names for locally complete LCS include **fast complete**, **Mackey complete**, and **boundedly complete**. In deference to topological purists, the adjective "locally" has been typically used for concepts such as in Definition 7.1.1 above, even though the sets in question need not be neighborhoods.

Relations between several completeness-related concepts we have seen up to now can be determined as follows. Observe first that Proposition 5.1.4 implies that sequentially complete LCS are locally complete. Locally complete LCS need not be sequentially complete; witness Exercise 4.5.4. The LCS of Exercise 4.5.4 is however, locally complete by Theorem 7.1.1 above. Proposition 5.4.1 indicates that infinite-dimensional normed spaces, equipped with their weak topologies, are incomplete. Next, by the corollary to Theorem 5.3.1 we have that every infinite-dimensional reflexive Banach space is weakly quasicomplete but not weakly complete. Meanwhile, recall that Equation (5.1) applies in the context of normed spaces. In other words, an incomplete normed space that is a Baire space is locally Baire and not locally complete, and a normed space that is barreled and not Baire represents a LCS that is locally barreled but not locally Baire. Putting together these observations, the display shown in (7.1) below summarizes the connections between these six types of LCS:

$$\text{Complete} \; \overset{\Rightarrow}{\nLeftarrow} \; \text{quasicomplete} \; \overset{\Rightarrow}{\nLeftarrow} \; \text{sequentially complete}$$

$$\overset{\Rightarrow}{\nLeftarrow} \; \text{locally complete} \; \overset{\Rightarrow}{\nLeftarrow} \; \text{locally Baire} \; \overset{\Rightarrow}{\nLeftarrow} \; \text{locally barreled.} \qquad (7.1)$$

In the case of local completeness, we have the following two useful facts.

Proposition 7.1.1. *A LCS is locally complete if and only if every closed disk is a Banach disk.*

Proof. ($\Rightarrow$): Assume E is locally complete and let A be a closed disk in E. There exists a Banach disk B that contains A by assumption. Because the topology of $\| \cdot \|_B$ is finer than the topology inherited from E, A is $\| \cdot \|_B$-closed. That $A \subset B$ implies that A is $\| \cdot \|_B$-bounded, hence, the topology $\| \cdot \|_A$ on E_A is finer than the topology inherited from $\| \cdot \|_B$. The conclusion now follows by Robertson's Theorem 2.5.2.

($\Leftarrow$): Immediate. □

Proposition 7.1.2. *A metrizable LCS is locally complete if and only if it is complete.*

Proof. ($\Rightarrow$): Assume E is metrizable and locally complete. It suffices to prove that E is sequentially complete. Let (x_n) be a Cauchy sequence in E. Let $A = \{x_n : n \in \mathbb{N}\}$. Then A is bounded in E. As E is metrizable, Corollary 6.3.5 implies that E satisfies the strict Mackey convergence condition (SMCC). Hence, there is a closed disk B such that the topology of $\| \cdot \|_B$ and the topology of E coincide on the bounded set A. We instantly observe that (x_n)

is Cauchy with respect to the norm $\|\cdot\|_B$. By assumption, E is locally complete, so B is a Banach disk by Proposition 7.1.1 and, therefore, (x_n) converges in the normed space E_B. The injection from E_B to E is continuous, which implies that (x_n) converges in E, which is thus sequentially complete.

$\quad$ ($\Leftarrow$): Obvious. $\qquad\qquad\qquad\qquad\qquad\qquad\qquad\qquad\qquad\qquad\qquad\quad\square$

In the sequel are three more results on equivalences to local completeness, the first of which contains Proposition 7.1.1.

Theorem 7.1.2. *The following are equivalent for a Hausdorff LCS $(E, \mathcal{T})$:*
(a) *E is locally compete.*
(b) *Every closed disk is a Banach disk.*
(c) *For every closed disk B, the Cauchy sequences in $(E_B, \|\cdot\|_B)$ are convergent with respect to $\mathcal{T}$.*
(d) *For every closed disk B, the Cauchy sequences in $(E_B, \|\cdot\|_B)$ are convergent with respect to $\sigma(E, E')$.*

Proof. (a) $\Leftrightarrow$ (b) is the statement of Proposition 7.1.1 and (b) $\Rightarrow$ (c) $\Rightarrow$ (d) can be deduced by comparing topologies, as you can check in Exercise 7.1.5. To complete the proof, observe that assuming (d) allows us to apply Robertson's Theorem 2.5.2 with respect to item (a). $\qquad\qquad\qquad\qquad\qquad\qquad\qquad\qquad\qquad\qquad\qquad\qquad\qquad\qquad\quad\square$

Theorem 7.1.3. *The following are equivalent for a Hausdorff LCS $(E, \mathcal{T})$:*
(a) *E is locally compete.*
(b) *The closed absolutely convex hull of each Mackey null sequence is compact.*
(c) *The closed absolutely convex hull of each Mackey null sequence is weakly compact.*
(d) *The closed absolutely convex hull of each null sequence is compact.*

Proof. See [95, Theorem 5.1.11, p. 153]. $\qquad\qquad\qquad\qquad\qquad\qquad\qquad\qquad\qquad\quad\square$

A more in-depth result is the following general characterization of local completeness.

Theorem 7.1.4. *Consider the dual pair (E, E') and let $\mathcal{T}$ be any Hausdorff topology on E consistent with the duality. The following are equivalent for $(E, \mathcal{T})$:*
(a) *E is locally compete.*
(b) *For some (every) p, $1 \le p \le \infty$, one has: If $B \subset E$ is a disk and (x_n) is a sequence in E_B such that $\{\|x_n\|_B : n \in \mathbb{N}\} \in l_p$ ($\in c_0$ for $p = \infty$), then $\overline{\mathrm{conv\,bal}}\{x_n : n \in \mathbb{N}\}$ is compact in $(E_C, \|\cdot\|_C)$, for some disk C in E.*
(c) *Under the assumptions of part (b), $\overline{\mathrm{conv\,bal}}\{x_n : n \in \mathbb{N}\}$ is compact in E.*
(d) *For some (every) p, $1 \le p \le \infty$, one has: If (x_n) is a sequence in E such that the sequence $(\langle f, x_n\rangle) \in l_p$ ($\in c_0$ for $p = \infty$) holds for every $f \in E'$, then $\overline{\mathrm{conv\,bal}}\{x_n : n \in \mathbb{N}\}$ is compact in $(E, \sigma(E, E'))$.*

(e) *For some (every) p, $1 < p < \infty$, one has: Every linear map $T : (E', \sigma(E', E)) \to (l_p, \sigma(l_p, l'_p))$ having a closed graph (See Definitions 8.1.1, 8.1.2), is continuous.*

Proof. Implications (a) $\Rightarrow$ (b) $\Rightarrow$ (c) and (a) $\Rightarrow$ (d) are straightforward and are left as Exercise 7.1.8. For a proof of the remaining implications, see [70, pp. 198–199]. $\qquad\square$

For locally complete LCS, we have the following hereditary properties.

Theorem 7.1.5. *The property of local completeness is preserved under the following formations:*
(a) *Closed subspaces*
(b) *Products*
(c) *Projective limits*
(d) *Direct sums*
(e) *Regular inductive limits.*

Proof. (a) Let F be a closed subspace of a locally complete LCS E and let B be any closed, absolutely convex, bounded subset of F. Then B is bounded in E and is also closed. Thus by Proposition 7.1.1, B is a Banach disk in E. As $B \subset F$, E_B is a subspace of F and B is thus a Banach disk in F.

(b) Consider $\{(E_\alpha, \mathcal{T}_\alpha) : \alpha \in I\}$ any collection of locally complete LCS. Put $E = \prod_{\alpha \in I} E_\alpha$, equipped of course, with the product topology. If A is any bounded subset of E, the continuity of each canonical projection $\pi_\alpha : E \to E_\alpha$ implies that A is bounded in E_α. So, $\pi_\alpha(A)$ is contained in a Banach disk $B_\alpha \subset E_\alpha$ for each $\alpha \in I$. Let $B = \prod_{\alpha \in I} B_\alpha$. As a product of absolutely convex sets, B is absolutely convex (see Theorem 1.1.1(g)). In addition, by Proposition 3.3.3, B is bounded in E. It remains to be shown that B is a Banach disk in E. Hence, let (x_n) be a Cauchy sequence in $(E_B, \|\cdot\|_B)$. Let $x_{\alpha,n} = \pi_\alpha(x_n)$, for each $\alpha \in I, n \in \mathbb{N}$. Using the zero neighborhood base of the topology of $\|\cdot\|_B$ on E_B, namely $\{k^{-1}B : k \in \mathbb{N}\}$, we have that, given any $k \in \mathbb{N}$, there is $N = N_k \in \mathbb{N}$ such that $(\forall m, n \geq N_k), x_m - x_n \in \frac{1}{k}B$. Thus, for each $\alpha \in I$,

$$x_{\alpha,m} - x_{\alpha,n} \in \pi_\alpha\left(\frac{1}{k}B\right) = \frac{1}{k}B_\alpha. \tag{7.2}$$

Define, for each $\alpha \in I$, $F_\alpha = \mathrm{span}(B_\alpha)$ as a vector space and equip each F_α with the normed topology generated by $\|\cdot\|_{B_\alpha}$, forming a Banach space $(F_\alpha, \|\cdot\|_{B_\alpha})$. The completeness of each F_α with respect to the corresponding normed topology $\|\cdot\|_{B_\alpha}$ implies that $(\forall \alpha \in I)(\exists x_\alpha \in F_\alpha)$ such that $x_{\alpha,n} \to x_\alpha$, as $n \to \infty$. Now define $x = (x_\alpha : \alpha \in I) \in E$. In Equation (7.2), fix n and let $m \to \infty$, obtaining that for any $k \in \mathbb{N}$ and sufficiently large n,

$$x_\alpha - x_{\alpha,n} \in \frac{1}{k}B_\alpha, \tag{7.3}$$

for every $\alpha \in I$. That is, $x - x_n \in \frac{1}{k}B$. We conclude that $x_n \to x$ with respect to the topology $\| \cdot \|_B$. To complete the proof of this part, we observe that for large enough n, $x - x_n \in B$, and $x_n \in E_B$ together imply that $x = (x - x_n) + x_n \in E_B$.

(c) By Theorem 3.4.1, a projective limit of LCS $\{(E_\alpha, \mathcal{T}_\alpha)\}$ (see Definition 3.4.3) is a closed subspace of the product $\prod_{\alpha \in I} E_\alpha$, therefore, the local completeness of the projective limit follows from parts (a) and (b). Notice that the result in this part generalizes Proposition 3.4.5.

(d) Consider the direct sum $E = \bigoplus_{\alpha \in I} E_\alpha$ of locally complete LCS E_α, $\alpha \in I$ (see Definition 3.5.2). If A is a bounded subset of E, then by Proposition 3.5.3, A is contained in and bounded in a finite sum of subsets of the spaces E_α. We denote this property as $A \subset \bigoplus_{j=1}^{n} E_{\alpha_j}$, for some $n \in \mathbb{N}$. Because a finite direct sum is equivalent to a finite product by Definition 3.5.2, we see that $\bigoplus_{j=1}^{n} E_{\alpha_j}$ is locally complete by part (b). Thus, there is a bounded set $B \subset \bigoplus_{j=1}^{n} E_{\alpha_j}$ such that B is a Banach disk, and $A \subset B$. Clearly, B is also bounded in the direct sum as well, and we conclude that the direct sum is locally complete.

(e) Let $E = \mathrm{ind}_n E_n$ such that each E_n is locally complete and E is regular. If A is any bounded subset of E then there is some $m \in \mathbb{N}$ such that $A \subset E_m$ and is bounded there. By assumption E_m is locally complete, so there is a Banach disk $B \subset E_m$ such that $A \subset B$. The continuity of $\mathrm{id}_m : E_m \to E$ implies that B is bounded in E, as well as the fact that B is absolutely convex in E. That is, B is a Banach disk in E and B contains A. $\qquad\square$

A related result is next.

Proposition 7.1.3. *Regular inductive limits of locally Baire LCS are locally Baire; regular inductive limits of locally barreled LCS are locally barreled.*

Proof. Exercise 7.1.6. $\qquad\square$

Another observation about local completeness is next.

Proposition 7.1.4. *If a LCS contains a dense locally complete hyperplane, then it is locally complete.*

Proof. Exercise 7.1.7. $\qquad\square$

You may have noticed that in Theorem 7.1.5 we assumed regularity of the inductive limits in part (e), and quotients of locally complete LCS were not on the list of items. It is because local completeness can fail in non-regular inductive limits and quotients. This issue will be discussed in Chapter 8. We could consider hereditary properties of locally Baire LCS, however, there is an obstacle: There exists an example of a product of two normed spaces that are Baire such that the product is not a Baire space (see [95, 1.4.7, p. 30]). Meanwhile, it seems hereditary properties of locally barreled LCS have not been considered. Nonetheless, locally Baire and locally barreled spaces will be useful to us in a variety of ways, such as in Section 7.2 of this chapter and in Chapter 8. Our next result connects LCS having property $\mathcal{K}$ (see Definition 6.1.2) to locally Baire LCS. See also [55].

Theorem 7.1.6. *Every LCS that satisfies property $\mathcal{K}$ is locally Baire.*

Proof. Let A be any bounded subset of a LCS E, where we assume E satisfies property $\mathcal{K}$. Let B denote the absolutely convex, closed hull of A. We intend to show that E_B, with the finer, normed topology generated by $\|\cdot\|_B$, satisfies property $\mathcal{K}$. Showing that will allow us to cite Theorem 6.1.4 to reach the conclusion. Let (x_n) be any null sequence in the normed space $(E_B, \|\cdot\|_B)$. Because the topology of the norm on E_B is finer than the relative topology on the vector space E_B, (x_{n_k}) converges to 0 in E. We choose a subsequence (x_{n_k}) such that for each $k \in \mathbb{N}$, $x_{n_k} \in 2^{-k}B$. Then the sequence of partial sums of the series $\sum_{k=1}^{\infty} x_{n_k}$ form a Cauchy sequence as we have seen earlier. By the assumption of property $\mathcal{K}$, there is a further subsequence of (x_{n_k}), which we write without loss of generality as (x_{n_k}), such that the series $\sum_{k=1}^{\infty} x_{n_k}$ converges in E. Denote the limit of this series by y. Now define the sequence (y_m) in E as the partial sums of $\sum_{k=1}^{\infty} x_{n_k}$; that is, $y_m = \sum_{k=1}^{m} x_{n_k}$. Observe that the sequence (y_m) converges in E by property $\mathcal{K}$, and also converges in the relative topology on E_B. Moreover, as B is closed in E, $\{\frac{1}{n}B : n \in \mathbb{N}\}$ is a base of closed zero neighborhoods for the norm on E_B, and this prompts us to apply Robertson's Theorem 2.5.2, concluding that $\sum_{k=1}^{\infty} x_{n_k}$ converges to y with respect to the normed topology on E_B. We now see that (y_m) is bounded in E_B, which implies that $y \in E_B$. We have shown that $(E_B, \|\cdot\|_B)$ satisfies property $\mathcal{K}$, which allows us to cite Theorem 6.1.4, from which we conclude that $(E_B, \|\cdot\|_B)$ is a Baire space. $\square$

Recall from Definition 6.2.2 that a null sequence is fast convergent if it is a null sequence in $(E_B, \|\cdot\|_B)$, for some Banach disk B. It turns out that the disk B can just as well be compact, as we see next. In fact, fast convergence is connected to local completeness as shown next.

Proposition 7.1.5. *The following are equivalent for a Hausdorff LCS E:*
(a) *E satisfies the fast convergence condition FCC (Definition 6.2.2).*
(b) *For every null sequence in E, there exists a compact disk K such that the sequence is null in $(E_K, \|\cdot\|_K)$.*
(c) *E is locally complete and satisfies the Mackey convergence condition MCC (Definition 6.2.1).*

Proof. (a) $\Leftrightarrow$ (b) is immediate by Proposition 6.2.5.

(b) $\Rightarrow$ (c): Assuming (b), we only need to prove that E is locally complete. To establish this, we observe that Theorem 7.1.3 is rather appealing. In fact, we will appeal to it by proving that the closed absolutely convex hull of any null sequence is compact. Let (x_n) be a null sequence in E. Then for some compact disk K, $x_n \to 0$ in $(E_K, \|\cdot\|_K)$. That K is a Banach disk implies that if A denotes the $\|\cdot\|_K$-closure of convbal$\{x_n : n \in \mathbb{N}\}$, it follows that A is $\|\cdot\|_K$-compact. The injection id : $E_K \to E$ is continuous, so A is compact in E as well and, therefore, A is closed in the topology of E. We now proceed to prove that A_0, the E-closure of convbal$\{x_n : n \in \mathbb{N}\}$, is equal to A. Now, A is bounded with respect to $\|\cdot\|_K$, so there exists $a > 0$ such that $A \subset aK$. Meanwhile, aK is closed in E, which tells us that $A_0 \subset aK \subset E_K$. Let $x \in A_0$ be arbitrary and suppose $\mathfrak{F}$ is a filter of subsets of A that converges to x. By way of the continuity of id $: E_k \to E$ and that $\{n^{-1}K : n \in \mathbb{N}\}$ is a $\|\cdot\|_K$-base of zero neighborhoods on E_K consisting of sets that are closed in E, we have that $\{n^{-1}K : n \in \mathbb{N}\}$ are closed in the linear span of K. This allows us to apply Robertson's Theorem 2.5.2, to conclude that $\mathfrak{F} \to x$ with respect to $\|\cdot\|_K$. Think back to a few minutes ago to recall that A is $\|\cdot\|_K$-closed, and we conclude that $x \in A$. Thus, $A_0 = A$.

(c) $\Rightarrow$ (a): Obvious. $\qquad\square$

Next, we compare some properties such as being locally complete, locally Baire, and locally barreled versus properties such as being Baire, barreled, and more.

> **Example 7.1.1.** There exist examples of LCS that satisfy the following:
> (a) A locally Baire LCS that does not satisfy property $\mathcal{K}$.
> (b) A locally Baire LCS that is not Baire.
> (c) A locally barreled LCS that is not barreled.
> (d) A sequence of LCS each of which satisfies property $\mathcal{K}$ and for which the inductive limit of that sequence also satisfies property $\mathcal{K}$.

We describe examples of such spaces here for parts (a) to (c). Part (d) is left as Exercise 7.1.10. Earlier examples involving property $\mathcal{K}$ can be reviewed in Section 6.4. Further distinguishing examples will be provided in Section 8.2.

(a) Let E be any Banach space having weakly convergent sequences that are not convergent in norm. By Theorem 7.1.1, $(E, \sigma(E, E'))$ is locally complete, hence locally Baire. On the other hand, every $\mathcal{K}$-convergent sequence with respect to $\sigma(E, E')$ would necessarily be convergent in norm by Theorem 6.1.2, which is not possible. Hence, this space does not satisfy property $\mathcal{K}$.

(b) Any proper strict (LB) space is locally complete by Theorem 7.1.5(e), hence locally Baire. Such a space is not Baire by Proposition 3.5.8.

(c) Consider any infinite-dimensional Banach space E, equipped with its weak topology. As in part (a) above, E is locally complete, hence locally barreled. Meanwhile, by Theorem 5.4.1, E is not barreled.

By the way, the author does not know of an example of a barreled LCS that is not locally barreled.

Proposition 7.1.6. *There exists a LCS that is locally complete and webbed but does not satisfy the MCC.*

Proof. Exercise 7.1.11. $\qquad\square$

Below are a few variations on the theme of "localness."

Definition 7.1.2. A LCS E is **locally reflexive** if each of its bounded sets is contained in a closed disk B such that $(E_B, \|\cdot\|_B)$ is a reflexive Banach space.

Clearly, every reflexive Banach space is locally reflexive. In Exercise 7.1.12, you can explore some properties of locally reflexive LCS, including the existence of locally reflexive LCS that are not metrizable. In [100], a generalization of the concept of local completeness was given by Qiu, as stated next.

Definition 7.1.3. A LCS $(E, \mathcal{T})$ is **quasilocally complete** (also **quasifast complete**) if given any of its bounded subsets, there exists a coarser locally convex topology ρ on E, independent of the bounded set, and such that the set is contained in a ρ-Banach disk.

Among other results, it is shown in [100] that if $E = \operatorname{ind}_n E_n$ is an inductive limit of strictly webbed LCS such that E is quasilocally complete, then E is regular. We can expand the concept of quasilocally complete with that given below.

Definition 7.1.4. A LCS $(E, \mathcal{T})$ is **quasilocally Baire** if given any of its bounded subsets, there exists a coarser locally convex topology ρ on E, independent of the bounded set, and such that the set is contained in a ρ-Baire disk.

Naturally, every locally complete LCS is quasi locally complete and every quasilocally complete LCS is quasilocally Baire. See also [58]. Below are a few distinguishing examples.

Example 7.1.2. A strictly webbed quasilocally Baire LCS that is not Baire.

Let E be any infinite-dimensional Banach space. Equip E with its weak topology $\sigma(E, E')$. Applying Proposition 6.3.3 and Theorem 6.3.3(b) leads to the observation that E with its weak topology has a strict web. The LCS E with its weak topology is locally complete by Theorem 7.1.1, hence is quasilocally Baire. We finish by observing that E with its weak topology is not Baire because it is not even barreled, courtesy of Theorem 5.4.1. By the way, the relevance of including the strictly webbed aspect in this example is so that it can be referred to in Corollary 8.2.3.

Example 7.1.3. A quasilocally Baire LCS that is not locally Baire.

Such spaces can be deduced from earlier results, the details of which are left as Exercise 7.1.13.

Example 7.1.4. A quasilocally Baire LCS that is not quasi locally complete.

An example of this kind of LCS can also be deduced from earlier results, the details of which are left as Exercise 7.1.14. Yet another version of local completeness is the following.

Definition 7.1.5. A LCS E is

(a) **Dual locally complete** if $(E', \sigma(E', E))$ is locally complete.
(b) **Dual locally barreled** if for every subset A of E' (with respect to a topology consistent with the duality), there exists a barreled disk $B \subset E'$ such that $A \subset B$.

For a sampling of results connected with dual locally complete LCS, check out [110] and the references therein. Some properties results regarding dual locally barreled LCS can be seen in [121] and references therein.

Exercises

7.1.1. Give a another proof that if D is an absolutely convex, bounded, and complete subset of E, then $(E_D, \|\cdot\|_D)$ is a Banach space; that is, prove this result without citing Robertson's Theorem 2.5.2.

7.1.2. Prove that if $U \in \mathcal{N}_0$ for a Hausdorff LCS E, then its polar U° is a Banach disk in $(E', \sigma(E', E))$.

7.1.3. Prove directly that any infinite-dimensional Banach space, equipped with its weak topology, is a non-normable locally complete LCS that is not complete.

7.1.4. Here is another example of a locally barreled LCS that is not barreled: Let E be a semireflexive LCS that is not reflexive. Prove that E is locally barreled (in fact, locally complete), but not barreled.

7.1.5. Prove implications (b) $\Rightarrow$ (c) $\Rightarrow$ (d) of Theorem 7.1.2.

7.1.6. Prove Proposition 7.1.3.

7.1.7. Prove Proposition 7.1.4.

7.1.8. Prove implications (a) $\Rightarrow$ (b) $\Rightarrow$ (c) and (a) $\Rightarrow$ (d) of Theorem 7.1.4. *Hint:* For (a) $\Rightarrow$ (d), apply Robertson's Theorem 2.5.2.

7.1.9. Prove that a regular inductive limit of a sequence of LCS that each satisfy property $\mathcal{K}$ is locally Baire.

7.1.10. Describe an example of an inductive limit of LCS, each of which satisfies property $\mathcal{K}$, such that the inductive limit also satisfies property $\mathcal{K}$.

7.1.11. Prove Proposition 7.1.6. *Hint:* Consider any Banach space E that has weakly convergent sequences that are not norm convergent (for example $l_p, 1 < p < \infty$). Let $\mathcal{W} = \{2^{-k}B : k \in \mathbb{N}\}$, where B is the closed unit ball of E. Prove that $\mathcal{W}$ is a completing, hence compatible web with respect to the weak topology $\sigma(E, E')$. Then explain why E is locally complete and does not satisfy the MCC.

7.1.12. Prove the following properties of locally reflexive LCS that were defined in Definition 7.1.2.

    (a) Every locally reflexive LCS is semireflexive.

    (b) The property of being locally reflexive is a property of the dual pair.

    (c) There exists an example of a locally complete LCS that is not locally reflexive.

    (d) A regular inductive limit of locally reflexive LCS is locally reflexive.

    (e) In any (reflexive) infinite dimensional Banach space E, there exists an absolutely convex, compact subset set B such that $(E_B\|\cdot\|_B)$ is a non-reflexive Banach space. *Hint:* You may wish to consult [122] about this.

    (f) In any infinite-dimensional Banach space E, there exists an absolutely convex, compact subset set B such that the linear span of B is not closed. *Hint:* Use part (e) of this exercise.

    (g) There exists a locally reflexive LCS that is neither metrizable nor complete.

7.1.13. Give an example of a quasi locally Baire space that is not locally Baire. *Hint:* Take a look at Theorem 5.4.2.

7.1.14. Give an example of a quasilocally Baire space that is not quasilocally complete. *Hint:* After Definition 5.1.4 it was indicated that there exists an incomplete Baire normed space. Apply Theorem 4.3.2 and Mackey's Theorem 5.1.5.

7.2 The uniform boundedness principle and Banach–Steinhaus theorem

In this section, we will see several versions of the Uniform Boundedness Principle, abbreviated as **PUB**, and the Banach–Steinhaus theorem. Motivated by material on Banach spaces, there is the celebrated Banach–Steinhaus theorem for Banach spaces (see Theorem A.5.4), which states that pointwise bounded collections of elements of bounded linear maps from a Banach space to a normed space are uniformly bounded. As boundedness and continuity coincide for normed spaces, we can replace "uniformly bounded" by "equicontinuous" in the last sentence. In fact, in the context of normed spaces, the Banach–Steinhaus theorem is sometimes also referred to as the uniform boundedness principle. Of course for general LCS, boundedness of linear maps does not imply continuity, as we saw in Example 5.1.3. This implies that we obtain two types of results, one for uniform boundedness and one for equicontinuity. For LCS even uniform boundedness does not imply equicontinuity, as we will see in Example 7.2.2. We have already seen the Banach–Steinhaus theorem for the dual space of a LCS in Corollary 5.1.3. In this section, we prove some versions of the principle of uniform boundedness and the Banach–Steinhaus theorem in the more general context of $\mathcal{L}(E, F)$, the collections of continuous linear maps E to F, where E and F are both LCS (see Definition 2.3.2). The process of creating topologies on $\mathcal{L}(E, F)$ is a generalization of the process of creating polar topologies that were discussed in Chapter 4. Let us get to it.

First, a more general definition of $\mathcal{S}$ topologies than what we saw in Section 4.3, appears next.

Definition 7.2.1. Let $\mathcal{S}$ denote a collection of bounded subsets of E, and let $\mathcal{N}_{0,F}$ be a base of absolutely convex zero neighborhoods of F. For each $B \in \mathcal{S}$ and each $V \in \mathcal{N}_{0,F}$, define

$$W_{B,V} = \{f \in \mathcal{L}(E, F) : f(B) \subset V\}. \tag{7.4}$$

By the linearity of each $f \in \mathcal{L}(E, F)$, each $W_{B,V}$ is absolutely convex. The continuity of each $f \in \mathcal{L}(E, F)$ ensures that $f(B)$ is bounded in F for each $B \in \mathcal{S}$, which means that given any $V \in \mathcal{N}_{0,F}$, there exists $a > 0$ such that $f(B) \subset aV$, so that $f \in aW_{B,V}$. That is, $W_{B,V}$ is absorbing in $\mathcal{L}(E, F)$. We conclude that $\mathcal{S}$ will generate a locally convex topology on $\mathcal{L}(E, F)$ if the two conditions of Corollary 4.3.1 are satisfied. Let us formalize this in the next definition.

Definition 7.2.2. Assume E and F are LCS, $\mathcal{S}$ is a collection of bounded subsets of E satisfying the conditions of Corollary 4.3.1, and $\mathcal{N}_{0,F}$ is a base of absolutely convex zero neighborhoods of F. The locally convex topology generated by

$$\{W_{B,V} : B \in \mathcal{S}, V \in \mathcal{N}_{0,F}\}, \tag{7.5}$$

is called the $\mathcal{S}$ **topology** on $\mathcal{L}(E, F)$.

Other names include the **topology of S convergence** and the **topology of uniform convergence on the sets of S**. Below are descriptions of specific collections S of bounded subsets of a LCS. You will notice that these collections are reminiscent of collections of sets that generate polar topologies, as discussed in Section 4.3.

Definition 7.2.3. On $\mathcal{L}(E,F)$, the following locally convex topologies on $\mathcal{L}(E,F)$ are generated by the indicated collection S of bounded subsets of E.

(a) The **topology of pointwise convergence**, when S consists of the finite subsets of E.
(b) The **topology of absolutely convex, compact convergence**, when S consists of the absolutely convex, compact subsets of E.
(c) The **topology of totally bounded convergence**, when S consists of the totally bounded subsets of E.
(d) The **topology of uniformly bounded convergence**, when S consists of all bounded subsets of E.

Read from top to bottom, the topologies of Definition 7.2.3 range from coarsest to finest. Furthermore, the collections of bounded subsets of Definition 7.2.3 all satisfy the property that E is the union of the sets of the collection. In general, we have the definition below.

Definition 7.2.4. A collection S of bounded subsets of a TVS E such that $E = \bigcup\{S : S \in S\}$ is said to **cover** E.

The next four propositions represent some observations about S topologies and the new types of LCS just defined.

Proposition 7.2.1. *For an S-topology on $\mathcal{L}(E,F)$, the following are equivalent regarding subsets H of $\mathcal{L}(E,F)$:*

(a) *H is bounded with respect to the S topology.*
(b) *For every $V \in \mathcal{N}_{0,F}$, $\bigcap\{f^{-1}(V) : f \in H\}$ absorbs every set $A \in S$.*
(c) *$\bigcup\{f(A) : f \in H\}$ is bounded in F for every $A \in S$.*

Proof. (a) $\Rightarrow$ (b): Let $V \in \mathcal{N}_{0,F}$. Without loss of generality, assume V is balanced. If $H \subset \mathcal{L}(E,F)$ is bounded with respect to an S topology, then H is absorbed by each $W_{B,V}$ of Equation (7.4). Thus, for each $f \in H$, and each $A \in S$, $f(A) \subset aV$ for a suitable $a > 0$. This implies that $A \subset a\bigcap\{f^{-1}(V) : f \in H\}$, and (b) holds.

(b) $\Rightarrow$ (c): Given any $A \in S$ and a balanced $V \in \mathcal{N}_{0,F}$, if for some $a > 0$ we have $A \subset a\bigcap\{f^{-1}(V) : f \in H\}$, then $f(A) \subset aV$ for all $f \in H$. We conclude that $\bigcup\{f(A) : f \in H\}$ is bounded in F for every $A \in S$.

(c) $\Rightarrow$ (a): Again, given $A \in S$ and a balanced $V \in \mathcal{N}_{0,F}$, if there exists $a > 0$ such that for all $f \in H$, $f(A) \subset aV$, it follows that $H \subset aW_{B,V}$, so H is bounded with respect to the S topology. $\qquad\square$

Proposition 7.2.2. *If $H \subset \mathcal{L}(E,F)$ is equicontinuous, then H is bounded with respect to all S-topologies on $\mathcal{L}(E,F)$.*

Proof. Assume H is an equicontinuous subset of $\mathcal{L}(E,F)$. Let S be any collection of bounded subsets of E that covers E. Given any $V \in \mathcal{N}_{0,F}$, there exists, by the equicontinuity of H, a zero neighborhood $U \subset E$ such that $f(U) \subset V$ for all $f \in H$. If $B \in S$, then there exists $a > 0$ for which $B \subset cU$ for all $c \in \mathbb{K}$ such that $|c| \geq a$. Hence, $f(B) \subset cV$ for all $f \in H$ and all such c. $\qquad\square$

Proposition 7.2.3. *An S topology on $\mathcal{L}(E,F)$ is Hausdorff if and only if F is a Hausdorff LCS.*

Proof. Exercise 7.2.1. $\qquad\square$

Proposition 7.2.4. *Suppose $\{q_\alpha : \alpha \in I\}$ is a collection of continuous seminorms that generate the topology of F. For an S topology on $\mathcal{L}(E,F)$, given $f \in \mathcal{L}(E,F)$, $B \in S$ and any q_α,*

$$q_B(f) = \sup\{q_\alpha(f(x)) : x \in B\}, \tag{7.6}$$

is a seminorm on $\mathcal{L}(E,F)$. Moreover, $\{q_B : B \in S\}$ generates the S topology on $\mathcal{L}(E,F)$.

Proof. Exercise 7.2.2. $\qquad\square$

In addition, we may give names to sets that are bounded with respect to the topologies of Definition 7.2.3. For what we hope to accomplish, the main items of Definition 7.2.3 are the first and last, as indicated below.

Definition 7.2.5. On $\mathcal{L}(E,F)$, the following are names of boundedness properties of subsets of E, according to the indicated collection S:
(a) $H \subset \mathcal{L}(E,F)$ is **pointwise bounded** if H is bounded with respect to the S topology, where S consists of the finite subsets of E.
(b) $H \subset \mathcal{L}(E,F)$ is **uniformly bounded** if H is bounded with respect to the S topology, where S consists of all bounded subsets of E.

We are now ready to prove a general version of the PUB.

Theorem 7.2.1. *(Principle of uniform boundedness)* *If E is locally barreled, then the collections of bounded subsets of $\mathcal{L}(E,F)$ are the same for any S topology, where S is a collection of bounded subsets of E that covers E.*

Proof. It suffices to prove that in this context every pointwise bounded subset of $\mathcal{L}(E,F)$ is uniformly bounded. Assume E is locally barreled and let $H \subset \mathcal{L}(E,F)$ be pointwise bounded. For a given closed absolutely convex neighborhood of zero V in F, put

$$D = \bigcap_{f \in H} f^{-1}(V). \tag{7.7}$$

The linearity and continuity of each f imply that D is absolutely convex and closed in E. As H is pointwise bounded, if $x \in E$, then there is $a > 0$ such that $f(x) \in aV$ for every $f \in H$, hence $x \in f^{-1}(V)$ for every $f \in H$. This implies that $x \in aD$. We have shown that D is absorbing in E. Because D is also closed and absolutely convex, we have that D is in fact, a barrel. The Banach–Mackey Theorem 5.1.4 now applies to conclude that D absorbs the barreled disks of E. Furthermore, the assumption that E is locally barreled implies that any bounded subset C of E is contained in a closed barreled disk; that is, C is absorbed by D. By Proposition 7.2.1(b), H is bounded with respect to the uniform S topology. Thus, H is uniformly bounded as stated. $\qquad\square$

With $F = \mathbb{K}$, we have the following.

Corollary 7.2.1. *Weak $*$ bounded subsets of the dual of a locally barreled LCS are strongly bounded.*

Corollary 7.2.2. *Every locally barreled LCS that is also quasi barreled (see Definition 5.1.5) must be barreled.*

Proof. Exercise 7.2.3. $\qquad\square$

By Definition 5.1.2, any LCS that satisfies Theorem 7.2.1 with respect to its dual is a Banach–Mackey space. There are a few more results that can be obtained regarding Banach–Mackey spaces, the first of which amounts to another corollary to Theorem 7.2.1.

Example 7.2.1. An example of a locally barreled, hence Banach–Mackey LCS, that is not barreled.

Indeed, there are many such spaces, specifically any infinite-dimensional Banach space that is equipped with its weak topology. Such a space is locally complete, hence, locally barreled. See Theorem 5.4.1 for more details.

Below are two more results regarding Banach–Mackey spaces. Recall the notation A° of the polar of a set A (see Notation 4.2.1).

Theorem 7.2.2. *For a dual pair (E, F) of LCS, if E is a Banach–Mackey space, then F is also a Banach–Mackey space.*

Proof. Suppose $A \subset F$ is bounded. Let U be a barrel in F. Then the polar U° of U is bounded in E by Proposition 4.2.5. Therefore, U° is absorbed by the polar A° of A, because of the assumption of E being a Banach–Mackey space. That is, U° is strongly bounded. Hence, U is a bornivorous barrel, and the result follows by Theorem 5.1.6. $\qquad\square$

Theorem 7.2.3. *A LCS is a Banach–Mackey space if and only if every barrel is a bornivore.*

Proof. This result is also a consequence of Theorem 5.1.6. You can provide the details in Exercise 7.2.7. $\qquad\square$

In the spirit of weakly bounded subsets being strongly bounded, here is another such result.

Theorem 7.2.4. *In any Hausdorff LCS, every weakly relatively compact convex subset of the dual is strongly bounded.*

Proof. Exercise 7.2.8. ☐

Next is another result that involves $\mathcal{L}(E,F)$. We consider the vector space defined below.

Definition 7.2.6. Define $\mathcal{LB}(E,F)$ to be the collection of linear maps from a LCS E to a LCS F that map a zero neighborhood to a bounded set.

It is clear that $\mathcal{LB}(E,F)$ is a vector space. Exercise 7.2.9 asks for a proof of the following.

Proposition 7.2.5. *$\mathcal{LB}(E,F)$ is a subspace of $\mathcal{L}(E,F)$; that is, every element of $\mathcal{LB}(E,F)$ is continuous.*

Proof. Exercise 7.2.9. ☐

The question arises as to when $\mathcal{LB}(E,F)$ is equal to $\mathcal{L}(E,F)$. In [1], Albanese proved several characterizations of pairs of LCS E and F for which $\mathcal{L}(E,F) = \mathcal{LB}(E,F)$ as well as describing motivations for studying this question and related references. One result from [1] that is of independent interest is stated below. Its proof, which employs concepts such as Mackey first countability (Definition 2.2.8), the topology of an inductive limit (Proposition 3.5.6), and the Mackey convergence condition MCC (Definition 6.2.1), is outlined in Exercises 7.2.10 and 7.2.11.

Theorem 7.2.5 ([1, Proposition 5, p. 16]). *Suppose $E = \mathrm{ind}_n E_n$ is an inductive limit of LCS and F is a Fréchet space. Then;*
(a) *If $\mathcal{L}(E_n,F) = \mathcal{LB}(E_n,F)$ for each $n \in \mathbb{N}$, then $\mathcal{L}(E,F) = \mathcal{LB}(E,F)$.*
(b) *If E is regular and $\mathcal{L}(F,E_n) = \mathcal{LB}(F,E_n)$ for each $n \in \mathbb{N}$, then $\mathcal{L}(F,E) = \mathcal{LB}(F,E)$.*

Proof. The proof of part (a) is outlined in Exercise 7.2.10. Regarding part (b), one proves a version of the **Grothendieck–Floret Factorization Theorem** (see [95, Proposition 8.5.38, p. 298]), which states that in this context if $T \in \mathcal{L}(E,F)$ then there exists $m \in \mathbb{N}$ such that

$$T(F) \subset E_m \quad \text{and} \quad T : F \to E_m \text{ is continuous.} \tag{7.8}$$

The steps of this proof are outlined in Exercise 7.2.11. ☐

Now we consider the Banach–Steinhaus Theorem for $\mathcal{L}(E,F)$. The first task is to make good on the claim that uniform boundedness and equicontinuity are distinct concepts in the context of general LCS. This good-making consists of two quick observations that appear as a proposition and example, as seen next.

Proposition 7.2.6. *In $\mathcal{L}(E, F)$, every equicontinuous subset is uniformly bounded.*

Proof. Exercise 7.2.5. $\qquad\square$

Example 7.2.2. A uniformly bounded set that is not equicontinuous.

Consider $E = (c_0, \sigma(c_0, l_1))$. Then $A \subset l_1 = E'$ is bounded with respect to the topology of uniform boundedness of Definition 7.2.5(b) (i. e., the topology of uniform convergence on the weakly bounded subsets of c_0) if and only if A is bounded with respect to the topology induced by the norm of l_1. The specific set $A = \{e^{(n)} : n \in \mathbb{N}\}$ of unit vectors (1 in the nth entry, zeros for all other entries) is bounded with respect to the l_1 norm, but is not equicontinuous. In Exercise 7.2.6, you can verify the details of these statements.

We may ask if there are conditions under which uniformly bounded subsets of $\mathcal{L}(E, F)$ are equicontinuous. The next proposition gives such a condition.

Proposition 7.2.7. *If E is quasibarreled, then every uniformly bounded subset of $\mathcal{L}(E, F)$ is equicontinuous.*

Proof. Let $H \subset \mathcal{L}(E, F)$ be uniformly bounded. If $V \in \mathcal{N}_{0,F}$ is absolutely convex and closed, it suffices to prove that $U = \bigcap\{f^{-1}(V) : f \in H\}$ is bornivorous in E, because U must then be a zero neighborhood in E. To this end, we note that as in the proof of Theorem 7.2.1, U is absolutely convex, and closed in E. Now let A be any bounded subset of E. By hypothesis, H is uniformly bounded on A, so there exists $a > 0$ such that for each $f \in H, f(A) \subset aV$. Thus, $A \subset aU$. Having shown that U is bornivorous, U is a zero neighborhood by the assumption that E is quasibarreled. We conclude that H is equicontinuous. $\qquad\square$

We now prove a general Banach–Steinhaus theorem.

Theorem 7.2.6 (Banach–Steinhaus theorem). *A pointwise bounded subset of $\mathcal{L}(E, F)$ is equicontinuous if E is barreled.*

Proof. Assume E is barreled and $H \subset \mathcal{L}(E, F)$ is pointwise bounded. For a given closed $V \in \mathcal{N}_{0,F}$, we seek a zero neighborhood U in E such that $\bigcup\{f(U) : f \in H\} \subset V$.

By Theorem 2.1.1, we may assume that V is a barrel. The linearity and continuity of each f imply that $f^{-1}(V)$ is absolutely convex and closed in E. Absolute convexity and closures are preserved by set intersection, and that implies that

$$U = \bigcap_{f \in H} f^{-1}(V) \tag{7.9}$$

is also absolutely convex and closed in E. To establish that U is a zero neighborhood in E we show that U is absorbing because that will imply that U is a barrel. If $x \in E$ is given, then for some sufficiently small scalar a, we have $ax \in U$ if and only if $a\{f(x) : f \in H\} \subset V$, but this is satisfied exactly by the assumption that H is pointwise bounded. □

Below is stated a significant converse to Theorem 7.2.6, which provides essentially a generalization of Corollary 5.1.3.

Theorem 7.2.7. *Suppose E and F are nontrivial LCS such that F does not carry the trivial topology. If every pointwise bounded subset of $\mathcal{L}(E,F)$ is equicontinuous, then E is barreled.*

Proof. See [91, Theorem 11.9.1, p. 400]. □

A sequential version of the Banach–Steinhaus theorem can be proven using property $\mathcal{K}$, as we see next. For the result below, recall from Definition 2.3.2 that $\mathcal{L}_S(E,F)$ represents the vector space of sequentially continuous linear maps from a TVS E to a TVS F.

Theorem 7.2.8 (Banach–Steinhaus theorem for Property $\mathcal{K}$). *Suppose E and F are TVS such that F is Hausdorff. Let $(T_n) \subset \mathcal{L}_S(E,F)$ such that $T_n \to T$ pointwise; that is, for every $x \in E$, $T_n(x) \to T(x)$. Then for a $\mathcal{K}$-convergent sequence (x_j) in E, the following hold:*
(a) $\lim_{j \to \infty} T(x_j) = 0$.
(b) For each $j \in \mathbb{N}$, $\lim_{n \to \infty} T_n(x_j) = T(x_j)$ uniformly.

Proof. Form the matrix $\mathcal{M}$ (see Definition 6.1.5) that has entries $T_n(x_j)$, for $n, j \in \mathbb{N}$; that is, $\mathcal{M} = [T_n, x_j]$. In Exercise 7.2.13, you can prove that the assumptions here satisfy the conditions of the Basic Matrix Theorem 6.1.1, from which items (a) and (b) follow. □

For more results involving property $\mathcal{K}$ regarding equicontinuity and uniform boundedness, see [117, Chapters 4, 5].

Exercises

7.2.1. Prove Proposition 7.2.3.

7.2.2. Prove Proposition 7.2.4.

7.2.3. Prove Corollary 7.2.2.

7.2.4. Prove directly that a subset H of $\mathcal{L}(E, F)$ is bounded if and only if $\bigcup\{f(A) : f \in H\}$ is bounded in F for every $A \in \mathcal{S}$.

7.2.5. Prove Proposition 7.2.6.

7.2.6. Verify the details of Example 7.2.2. *Hint:* For the last statement, consider the set $\{e^{(n)} : n \in \mathbb{N}\}$ also as elements of c_0. Prove that as a sequence, $e^{(n)} \to 0$ with respect to $\sigma(c_0, l_1)$ but the evaluation $e^{(n)}$ of $e^{(n)}$ (i. e., $e^{(n)}$ as an element of l_1 acting on $e^{(n)}$ as an element of c_0), is 1 for each $n \in \mathbb{N}$.

7.2.7. Prove Theorem 7.2.3. *Hint:* Apply Theorem 5.1.6.

7.2.8. Prove Theorem 7.2.4. *Hint:* First, apply Proposition 2.1.2 to any $H \subset E'$ that is convex and relatively $\sigma(E', E)$-compact. Next, apply Theorem 2.1.3 to H_1, where H_1 is the $\sigma(E', E)$-closure of convbal(H). Then show that H_1 is absorbed by the polar of any $\sigma(E, E')$-bounded subset of E, so as to apply the result of Exercise 5.3.2. More details can be seen in [91, Theorem 11.11.5, p. 407].

7.2.9. Prove Proposition 7.2.5.

7.2.10. Prove part (a) of Theorem 7.2.5. *Hint:* First consider $T \circ \mathrm{id}_n$, where $\mathrm{id}_n : E_n \to E$ is the canonical injection for each $n \in \mathbb{N}$, and find zero neighborhoods $U_n \subset E_n$ such that $T(\mathrm{id}_n(U_n))$ is bounded in F. Then apply Theorem 2.2.2 to F to find scalars $a_n, n \in \mathbb{N}$, for which $\bigcup\{a_n T(\mathrm{id}_n(U_n))\}$ is bounded in F. Verify that the result then follows by Proposition 3.5.6.

7.2.11. Prove the statement shown in Expression (7.8) regarding Theorem 7.2.5. *Hint:* Let $\mathfrak{F}$ be the filter generated by $\{U_k : k \in \mathbb{N}\}$, where $\{U_k : k \in \mathbb{N}\}$ is a decreasing base of zero neighborhoods of F. Let $\mathfrak{F}_n$ denote the filter generated by a base of zero neighborhoods of E_n, for each $n \in \mathbb{N}$. Show that given a convergent sequence (x_k) with respect to $\mathfrak{F}$, there exists for each $k \in \mathbb{N}, y_k \in F$ such that $T(y_k) = x_k$ and (y_k) converges in F. Then explain the property of F that guarantees the existence of a sequence (b_k) of scalars such that $b_k \to \infty$ as $k \to \infty$ and satisfies: $B = \{b_k x_k : k \in \mathbb{N}\} = \{T(b_k y_k) : k \in \mathbb{N}\}$ is bounded in E. Then explain why there exists $m \in \mathbb{N}$ such that $B \subset E_m$ and is bounded in E_m. Finish by applying the property that there exists $l \in \mathbb{N}$ for which $\mathfrak{F}_l$ is coarser than $\mathfrak{F}$.

7.2.12. Prove that part (a) of Theorem 7.2.5 remains true if F is only assumed to be Mackey first countable (Definition 2.2.8).

7.2.13. Prove Theorem 7.2.8.

8 The closed graph theorem

From the point of view of analysis and topology, any theorem that can be used to show continuity of a function catches our attention. In this chapter, we will study such a theorem—the closed graph theorem. It is fundamentally important in the world of TVS. In Section 8.1, we prove the main result of this chapter, Theorem 8.1.2, which consists of a version of a closed graph theorem in which the domain LCS is a metrizable Baire space and the range LCS has a completing web. We also prove a sort of dual version of the closed graph theorem, the Open Mapping Theorem 8.1.6. In Section 8.2, we study a significant consequence of the closed graph theorem, called the Localization Theorem 8.2.1. Some applications of the results of Sections 8.1 and 8.2 are indicated in Section 8.3. Like every deep result, there are numerous versions of the closed graph theorem. Several of the many other versions of the closed graph theorem are briefly described in Subsection 8.3.1. Let us get started!

8.1 A closed graph theorem for webbed spaces

Naturally, we begin with the definition of a closed graph. Note that we use the terms "map" and "mapping" interchangeably.

Definition 8.1.1. Consider Hausdorff TVS E and F, and a linear map $T : E \to F$. The **graph of** T is $G_T = \{(x, T(x)) : x \in E\}$.

Definition 8.1.2. For E, F, and T as above, we say that the graph of T is **(sequentially) closed** (or T is a **(sequentially) closed linear map**) if G_T is a (sequentially) closed subset of the product space $E \times F$.

In Exercise 8.1.1(a), you can prove that if we already know a linear map T is continuous, then its graph must be closed. In Exercise 8.1.1(b), you can verify that the converse does not hold, concluding the following.

Example 8.1.1. If E is any infinite-dimensional Banach space, then id : $(E, \sigma(E, E')) \to (E, \| \cdot \|)$ is a linear, discontinuous map having a closed graph.

Another example of a discontinuous linear map having a closed graph is next.

Example 8.1.2. The identity map from the continuously differentiable functions to the continuous functions on $[0, 1]$ has a closed graph and is discontinuous.

To verify this statement, let F denote the Banach space of continuous real valued functions on $[0, 1]$, and let E denote the normed space of continuously differentiable functions on $[0, 1]$, both with the sup norm. We consider the map $T : E \to F$ given by the

https://doi.org/10.1515/9783111392868-008

derivative; that is, $T(f) = f'$, for each $f \in E$, E being viewed as a subspace of F. By calculus, T is linear. To see that the graph of T is closed, assume $f_n \to f$ in E and $T(f_n) \to y$ in F. By uniform convergence and advanced calculus, we conclude that

$$T(f) = f' = \lim_{n \to \infty} f_n' = y, \tag{8.1}$$

implying that the graph of T is closed. Meanwhile, consider $f_n(x) = x^n$ for each $n \in \mathbb{N}$, $x \in [0,1]$. Then $\{T(f_n) : n \in \mathbb{N}\} = \{nx^{n-1} : n \in \mathbb{N}\}$, which is unbounded. Thus, T maps the bounded set $\{x^n : n \in \mathbb{N}, x \in [0,1]\} \subset E$ to an unbounded set. By Proposition 2.3.3, this means T cannot be continuous.

Below is a conclusion that can be made if the graph of a linear function is closed.

Proposition 8.1.1. *Suppose E and F are TVS such that F is Hausdorff. For a linear map $T : E \to F$, if the graph of T is closed, then $T^{-1}(\{0\})$ is closed in E.*

Proof. The assumption that F is Hausdorff implies that $E \times \{0\}$ is closed in $E \times F$. If the graph G_T of T is closed, then $T^{-1}(\{0\}) \times \{0\} = G_T \cap (E \times \{0\})$ is closed in $E \times F$. The map $f : E \to E \times F$ given by $f(x) = (x, 0)$ is continuous. The last statement implies that $f^{-1}(T^{-1}(\{0\}) \times \{0\}) = T^{-1}(\{0\})$ is closed in E. $\qquad\square$

The big question that arises consists of asking under which conditions can we conclude that a linear map is continuous if its graph is closed? Attempts to answer this question for rather general contexts is the focus of this section. Whenever we obtain a result that gives a positive answer, we say that we have obtained a closed graph theorem. Banach [8] (see [9] for an updated printing of this classic) proved an early version of the closed graph theorem, which reads as follows.

Theorem 8.1.1 (Banach's closed graph theorem). *Suppose $T : E \to F$ is a linear map, where E and F are complete, metrizable TVS. If the graph of T is closed, then T is continuous.*

The above version of the closed graph theorem was proved for a variety of special cases of types of spaces E and F until the 1960s; see [68]. An important motivation for continuing to seek generalizations of Banach's theorem came from Grothendieck [63] who in 1955, conjectured that if the domain would be a Fréchet space, then the range for a closed graph theorem should have good permanence properties, implying that the theorem would apply to large collections of LCS. Grothendieck's conjecture was verified by several people, the first version of which was proved in 1969 by De Wilde [35]. Several other versions of a solution to Grothendieck's conjecture have since been obtained.

Banach's original proof of Theorem 8.1.1 in [8] makes use of the completeness property in the range space and that of a complete metric space being of second category via the Baire Category Theorem A.3.9 in the domain space. One kind of structure for the range space that satisfies Grothendieck's conjecture, turns out to be that of a completing web (see Definition 6.3.5). For the range space, the idea of what we have defined as a (Robertson) web (see Definition 6.3.1) goes back to De Wilde's original definition [35], called **réseaux absorbents**. The concept of a réseaux web is stated in Definition 8.1.3 later in this section.

The following property of strands of a web in a Baire space will be useful in the sequel.

Proposition 8.1.2. *If E is a Baire space with web $\mathcal{W}$, then there exists a strand (W_k) of $\mathcal{W}$ such that for every $k \in \mathbb{N}$, $\overline{(W_k)} \in \mathcal{N}_0$.*

Proof. In Exercise 8.1.2, you can prove that in the first layer of $\mathcal{W}$ there exists at least one set W_1 such that $\overline{W_1}$ is a zero neighborhood of E. By way of induction, assume there are sets $W_k \in \mathcal{W}$, $k = 2,\ldots,n$ of a strand (W_k) of $\mathcal{W}$ for which $\overline{W_k} \in \mathcal{N}_0$. Consider $\{A_j : j \in \mathbb{N}\}$, the set of all subsets of W_n that are in the $(n+1)$st layer of $\mathcal{W}$; that is, sets determined by W_n. Then $\bigcup\{A_j : j \in \mathbb{N}\}$ is absorbing in W_n. This can be expressed as

$$W_n \subset \bigcup_{j \in \mathbb{N}} \bigcup_{m \in \mathbb{N}} mA_j. \tag{8.2}$$

$\overline{W_n} \in \mathcal{N}_0$ implies that for some j_0, $\overline{A_{j_0}} \in \mathcal{N}_0$. Choose $W_{n+1} = A_{j_0}$. Then $W_{n+1} \subset W_n$ and $\overline{W_{n+1}} \in \mathcal{N}_0$. $\qquad\square$

Corollary 8.1.1. *Every Fréchet LCS, in particular every Banach space, satisfies the conclusion of Proposition 8.1.2.*

Proof. This follows by the Baire Category Theorem A.3.9. $\qquad\square$

In preparation for the next result, assume E and F are Hausdorff LCS and $T : E \to F$ is linear. In Exercise 8.1.3, you can verify that if $\mathcal{W}$ is a web on F, then $\{T^{-1}(W) : W \in \mathcal{W}\}$ is a web on E. As necessary, refer to Section 6.3 regarding webs. The main result of this section is next, the proof of which follows that of [105, Theorem 14, p. 716].

Theorem 8.1.2 (Closed graph theorem for webbed spaces). *Suppose E is a metrizable Baire LCS, F is a LCS that has a completing web, and the graph of T is closed. Then T is continuous.*

Proof. Let $\mathcal{W}$ be a completing web on F. By Proposition 8.1.2, there is a strand (W_k) of $\mathcal{W}$ such that for every $k \in \mathbb{N}$,

$$\overline{T^{-1}(W_k)} \in \mathcal{N}_{0,E}. \tag{8.3}$$

Therefore, there exists a sequence (U_n) of zero neighborhoods in E such that for every $n \in \mathbb{N}$, $U_n \subset \overline{T^{-1}(W_n)}$. Without loss of generality, we can take $\{U_n : n \in \mathbb{N}\}$ to be a zero neighborhood base for E, which you can check in Exercise 8.1.4. By Equation (1.9), $U_n \subset \overline{T^{-1}(W_n)}$ is equivalent to: for every $U \in \mathcal{N}_{0,E}$, $U_n \subset T^{-1}(W_n) + U$. Consequently, for every $n \in \mathbb{N}$,

$$T(U_n) \subset W_n + T(U_{n+1}). \tag{8.4}$$

Let V be any zero neighborhood in F, which we may assume to be closed. The wish is to find a zero neighborhood U in E such that $T(U) \subset V$. There is already a set relation that draws our attention: Applying the compatibility of $\mathcal{W}$ to the strand (W_k) above, there exists $N \in \mathbb{N}$ such that $W_{N-1} \subset V$. If we can show that $T(U_N) \subset V$, we will have completed the proof. We start by letting $x_0 \in U_N$ be arbitrary. The rest of the proof will consist of finding elements $x_k \in U_{N+k}$ for integers $k \geq 1$ that lead us to conclude that $T(U_N) \subset V$. By rearranging Equation (8.4), there is $x_1 \in U_{N+1}$ such that $T(x_0) - T(x_1) \in W_N$. Next, there is $x_2 \in U_{N+2}$ with $T(x_1) - T(x_2) \in W_{N+1}$. This pattern continues and, in general, there is, for each $k \in \mathbb{N}$, $x_k \in U_{N+k}$ such that

$$T(x_k) - T(x_{k+1}) \in W_{N+k}. \tag{8.5}$$

By the assumption that the web $\mathcal{W}$ is completing, the series

$$\sum_{k=1}^{\infty} (T(x_k) - T(x_{k+1})) \tag{8.6}$$

converges to some element $y \in F$. Moreover, that $\{U_k : k \in \mathbb{N}\}$ is a zero neighborhood base for E implies that $x_k \to 0$ in E. We now apply the closed graph assumption: $x_k \to 0$ in E and $(T(x_k)) \to y$ in F tells us that $y = T(0) = 0$. Hence, $T(x_k) \to 0$ as $k \to \infty$. To complete the proof, we determine what happens to $T(x_0)$ by telescoping as follows:

$$T(x_0) - T(x_k) = \sum_{j=1}^{k-1} (T(x_j) - T(x_{j+1})) \in W_N + W_{N+1} + \cdots + W_{N+k-1} \subset W_{N-1}, \tag{8.7}$$

by properties of webs. Letting $k \to \infty$, $T(x_k) \to 0$, so the sum above becomes $T(x_0) \in \overline{W_{N-1}}$. Recall that by the compatibility of $\mathcal{W}$, $W_{N-1} \subset V$. Thus,

$$T(x_0) \in \overline{W_{N-1}} \subset \overline{V} = V, \tag{8.8}$$

because we assumed V is closed. As $x_0 \in U_N$ was arbitrary, we have shown that $T(U_N) \subset V$, and T is therefore continuous. $\qquad\square$

Observe that we only needed to make use of the graph of T being sequentially closed. There is a certain calculational advantage to sequential closedness in that it is often easier to verify than general topological closedness. Recall from Proposition 3.5.10 that no proper (LB) space is metrizable. Thus, if E is a Fréchet space, F is a proper (LB) space and $T : E \to F$ is linear with a closed graph, then Theorem 8.1.2 applies, and this is a positive response to Grothendieck's conjecture.

Theorem 8.1.3. *Assume E carries the final topology of a collection $(E_\alpha, \mathcal{T}_\alpha, f_\alpha : \alpha \in I)$ of metrizable Baire LCS, and corresponding continuous linear maps $f_\alpha : E_\alpha \to E$. Assume F is webbed, and the linear map $T : E \to F$ has a closed graph. Then T is continuous.*

Proof. The conclusion will hold if Theorem 8.1.2 can be applied to each of the maps $T \circ f_\alpha$. To this end, suppose $x_k \to x$ in E_α and $T(f(x_k)) \to y$ in F. The continuity of each f_α ensures that $f_\alpha(x_k) \to f_\alpha(x)$. The graph of T is closed, so $y = T(f_\alpha(x))$ and Theorem 8.1.2 applies: Each $T \circ f_\alpha$ is continuous. It only remains to cite Proposition 3.5.4 to conclude that $T : E \to F$ is continuous. $\qquad\square$

Corollary 8.1.2 (Closed graph theorem for ultrabornological spaces). *A closed linear map from an ultrabornological LCS to a webbed LCS is continuous.*

Proof. Apply the Baire Category Theorem A.3.9 to the relevant Banach spaces of the ultrabornological LCS, and the result then follows immediately. $\qquad\square$

Next, no discussion of webbed spaces would be complete (pun intended) without including De Wilde's classic construction; see [36] for a detailed discussion of this approach. The definition follows.

Definition 8.1.3. A **réseaux (absorbents) web** (or **De Wilde web**) on a LCS F is a countable collection $\mathcal{R}$ of sets, denoted by $\{C_{n_1 n_2 \cdots n_k} : k, n_1, n_2, \ldots, n_k \in \mathbb{N}\}$, satisfying the following properties:

(a) $\bigcup \{C_{n_1} : n_1 \in \mathbb{N}\} = F$.

(b) For each $k \in \mathbb{N}$, $\bigcup \{C_{n_1 n_2 \cdots n_{k+1}} : n_{k+1} \in \mathbb{N}\} = C_{n_1 n_2 \cdots n_k}$.

The concept of a strand is given by indicating a specific sequence $(n_k : k \in \mathbb{N})$ of positive integers. In Exercise 8.3.2, you can experience a particular comparison of this Definition 8.1.3 of a réseaux web, and Definition 6.3.1. For a more detailed discussion of connections between these two definitions, see [105, pp. 723–726]. The (strict) completeness property reads as follows.

Definition 8.1.4. A **réseaux web** $\mathcal{R}$ on a LCS F is of **type** $\mathcal{C}$ if given any fixed sequence (n_k) of positive integers, there exists a sequence (λ_k) of positive numbers, such that the series $\sum_{k=1}^\infty \mu_k x_k$ converges in F for μ_k such that $0 < |\mu_k| \le \lambda_k$, $k \in \mathbb{N}$, and $x_k \in C_{n_1 n_2 \cdots n_k}$ for each $k \in \mathbb{N}$. If in addition the convergent series satisfy $\sum_{k=p}^\infty \mu_k x_k \in C_{n_1 n_2 \cdots n_p}$, then $\mathcal{R}$ is **strict**.

One more definition is needed before we state De Wilde's closed graph theorem.

Definition 8.1.5. A subset A of a LCS E is **fast sequentially closed** if it contains the limits of all sequences of its elements that are fast convergent (see Definition 6.2.2).

Below is De Wilde's closed graph theorem. Steps of the proof are similar to those of the proof of Theorem 8.1.2.

Theorem 8.1.4 (De Wilde's closed graph theorem). *Suppose E, F are LCS such that E is a Fréchet space and F is strictly webbed. Let $T : E \to F$ be a linear map. If the graph of T is fast sequentially closed, then T is continuous.*

Proof. See [36, Section IV.5, pp. 63–69]. $\qquad\square$

8.1.1 Open mapping theorems

Our next goal is to obtain an open mapping theorem from the closed graph theorem. The first step is to define an open map.

Definition 8.1.6. A linear map is **open** if it sends open sets to open sets.

Notice that if we already know a linear map is open, then it must be surjective. By essentially working backward from the closed graph theorem, we can obtain a result that can be used to show that a linear map is open. Theorems 8.1.5 and 8.1.6 below are of the same fundamental importance as the corresponding closed graph theorems, and in fact, one can first prove an open mapping theorem and then deduce the corresponding closed graph theorem from that open mapping theorem. Such a route is taken for example in [67, Section 3.17] for the domain space being a Fréchet space and range being a Baire LCS.

Theorem 8.1.5 (Banach's open mapping theorem). *Suppose $T : F \to E$ is a continuous, surjective linear map, where E and F are complete, metrizable TVS. Then T is an open map.*

Proof. See [67, Theorem 3.17.1, pp. 294–296] for a direct proof. Note that Banach's open mapping theorem is traditionally referred to as **(Banach's homomorphism theorem)**. □

The next theorem is in the context of spaces with completing webs. Its proof relies on Theorem 8.1.2.

Theorem 8.1.6 (Open mapping theorem for webbed spaces). *Suppose E is a Fréchet LCS, F has a completing web, and $T : F \to E$ is linear, continuous, and surjective. Then T is an open map.*

Proof. The proof involves a typical factoring through the quotient of the null space $T^{-1}(\{0\})$ of T. Let $\varphi : F \to F/T^{-1}(\{0\})$ be the canonical quotient map onto $T^{-1}(\{0\})$. Recall that by Proposition 1.1.5, φ is linear, surjective, and open. Think back to the Diagram 1.4 regarding the quotient map. A modified version of that diagram is displayed in Diagram 8.9 below. Observe that the inverse $S^{-1} : F/T^{-1}(\{0\}) \to E$ of S has been included in Diagram 8.9:

$$\begin{array}{ccc}
F & \xrightarrow{\ \ T = S \circ \varphi\ \ } & E \\
 & \searrow^{\varphi} & \big\uparrow S \big\downarrow S^{-1} \\
 & & F/T^{-1}(\{0\})
\end{array} \tag{8.9}$$

Here, $S : F/T^{-1}(\{0\}) \to E$ represents a bijective linear map such that $T = S \circ \varphi$. The continuity of T implies that S is continuous, and hence, the graph of S^{-1} is closed. Meanwhile, by Theorem 6.3.3(c), the quotient LCS $F/T^{-1}(\{0\})$ has a completing web because F does. Thus, Theorem 8.1.2 applies to S^{-1} so as to conclude that S^{-1} is continuous. S is therefore a homeomorphism. Theorem A.3.7 now tells us that S is an open map. The conclusion follows by noting that a composition of open maps is open (which you can verify in Exercise 8.1.13). $\qquad\square$

An open mapping theorem can be obtained independently of assuming a closed graph theorem and visa versa. See [94, Section 4.1, 4.5] for more results and details.

Exercises

8.1.1. On when the graph of a linear map is closed.
   (a) Prove that the graph of a continuous linear map between two Hausdorff LCS is closed.
   (b) Suppose that on a vector space E, two Hausdorff linear topologies $\mathcal{T}_1$ and $\mathcal{T}_2$ are such that $\mathcal{T}_1$ is strictly coarser than $\mathcal{T}_2$. Prove that id $: (E, \mathcal{T}_1) \to (E, \mathcal{T}_2)$ is a closed, discontinuous map. *Hint:* What can be said about the graph of the inverse of id?

8.1.2. Verify the first line of the proof of Proposition 8.1.2. *Hint:* Let $\{W_k : k \in \mathbb{N}\}$ represent the first layer of a web on a Baire LCS. Define a collection $\mathcal{V} = \{V_n : n \in \mathbb{N}\}$ by $V_n = nW_1 \cup nW_2 \cup \cdots \cup nW_n$.

8.1.3. Suppose E and F are Hausdorff LCS and $T : E \to F$ is linear. Prove that if $\mathcal{W}$ is a web on F, then $\{T^{-1}(W) : W \in \mathcal{W}\}$ is a web on E.

8.1.4. Verify the statement "Without loss of generality, we can take $\{U_n : n \in \mathbb{N}\}$ to be a zero neighborhood base for E" from the proof of Theorem 8.1.2.

8.1.5. Suppose $\mathcal{W}$ is a compatible web on E, and that (a_k) is any sequence of nonzero scalars. Prove that the set $S = \bigcap_{k=1}^{\infty} a_k W_k$ is bounded for any strand (W_k) of $\mathcal{W}$.

8.1.6. Prove that if E is locally complete and each set in $\mathcal{W}$ is closed, then $S = \bigcap_{k=1}^{\infty} a_k W_k$ is a Banach disk, for any strand (W_k) of $\mathcal{W}$.

8.1.7. Prove that on a fixed vector space it is impossible for two distinct Hausdorff compatible (with respect to the linear structure, that is) topologies to be comparable as inductive limits of Fréchet spaces. *Hint:* Use Theorem 8.1.3.

8.1.8. Prove that a continuous linear map from a webbed LCS onto an ultrabornological LCS is open.

8.1.9. Suppose E is a nonreflexive Fréchet space. Prove that E as a subspace of E'' is a countable union of nowhere dense subsets (Definition A.3.24). *Hint:* Prove that if the conclusion fails, then the identity map from E to E'' would be surjective. Then apply the Open Mapping Theorem 8.1.5 to reach a contradiction.

8.1.10. Use Definition 8.1.4 to prove that a réseaux web $\mathcal{W}$ on a LCS E is of type $\mathcal{C}$ if for every fixed sequence (n_k) of positive integers, there exists a sequence (a_k) of positive numbers, such that every sequence $(a_k x_k)$ for which $x_k \in C_{n_1 n_2 \cdots n_k}$, is contained in a Banach disk.

8.1.11. Prove that if E is a Baire LCS, F is webbed, and $T : E \to F$ is linear and has a closed graph (rather than sequentially closed), then T is continuous. *Hint:* The proof generally follows that of Theorem 8.1.2 except that the collection $\{U_k : k \in \mathbb{N}\}$ need not be a base of zero neighborhoods in E, which means the sequence (x_k) need not be a null sequence in E. To deal with this obstacle, work with the closure of the graph (which is assumed closed).

8.1.12. Describe examples of functions that satisfy the conditions below:
   (a) A discontinuous function that maps open sets to open sets.
   (b) A discontinuous function that maps closed sets to closed sets.
   (c) A continuous function that neither maps open sets to open sets nor maps closed sets to closed sets.
   (d) A continuous function that maps open sets to open sets but does not map closed sets to closed sets.
   (e) A continuous function that maps closed sets to closed sets but does not map open sets to open sets.

8.1.13. Prove that a composition of open maps is an open map.

8.2 The localization theorem

So far, the property of a web being strict has not been highlighted—until now. Such highlighting is the topic of this section, in which we obtain significant properties of inductive limits and also observe some new examples and counterexamples.

Theorem 8.2.1 (Localization theorem). *Assume E and F are LCS such that E is a Baire LCS, F has a completing web $\mathcal{W}$, and the linear map $T : E \to F$ has a closed graph. Then for each $k \in \mathbb{N}$, there is a zero neighborhood U_k in E such that $T(U_k) \subset \overline{W_k}$ in F. If the web is strict, then $T(U_k) \subset W_k$ holds.*

Proof. By Proposition 8.1.2, there is a strand (W_k) of $\mathcal{W}$ such that $\overline{T^{-1}(W_k)} \in \mathcal{N}_0$, for each $k \in \mathbb{N}$. Next, as in the proof of Theorem 8.1.2, for every $U \in \mathcal{N}_{0,E}$, the closure of the previous statement is expressed as $U_k \subset T^{-1}(W_k) + U$. In other words,

$$T(U_k) \subset W_k + T(U). \tag{8.10}$$

T is continuous by Theorem 8.1.2, so given any zero neighborhood V in F, there exists a zero neighborhood U_V in E such that $T(U_V) \subset V$. If $U \subset U_V$ in Equation (8.10), then

$$T(U_k) \subset W_k + T(U) \subset W_k + T(U_V) \subset W_k + V. \tag{8.11}$$

As V was arbitrary, we have $T(U_k) \subset \overline{W_k}$.

Now suppose $\mathcal{W}$ is strict. Because $\overline{T^{-1}(W_k)}$ is a zero neighborhood in E, the strictness of $\mathcal{W}$ implies that $\overline{T^{-1}(W_{k+2})} \subset T^{-1}(W_k)$, for each $k \in \mathbb{N}$. We then observe that $T(U_{k+3}) \subset W_k$; that is, $U_{k+3} \subset T^{-1}(W_k)$. Because U_{k+3} is a zero neighborhood in E, the proof is completed. $\qquad\square$

A few useful corollaries result from the localization theorem, some of which are left as exercises.

Corollary 8.2.1. *Under the same hypotheses as Theorem 8.2.1, if $\mathcal{W}$ is a strict web on F, then there exists a strand (W_k) in $\mathcal{W}$ such that for each $k \in \mathbb{N}$, $T^{-1}(W_k)$ is a zero neighborhood in E.*

Proof. Exercise 8.2.1. $\qquad\square$

Corollary 8.2.2. *Consider E to be any Hausdorff LCS having a completing web $\mathcal{W}$. Suppose B is a Baire disk. Then there exists a strand (W_k) in $\mathcal{W}$ such that for each $k \in \mathbb{N}$ there exists a scalar a_k such that $T(B) \subset a_k \overline{W_k}$. If $\mathcal{W}$ is strict, then $T(B) \subset a_k W_k$.*

Proof. Exercise 8.2.2. $\qquad\square$

Corollary 8.2.3. *Any webbed Baire LCS is metrizable. Any strictly webbed Baire LCS is a Fréchet space.*

Proof. Apply Theorem 8.2.1 to the identity, id : $E \to E$, on the assumed Baire LCS E. Let $\mathcal{W}$ denote the completing web on E. There exists a strand (W_k) of $\mathcal{W}$ such that to each $k \in \mathbb{N}$ there corresponds a zero neighborhood U_k such that $\mathrm{id}(U_k) = U_k \subset \overline{W_k}$. Meanwhile, the compatibility of the web implies that for any closed zero neighborhood U, there is a $k \in \mathbb{N}$ such that $\overline{W_k} \subset U$. We conclude that $\{\overline{W_k} : k \in \mathbb{N}\}$ is a base of zero neighborhoods that generates the topology of E. The countability of this collection indicates that E is metrizable. In addition, if $\mathcal{W}$ is assumed to be strict, then any Cauchy sequence will be convergent to a point of one of the sets in the strand (W_k) in this metrizable topology. That is, E will be complete, hence a Fréchet space. $\qquad\square$

Corollary 8.2.4. *Let E be a Baire LCS and F an inductive limit of strictly webbed LCS F_n. Suppose $T : E \to F$ is linear with a closed graph. Then there exists $m \in \mathbb{N}$ such that $T(E) \subset F_m$ and the linear map from E to F_m induced by T is continuous.*

Proof. Exercise 8.2.8. $\qquad\square$

One of the main applications of the Localization Theorem 8.2.1 has to do with the regularity of inductive limits. We can obtain the following regularity result for locally Baire spaces, which include those that have property $\mathcal{K}$.

Theorem 8.2.2. *Let $E = \mathrm{ind}_n E_n$ be an inductive limit of strictly webbed locally Baire LCS. Then E is regular if and only if E is locally Baire.*

Proof. ($\Rightarrow$): Assume E is regular, and let $A \subset E$ be any bounded set. Then for some $m \in \mathbb{N}$, $A \subset E_m$ and is bounded in the topology of E_m. As E_m is locally Baire, it contains a closed Baire disk B such that $B \supset A$. The topology on the span of B generated by $\{n^{-1}B : n \in \mathbb{N}\}$ is the same in both E_m and E; that is, B is a Baire disk in E that contains A. Of course, B is bounded in E by the continuity of the identity map from E_m to E. Hence, E is locally Baire.

($\Leftarrow$): Now suppose E is locally Baire and let A be any bounded subset of E. Then there is a closed Baire disk B, this time in E, such that $A \subset B$. Invoking the hereditary property of Theorem 6.3.1(d) for inductive limits, we observe that a strict web can be defined on the inductive limit so that any strand of the strict web on E is a strand of the strict web on some step E_m. Consider the identity id : $E_B \to E$. It is a continuous linear map by the boundedness of B in E. Moreover, this map is from a metrizable Baire space to a strictly webbed space, so Corollary 8.2.2 applies to the bounded set B: There is a strand (W_k) of the web on E such that for each $k \in \mathbb{N}$, there exists a scalar a_k such that $\mathrm{id}(B) \subset a_k W_k$, where the collection $\{W_k : k \in \mathbb{N}\}$ represents a strand of a strict web in some step E_m. We conclude that $A \subset B \subset E_m$, but we seek more, namely to show that A is bounded in E_m. In order to conclude this, let V be any zero neighborhood in E_m. The compatibility of the web on E_m implies that there is some $K \in \mathbb{N}$ such that $W_K \subset V$. It now follows that

$$A \subset B = \mathrm{id}(B) \subset a_K W_K \subset a_K V, \tag{8.12}$$

and A is indeed bounded in E_m. $\qquad\square$

Corollary 8.2.5. *An (LB) space is regular if and only if it is locally Baire.*

Proof. Indeed, every Banach space has a strict web, which allows us to construct a strict web $\mathcal{W}$ on an (LB) space such that each strand of $\mathcal{W}$ is a strand of the strict web on one of the steps. Theorem 8.2.2 now applies. $\qquad\square$

The next two examples show that surprisingly, there is no connection between having a completing web and any kind of completeness.

Example 8.2.1. A complete LCS that is not webbed.

To verify this statement, we show: No uncountable product of Banach spaces can be webbed. We know that an arbitrary product of complete LCS is complete by Theorem 3.3.2. An uncountable product of Banach spaces is not only complete, it is a Baire space (see [28]). If such a product were to have a completing web, it would be metrizable by Corollary 8.2.3, an impossibility in light of the product's uncountability. To be sure, any uncountable product of Banach spaces is a complete LCS that has no completing web.

Example 8.2.2. A (strictly) webbed LCS that is not locally Baire.

Exercise 8.2.4 involves citing Corollary 8.2.5 to prove that there exist LCS that are strictly webbed and not even locally Baire (from Köthe's example of a nonregular (LB) space; see Theorem 3.5.6). In fact, any nonregular (LB) space represents such an example. The completeness properties of webbed and strictly webbed LCS are in fact rather unusual in the sense that they provide series convergence for only certain sequences. These two rather drastic distinctions of concepts of completeness are shown in Display 8.13 below:

$$\text{(strictly) webbed} \;\not\Rightarrow\; \text{locally Baire}, \quad \text{(locally) complete} \;\not\Rightarrow\; \text{(strictly) webbed}. \quad (8.13)$$

The next task is to make use of the Localization Theorem 8.2.1 to obtain sufficient conditions for the Mackey convergence condition MCC of Section 6.2. Recall Definition 6.3.4 of a sequential web.

Proposition 8.2.1. *A Hausdorff locally Baire LCS with a completing web that satisfies the MCC must have a sequential web.*

Proof. Let $\mathcal{W}$ be a completing web on the LCS E, where we assume E satisfies the MCC. We plan to prove that $\mathcal{W}$ is sequential. Let (x_n) be a null sequence in E. Per Proposition 6.2.1, there exists a sequence (a_n) of positive scalars such that $a_n \to \infty$ and $a_n x_n \to 0$ in E. Let $A = \{a_n x_n : n \in \mathbb{N}\}$. Then A is bounded in E, and consequently, is contained in a closed disk B such that $(E_B, \|\cdot\|_B)$ is a normed Baire space. The injection id $: E_B \to E$ is continuous and as such has a closed graph. By the Localization Theorem 8.2.1 there exists a strand (W_k) of $\mathcal{W}$ and a sequence (r_k) of scalars such that $\mathrm{id}(B) = B \subset r_k \overline{W_k}$, for each $k \in \mathbb{N}$. This implies that for each $n \in \mathbb{N}$, $a_n x_n \in r_k \overline{W_k}$. For a fixed $k \in \mathbb{N}$, we may find $N_k \in \mathbb{N}$ such that for each $n \geq N_k$, $\frac{|r_k|}{a_n} < 1$. Therefore, for all $n \geq N_k$,

$$x_n \in \frac{|r_k|}{a_n} \overline{W_k} \subset \overline{W_k}, \tag{8.14}$$

because each $\overline{W_k}$ is balanced. We conclude that $\mathcal{W}$ is sequential. $\qquad\square$

Corollary 8.2.6. *A Hausdorff locally Baire LCS that has a completing web satisfies the MCC if and only if the web is sequential.*

Proof. If such a space has a sequential web, then it satisfies the MCC by Theorem 6.3.2. The other implication is immediate from Proposition 8.2.1 above. $\qquad\square$

Corollary 8.2.7. *The strong dual of a Fréchet space satisfies the MCC if and only it has a sequential web.*

Proof. Exercise 8.2.7. $\qquad\square$

For yet another example, Exercise 8.2.8 involves using Proposition 8.2.1 above to find a complete LCS that is not strictly barreled but has a sequential web.

Proposition 8.2.2. *A Hausdorff locally complete LCS with a strictly completing web that satisfies the strict Mackey convergence condition SMCC, has a boundedly compatible web.*

Proof. Exercise 8.2.9. $\qquad\square$

Exercises

8.2.1. Prove Corollary 8.2.1.

8.2.2. Prove Corollary 8.2.2.

8.2.3. Let $E = \text{ind}_n E_n$ be a regular (LF) space. Use Proposition 8.2.1 to prove that the following are equivalent:
   (a) E is sequentially retractive.
   (b) E satisfies the Mackey convergence condition (MCC).
   (c) E satisfies the condition of Equation 8.14.

8.2.4. Apply the fact that an (LB) space is regular if and only if it is locally Baire as follows: Deduce from Köthe's example of a nonregular (LB) space (see Theorem 3.5.6) that there exists a strictly webbed LCS that is not locally Baire, thus not even locally complete.

8.2.5. Prove that an uncountable product of Fréchet spaces does not have a compatible web. Note that such a product is Baire (see [28]).

8.2.6. Prove Corollary 8.2.4.

8.2.7. Prove Corollary 8.2.7.

8.2.8. Describe a complete LCS that is not strictly barreled but has a sequential web as follows: There exists a Fréchet space E whose strong dual satisfies the MCC (see [70, Corollary 12.5.9, p. 266]). Using the construction from Theorem 6.3.1(g), form a web on the strong dual. Use relevant properties of polars and of strong duals of metrizable LCS to conclude the desired properties.

8.2.9. Prove Proposition 8.2.2. *Hint:* One implication follows from Theorem 6.3.5. For the other implication, modify the steps of Proposition 8.2.1 accordingly.

8.3 A few applications of the closed graph theorem

A few examples here will indicate some of the utility of the closed graph theorem.

Example 8.3.1. Suppose E and F are Fréchet spaces and $T : E \to F$ is linear. Assume that whenever (x_n) is a null sequence in E and $(T(x_n))$ converges to y in F, then $y = 0$. Then T is continuous.

We only need to show that the graph of T is closed, and then Theorem 8.1.1 applies. As E and F are metrizable, if (u, v) is a closure point of the graph of T, then there exists a sequence (x_n) in E that converges to u and $T(x_n) \to v$. Linearity leads to the calculation that $x_n - u \to 0$ in E, so $T(x_n - u) = T(x_n) - T(u) \to v - T(u)$. By assumption, $v - T(u) = 0$, which leads to $T(u) = v$, and the graph of T is closed. $\square$

Example 8.3.2. Let $H = (H, \langle \cdot \rangle)$ be a Hilbert space and let $T : H \to H$ be a linear operator, with adjoint T^*, appropriately defined. If T is self-adjoint, then T is continuous.

Having H be the domain and range space certainly satisfies the conditions the spaces for Theorem 8.1.1. Let us show that the graph of T is closed. Suppose $x_n \to x$ and $T(x_n) \to y$. The assumption that T is self-adjoint implies that for each $n \in \mathbb{N}$, and for each $h \in H$, $\langle T(h), x_n \rangle = \langle h, T(x_n) \rangle$. By the continuity of the inner product, we have that, for every $h \in H$, $\langle T(h), x \rangle = \langle h, T(x) \rangle = \langle h, y \rangle$, which implies $T(x) = y$.

Proposition 8.3.1. *Equipped with the sup norm, the space of continuously differentiable functions on $[0, 1]$ is not a closed subspace of the space of continuous functions on $[0, 1]$.*

Proof. Exercise 8.3.1. $\square$

8.3.1 Other versions of the closed graph theorem

Like a great piece of music that gets redone in different ways and in different contexts, the closed graph theorem (and resulting open mapping theorem) have been redone in different ways and in different contexts. Indeed, in 1965, Husain [68] wrote a survey of variations of the closed graph theorem, and that was even before the webbed spaces versions appeared. This section contains a fleeting glimpse of the many different closed graph theorems that have been proven.

Using De Wilde's construction (see Definition 8.1.3), Valdivia [124] proved a significant version of the closed graph theorem by using the concept of an ordered web, which we saw in Definition 6.3.7. By way of having only one strand, every Fréchet space has an ordered, strict web. In Example 6.3.2, we saw that every proper (LB) space has an ordered web, which is also a strict web via Theorem 6.3.3(d). In [124], it is assumed that the range space of the closed graph theorem (Theorem 8.3.1 below) is a LCS having an ordered, strict web, called a **quasi-(LB) space**, shown in Definition 8.3.2 below. It is worth noting that the original definition of a quasi-(LB) LCS was given with different but equivalent terminology, which is the gist of Definition 8.3.2. Preceding that definition is a notational one.

Definition 8.3.1. The symbol $\mathbb{N}^{\mathbb{N}}$ refers to the collection of all sequences of positive integers.

Definition 8.3.2. Given $\alpha = (a_n), \beta = (b_n) \in \mathbb{N}^{\mathbb{N}}$, one puts $\alpha \le \beta$ if $a_n \le b_n$ for every $n \in \mathbb{N}$. A **quasi-(LB) representation** on a LCS F is a family of Banach disks $\{A_\alpha : \alpha = (a_n) \in \mathbb{N}^{\mathbb{N}}\}$ such that
(a) $\bigcup\{A_\alpha : \alpha = (a_n) \in \mathbb{N}^{\mathbb{N}}\} = F$;
(b) For any α, β as defined above, if $\alpha \le \beta$, then $A_\alpha \subset A_\beta$.

In this case, the LCS F is called a **quasi-(LB) space.**

In [124, Theorem 4.1, p. 153], the following is proved.

Theorem 8.3.1. *A LCS E is a quasi-(LB) space if and only if, it has an ordered, strict web.*

Corollary 8.3.1. *Every Fréchet space, in particular every Banach space, is quasi-(LB).*

Corollary 8.3.2. *Every (LB) space is quasi-(LB).*

Proof. Let $E = \mathrm{ind}_n(E_n)$, where, for each $n \in \mathbb{N}$, B_n denotes the closed unit ball of the Banach space E_n. Multiplying by appropriate scalars if necessary, one may assume that for all $n \in \mathbb{N}$, one has $B_n \subset B_{n+1}$. The details of verifying that such a structure creates an ordered, strict web are straightforward and are left as Exercise 8.3.2.   □

The resulting closed graph theorem is stated next.

Theorem 8.3.2 (Valdivia's closed graph theorem). *Suppose a linear map $T : E \to F$ from a strictly barreled LCS E to a quasi-(LB) space F has a closed graph. Then T is continuous.*

The main idea of the proof of Theorem 8.3.2 is to consider the following collection: For $k, m_1, m_2, \ldots, m_k \in \mathbb{N}$, let

$$U_{m_1 m_2 \cdots m_k} = T^{-1}\left(\bigcup\{A_\alpha : \alpha = (a_n) \in \mathbb{N}^{\mathbb{N}}, a_n = m_n, n = 1, 2, \ldots, k\}\right). \tag{8.15}$$

One then shows (and you can verify in Exercise 8.3.4) that this collection forms an ordered, absolutely convex strict web on E, and hence, the assumption that E is strictly barreled applies to conclude that there exists a sequence (r_k) of positive integers such that

$$\overline{U_{r_1 r_2 \cdots r_k}} \in \mathcal{N}_{0,E} \tag{8.16}$$

for each $n \in \mathbb{N}$. Full details can be see in [124].

In [124], it is shown that quasi-(LB) spaces have good permanence properties comparable to those of (strictly) webbed spaces, as stated next.

Theorem 8.3.3. *The property of a LCS being quasi-(LB) is preserved under the following formations:*
(a) *Closed subspaces.*
(b) *Continuous linear images.*
(c) *Countable products.*
(d) *Countable direct sums.*
(e) *Every (LF) space is a quasi-(LB) space.*
(f) *The strong dual of an (LF) space is a quasi-(LB) space.*

Proof. Proofs of parts (a) and (b) are straightforward and left as Exercise 8.3.3. For the rest of the parts, please see [124, pp. 151–152]. $\qquad\square$

Another observation is that no specific type of LCS E is assumed for the domain; only the condition shown in Equation (8.15) needs to be satisfied as part of the hypotheses. We have seen a tangible collection of spaces that can serve as the domain space, namely the strictly barreled LCS of Definition 6.3.8. More general LCS as the domain space are considered in [59], for instance. Also, the basic structure of quasi-(LB) spaces can be applied in more general contexts, called **resolutions**. See [40, 43], and [44] for more details and further results regarding resolutions.

At one moment in the proof of Theorem 8.1.2, an expression of the form $U_{K-1} = \overline{T^{-1}(W_{K-1})} \in \mathcal{N}_{0,E}$, where W_{K-1} is a subset of an arbitrary zero neighborhood in F, seemed to imply that the continuity of T had *nearly* been shown. In fact, a generalized version of this statement is part of a separate line of closed graph theorems that will be outlined below. The definition follows.

Definition 8.3.3. Let E and F be TVS, and $T : E \to F$ a linear map. Then T is **nearly continuous** (or, **almost continuous**) if the closure of the inverse image of every zero neighborhood in F is a zero neighborhood in E. That is, for every $V \in \mathcal{N}_{0,F}$, $\overline{T^{-1}(V)} \in \mathcal{N}_{0,E}$.

The next result shows that there is a large collection of domain spaces for which linear maps are nearly continuous.

Theorem 8.3.4. *If E is barreled, then every linear map $T : E \to F$ where F is any LCS, is nearly continuous.*

Proof. If V is any barrel in F, then clearly $\overline{T^{-1}(V)}$ is a barrel in E. By the assumption that E is barreled, we have $\overline{T^{-1}(V)} \in \mathcal{N}_{0,E}$. We finish by noting that F has a base of zero neighborhoods consisting of barrels by way of Theorem 2.1.1. $\qquad\square$

Theorem 8.3.4 above leads to the following closed graph theorem.

Theorem 8.3.5. *If T is a linear map from a barreled LCS to a Fréchet LCS such that T has a closed graph, then T is continuous.*

Proof. A detailed proof is given in [91, Theorem 14.3.4, p. 465]. □

In terms of moving in the opposite direction toward open mappings, there is the following definition.

Definition 8.3.4. Let E and F be TVS and $T : E \to F$ a linear map. Then T is **nearly open** (or, **almost open**) if the closure of the image of every zero neighborhood in E is a zero neighborhood in the subspace $T(E)$ of F. That is, for every $U \in \mathcal{N}_{0,E}$, $\overline{T(U)} \in \mathcal{N}_{0,T(E)}$.

Below is a property of almost open linear maps.

Proposition 8.3.2. *Suppose F is a barreled LCS. Then a linear map from any LCS E onto F is nearly open.*

Proof. Because any LCS has a base consisting of barrels by Theorem 2.1.1, if U is any barreled neighborhood in E and $T : E \to F$ is surjective, then the closure of $T(U)$ is a barrel in F, and consequently, a zero neighborhood. □

In [94, pp. 112–118], Osborne gives proofs of some versions of the closed graph and open mapping theorems. The order of proof in one instance is to first prove an open mapping theorem, then deduce a closed graph theorem from the open mapping theorem. The open mapping theorem obtained is stated below.

Theorem 8.3.6. *[94, Theorem 4.35, p. 112] Suppose E and F are Hausdorff LCS and $T : E \to F$ is linear and continuous. If T is nearly open and E is a Fréchet space, then T is open (hence, surjective).*

The closed graph theorem that follows comes from the work of Pták, [99] (see also [67, Section 3.17]), and involves the following property from duality.

Definition 8.3.5. Suppose that for a linear subspace L of E', whenever M is an absolutely convex, $\sigma(E', E)$-closed equicontinuous subset of E' for which $L \cap M$ is $\sigma(E', E)$-closed, then L itself must be $\sigma(E', E)$-closed. Such a space E is called a **Pták space** (also **fully complete, B complete**).

In [67, pp. 299–300], it is shown that Pták LCS satisfy some general and hereditary properties, for example, that every Fréchet space as well as the strong dual of a Fréchet space are Pták spaces, every Pták space is complete, and that being a Pták space is preserved under the formations of closed subspaces and Hausdorff quotients. On the other hand, there exists a complete LCS that is not a Pták space, and a product of two Pták spaces can fail to be a Pták space, as described in [80, pp. 29–31]. For more details about Pták spaces and their generalization (called **infra-Pták** spaces), see [103, pp. 111–118] or [67, pp. 299–307]. The result below is a version of Pták's closed graph theorem.

Theorem 8.3.7 (Pták's closed graph theorem). *Let E and F be LCS and $T : E \to F$ a linear map. If T is nearly continuous and has a closed graph, and F is a Pták space, then T is continuous.*

One context that can be considered is that of nonlocally convex TVS. One such result for this context comes from W. Robertson's version of webbed spaces, in which the sets of a web are assumed only to be balanced; see [105].

Theorem 8.3.8 (W. Robertson's closed graph theorem). *Suppose E is an inductive limit of Baire TVS, F has a completing web using Equation (6.10) in Definition 6.3.1, and $T : E \to F$ is linear and has a closed graph. Then T is continuous.*

Proof. Exercise 8.3.7. $\qquad\square$

Regarding another closed graph theorem (stated below), a proof is obtained directly, not via an open mapping theorem. In fact, an enlightening discussion of deriving one of these theorems by way of the other can be found in [94, p. 114].

Theorem 8.3.9 ([94, Theorem 4.37, p. 114]). *Suppose E and F are Hausdorff LCS such that E is barreled and F is a Fréchet space. Let $T : E \to F$ be linear, with a closed graph. Then T is continuous.*

In [94, Appendix B, pp. 187–191] proofs of some closed graph theorems in the context of topological groups are given (see Definition A.3.2). Topological groups are more general than LCS, and you can learn more about them in [69].

Theorems 9.3.2 and 9.3.4 of Chapter 9 represent closed graph theorems in the context of a convergence vector space. See Section 9.3 for more details. Closed graph theorems can also be obtained in the context of non-Archimedean LCS, as seen in [96, Theorem 11.1.10, p. 413], [126], and [60]. See Section 9.2 for the definition of a non-Archimedean LCS. We finish this section with a closed graph theorem for quasibarreled LCS (see Definition 5.1.5).

Theorem 8.3.10. *Suppose E is quasibarreled, F is a Fréchet space, and $T : E \to F$ is linear with a closed graph. If T maps bounded sets to bounded sets, then T is continuous.*

Proof. Exercise 8.3.8. $\qquad\square$

Exercises

8.3.1. Prove Proposition 8.3.1. *Hint:* Using the notation of E and F from Example 8.1.2, apply Theorem 8.1.1 and the result of Example 8.1.2.

8.3.2. Strict ordered webs and strict ordered réseaux webs on any nontrivial (LB)-space. Let $E = \mathrm{ind}_n(E_n)$, where for each $n \in \mathbb{N}$, B_n denotes the closed unit ball of the Banach space E_n. Multiplying by appropriate scalars if necessary, assume that for all $n \in \mathbb{N}$, one has $B_n \subset B_{n+1}$.

   (a) Construct an ordered, absolutely convex web on E corresponding to Definition 6.3.1.

   (b) Verify that the following is an ordered, strict réseaux web as follows: Let $E = \bigcup_{n=1}^{\infty} nB_n$, be the first layer. For the second layer, write nB_n as an increasing countable union of Banach disks. Subsequent layers are obtained by repeating this process.

   (c) Prove that each of the constructions of parts (a) and (b) above can be obtained from the other.

8.3.3. Prove parts (a) and (b) of Theorem 8.3.3. *Hint:* Use Theorem 8.3.1.

8.3.4. Prove that the collection $\{U_{m_1 m_2 \cdots m_k} : k, m_1, m_2, \ldots, m_k \in \mathbb{N}\}$ shown in Equation (8.15) forms an ordered, absolutely convex strict web on the LCS E, under the hypotheses of Theorem 8.3.2.

8.3.5. Prove that if E is a barreled LCS and $T : E \to F$ is linear, where F is any LCS, then T is nearly continuous.

8.3.6. Prove that a Pták space remains a Pták space under any finer topology that is consistent with its duality.

8.3.7. Prove Theorem 8.3.8. *Hint:* The proof amounts to adjusting the proof of Theorem 8.1.2 by making use of Equation 6.10.

8.3.8. Prove Theorem 8.3.10. *Hint:* Prove that T is nearly continuous, then apply Theorem 8.3.5.

9 Applying concepts of locally convex spaces

This chapter is about some contexts in which concepts of LCS play an important role. Although there are many possibilities for such topics, we touch on just a few in order to get a cursory look at how the theory of LCS can be applied. The first section includes a look at the space $\mathcal{D}$ of test functions and its dual $\mathcal{D}'$, the classical space of distributions. The theory of distributions is historically important yet still relevant. The construction of distributions presented here is from early developments and is considered to be the basic background for further advancements. Another reason for considering $\mathcal{D}$ and its dual is that the area of locally convex spaces began to draw significant attention due to the impact of such early work on distribution theory, done notably by L. Schwartz and S. Sobolev. The space $\mathcal{D}$ of test functions turns out to be nonmetrizable, implying that $\mathcal{D}$ is a LCS that is strictly outside the realm of normed spaces. The rest of this chapter amounts to taking a step back and contemplating the overall concept of a (locally convex) topological vector space. Indeed, a TVS as discussed in Chapter 1 is nothing less than a marvelous interconnection between linear algebra and topology. We can therefore ask what happens when we change one or the other of these two contexts. In Section 9.2, we look at an example from the linear algebra aspect of LCS, specifically of what happens when we replace the scalar field $\mathbb{K}$ with some other kind of scalar field. In this case, one that does not satisfy the Archimedean property. In Section 9.3, an example from the topological aspect of LCS is considered, in which we replace the topology with something more general, that being of a convergence structure. In each of these two sections, we encounter some aspects that are essentially the same as those we have seen in earlier chapters, and we also encounter some aspects that are strikingly different than what we have previously seen.

9.1 An overview of distribution theory

In this section, we examine the LCS $\mathcal{D}$ and its dual, and present a brief summary of distribution theory. Detailed discussions of distribution theory and its applications can be seen in [20, Chapter 6], [108], [67, Chapter 4], [107, 119], and others. In general, $\mathcal{D}$ and its dual $\mathcal{D}'$ are defined over $\mathbb{C}^n$, though these LCS can be defined over other fields. We will use $\mathbb{R}$ as our field in this section. Recall from Example 3.5.5 and Definition 3.5.4 that $\mathcal{D} = \mathrm{ind}_n \mathcal{D}(K_n)$, where $\mathcal{D}(K_n) = \{f \in C^\infty(\mathbb{R}) : \mathrm{supp}(f) \subset K_n\}$, where we use $K_n = [-n, n]$. We have seen that certain LCS such as $(l_1, \sigma(l_1, l_\infty))$ have intriguing properties. In fact, $\mathcal{D}$ and its strong dual are also fascinating LCS, as summarized in Theorems 9.1.1 and 9.1.2 below.

Theorem 9.1.1. *The space $\mathcal{D}$ of test functions is:*
(a) *Sequentially retractive, hence is regular as an inductive limit. In fact, $\mathcal{D}$ is Hausdorff. Additionally, $\mathcal{D}$ is a hyperstrict (LF) space.*

https://doi.org/10.1515/9783111392868-009

(b) *Complete.*
(c) *Nonmetrizable.*
(d) *Barreled.*
(e) *Bornological.*
(f) *Montel, hence reflexive.*
(g) *A LCS that satisfies property $\mathcal{K}$.*
(h) *A LCS that satisfies the SMCC, hence also the MCC.*
(i) *A LCS that satisfies Schur's property.*
(j) *Strictly webbed.*

Proof. The proof of all these properties can be described by you in Exercise 9.1.3, by citing relevant results we have seen thus far. $\qquad\qquad\square$

For even more observations about $\mathcal{D}$, see [18, Section 2.10, pp. 139–143]. Meanwhile, we ask ourselves the question: As interesting as $\mathcal{D}$ is, what would be the advantage to equipping $\mathcal{D}$ with the nonmetrizable inductive limit topology of Definition 3.5.4? Indeed, an alternative locally convex topology on $\mathcal{D}$ turns out to be metrizable, as described next.

Definition 9.1.1. For each $\varphi \in \mathcal{D}$ and $n \in \mathbb{N}$, define

$$\|\varphi\|_n = \max\{|\varphi^{(j)}(x)| : x \in \mathbb{R}, j = 0, 1, \ldots, n\}, \tag{9.1}$$

where $\varphi^{(j)}$ represents the jth derivative of φ.

Because $\operatorname{supp}(\varphi)$ is compact for every $\varphi \in \mathcal{D}$, it is easy to see that $\|\cdot\|_n$ defines a norm on $\mathcal{D}$, for each $n \in \mathbb{N}$. Applying Theorem 2.2.1 to the countable collection $\{\|\cdot\|_n : n \in \mathbb{N}\}$, we observe that $\mathcal{D}$ becomes a metrizable, nonnormable LCS under the topology generated by this collection of norms. Is this metric "better" than the inductive limit topology? Well, it turns out that with this metric, $\mathcal{D}$ is not complete, as you can verify in Exercise 9.1.2. An observation answers the question: *In general it is preferable to have the predictability of completeness than the convenience of metrizability.* Thus, the inductive limit topology is used on $\mathcal{D}$, rather than the metrizable topology of Definition 9.1.1. Notice that this conclusion represents another example of the utility of locally convex spaces.

The dual of $\mathcal{D}$ is considered next.

Definition 9.1.2. A **(Schwartz–Sobolev) distribution** is an element of the continuous dual of the space of test functions, and is denoted by $\mathcal{D}'$.

The LCS $\mathcal{D}'$ is typically equipped with the strong topology, which we clarify below.

Notation 9.1.1. The symbol $\mathcal{D}'$ denotes the dual of $\mathcal{D}$ equipped with the strong topology; that is, $\beta(\mathcal{D}', \mathcal{D})$.

As good (mathematical) luck would have it, $\mathcal{D}'$ is also pleasantly intriguing, as summarized below.

Theorem 9.1.2. *The LCS $\mathcal{D}'$ with its strong topology is:*

(a) *Complete*

(b) *Strictly webbed*

(c) *Barreled*

(d) *Bornological*

(e) *Montel, hence reflexive*

(f) *A LCS that satisfies Schur's property.*

Proof. Properties (a)–(c) and (f) can be proven by citing relevant results that were developed in earlier parts of this book, as you can detail in Exercise 9.1.4. A proof of property (d) can be deduced from a result of Grothendieck, as seen in [67, Theorem 3.16.2, pp. 290–291], and a proof of property (e) can be seen in [67, Proposition 3.9.9, p. 236]. Property (f) can be deduced from a slightly stronger property, stated below. $\qquad\square$

Below is the first example of a distribution. It is perhaps historically the most important example because it motivated mathematicians to think deeply about what constitutes a function.

Definition 9.1.3. For a fixed $a \in \mathbb{R}$, define the **Dirac delta function** δ_a by

$$\delta_a(x) = \begin{cases} 0, & \text{if } x \neq a \\ \infty, & \text{if } x = a \end{cases}. \tag{9.2}$$

Physicists had used δ_a as if it were a function having the property that $\int_{\mathbb{R}} \delta(x)\, dx = 1$ to describe phenomena such as point mass or point charges. Mathematically speaking, δ_a is certainly not a function in the algebraic sense. However, the development of distribution theory by way of LCS structures of $\mathcal{D}$ and $\mathcal{D}'$ allows us to express δ_a as an element of $\mathcal{D}'$, as follows.

Definition 9.1.4. Let Ω be an open subset of $\mathbb{R}$ such that $a \in \Omega$. Define the distribution δ_a by: For each $\varphi \in \mathcal{D}$,

$$\delta_a(\varphi) = \langle \varphi, a \rangle = \varphi(a), \tag{9.3}$$

where we use the notation $< \cdot, \cdot >$ of an evaluation map (see Definition 4.2.2) from duality and in general write $T_f(\varphi) = \langle \varphi, f \rangle$. For $a = 0$, one writes δ in place of δ_0.

Clearly, δ_a is linear. In Exercise 9.1.13, you can provide a proof that δ_a is continuous; that is, $\delta_a \in \mathcal{D}'$.

Now that we have defined a distribution, we can involve continuous functions in calculations of distributions. That is the gist of the next example.

Example 9.1.1. Applying distributions to continuous functions.

Given a real valued continuous function f, we wish to apply an element of $\mathcal{D}'$ to f in a way that makes mathematical sense. To this end, we observe the following. For a function $\varphi \in \mathcal{D}$ such that $\mathrm{supp}(\varphi) \subset [-n, n]$, clearly $\mathrm{supp}(f\varphi) \subset [-n, n]$, which allows us to define $f\varphi = 0$ on the complement of $[-n, n]$ in $\mathbb{R}$. Because we assume f is continuous we may extend $f\varphi$ to a continuous function on $\mathbb{R}$. The integral $\int_{\mathbb{R}} f(x)\varphi(x)\, dx$ therefore exists. We now define a linear functional as follows: Given a continuous function f on $\mathbb{R}$, define $T_f : \mathcal{D} \to \mathbb{R}$ by: For every $\varphi \in \mathcal{D}$,

$$T_f(\varphi) = \langle \varphi, f \rangle = \int_{\mathbb{R}} f(x)\varphi(x)\, dx. \tag{9.4}$$

By definition, T_f is linear, and in Exercise 9.1.12 you will have an opportunity to prove that T_f is, in fact, continuous; that is, for any continuous function f on $\mathbb{R}$, $T_f \in \mathcal{D}'$. We give a formal name to T_f as stated below.

Definition 9.1.5. The element T_f of $\mathcal{D}'$, evaluated as in Equation (9.4), is called the **distribution associated with the function** f.

In general, one may define distributions for functions that are locally integrable, as defined below. The definition assumes some facts from Lebesgue measure and integration.

Definition 9.1.6. A function f is **locally integrable** on a nonempty open subset Ω of $\mathbb{K}^n$ if it is Lebesgue integrable on every compact subset of Ω.

Our next example requires a preliminary definition.

Definition 9.1.7. The **Cauchy principal value** T_C is defined by ($\forall \varphi \in \mathcal{D}$)

$$\langle T_C, \varphi \rangle = \lim_{\varepsilon \to 0} \int_{|x| \geq \varepsilon} \frac{\varphi(x)}{x}\, dx. \tag{9.5}$$

Example 9.1.2. The Cauchy principal value as a distribution.

The question arises as to whether this does indeed define a distribution. Thus, this example depends on proving the existence of the limit stated in Equation (9.5), which will be done via Proposition 9.1.1. The proof below establishes weak convergence. That the limit converges in the strong topology of $\mathcal{D}'$ follows by applying Schur's property to $\mathcal{D}'$ via Theorem 9.1.2(f).

Proposition 9.1.1. *For each* $\varphi \in \mathcal{D}$,

$$\lim_{\varepsilon \to 0} \int_{|x| \geq \varepsilon} \frac{\varphi(x)}{x}\, dx \tag{9.6}$$

exists.

Proof. One uses a continuous function η such that $\varphi(x) = \varphi(0) + x\eta(x)$ and $\eta(0) = \varphi'(0)$, and then uses η to calculate the limit shown in Equation (9.6). The rest of the details are left as Exercise 9.1.6. □

The next several results reveal that we can perform operations from calculus in the context of distributions.

To differentiate a function using distributions, we make a preliminary calculation as follows.

Proposition 9.1.2. *Suppose f is a real valued differentiable function. Applying integration by parts to the expression $\langle f', \varphi \rangle = \int_{\mathbb{R}} f'(x)\varphi(x)\, dx$, we obtain*

$$\langle f', \varphi \rangle = -\int_{\mathbb{R}} f(x)\varphi'(x)\, dx. \tag{9.7}$$

Proof. Exercise 9.1.14. □

Observe that we may write this differentiation symbolically as $(T_f)'(\varphi) = -T_f(\varphi')$. Because the right-hand side of this expression is independent of the differentiability of the function f, we may make this calculation for general functions defined on $\mathbb{R}$. Astoundingly, this applies even when f is not differentiable. In fact, we may repeat this process for all orders n. We summarize this in the next definition.

Definition 9.1.8. Let $T \in \mathcal{D}'$ be any distribution. The **distributional derivative of T** is defined by: For all $\varphi \in \mathcal{D}$,

$$\langle T', \varphi \rangle = -\langle T, \varphi' \rangle. \tag{9.8}$$

The nth **distributional derivative** $T^{(n)}$ is determined by: For all $\varphi \in \mathcal{D}$,

$$\langle T^{(n)}, \varphi \rangle = (-1)^n \langle T, \varphi^{(n)} \rangle. \tag{9.9}$$

Distributional differentiation satisfies the typical linearity properties as seen in calculus. The details are next.

Proposition 9.1.3. *Given any $S, T \in \mathcal{D}'$, $(S + T)' = S' + T'$, and for any $a \in \mathbb{K}$, $(aT)' = aT'$.*

Proof. Exercise 9.1.17. □

Regarding integration, we may think in terms of the inverse process of differentiation. As an example, consider the function defined below.

Definition 9.1.9. The **Heaviside function H** is defined by

$$H(x) = \begin{cases} 1, & \text{if } x \geq 0 \\ 0, & \text{if } x < 0 \end{cases}. \tag{9.10}$$

Of course, we can translate the Heaviside function to obtain $H_a(x) = H(x - a)$, for any $a \neq 0$ in $\mathbb{R}$.

Example 9.1.3. The Heaviside function as a distribution.

What needs to be established is that $H' = \delta$ in terms of derivatives of distributions per Definition 9.1.8. You can verify this as Exercise 9.1.16.

Third, we consider convergence of a series in $\mathcal{D}$.

Example 9.1.4. A convergent series of distributions.

Let (a_n) be any series whatsoever of real (or complex) numbers. Then the series $\sum_{-\infty}^{\infty} a_n \delta_n$ converges in $\mathcal{D}'$. To see this, note that any $\varphi \in \mathcal{D}$ has finite support, which implies that the series reduces to a finite sum. Thus, the series converges in $\mathcal{D}'$. Observe that this holds even if the series $\sum_{-\infty}^{\infty} a_n$ diverges.

It is possible to determine distributions as limits, as seen next.

Example 9.1.5. Approximating the delta distribution.

Using $a = 1$ in Example 3.5.6 leads to an element φ of $\mathcal{D}$ such that $\mathrm{supp}(\varphi) \subset [-1, 1]$. Dividing φ by the constant $\int_{\mathbb{R}} \varphi(x)dx$ produces an element ψ that satisfies:
(i) ψ is infinitely differentiable on $\mathbb{R}$.
(ii)   The support of ψ is contained in $[-1, 1]$, hence $\psi \in \mathcal{D}$.
(iii) $\psi(x) > 0$ on $(-1, 1)$, and in general, $\psi(x) \geq 0$ on $\mathbb{R}$.
(iv) $\int_{\mathbb{R}} \psi(x)dx = 1$.

Let $\varepsilon > 0$. If we define $\psi_\varepsilon(x) = \frac{1}{\varepsilon}\psi(\frac{x}{\varepsilon})$, we see that the support of ψ_ε is contained in $[-\varepsilon, \varepsilon]$ and $\psi_\varepsilon(x) > 0$ on $(-\varepsilon, \varepsilon)$. We now show that this set of functions converges to δ as $\varepsilon \to 0$, in the next item.

Proposition 9.1.4. *For ψ_ε as constructed above,* $\lim_{\varepsilon \to 0} \psi_\varepsilon = \delta$.

Proof. Let $\varphi \in \mathcal{D}$ be given. Apply item (iv) above to obtain the following expression:

$$\langle \psi_\varepsilon, \varphi \rangle - \langle \delta, \varphi \rangle = \langle \psi_\varepsilon, \varphi \rangle - \varphi(0) = \int_{\mathbb{R}} \psi_\varepsilon(x)(\varphi(x) - \varphi(0))\, dx. \tag{9.11}$$

Using other properties of ψ_ε, you can verify in Exercise 9.1.19 that Equation (9.11) leads to

$$\left| \int_{\mathbb{R}} \psi_\varepsilon(x)\varphi(x)\, dx - \varphi(0) \right| \leq \max\{|\varphi(x) - \varphi(0)| : |x| \leq \varepsilon\}. \tag{9.12}$$

By the continuity of φ, the limit as $\varepsilon \to 0$ of the maximum shown in Equation (9.12) is 0, which implies that $\psi_\varepsilon \to \delta$ as $\varepsilon \to 0$ with respect to $\sigma(\mathcal{D}', \mathcal{D})$. Observing Theorem 9.1.2(f), we conclude that $\lim_{\varepsilon \to 0} \psi_\varepsilon = \delta$ with respect to $\beta(\mathcal{D}', \mathcal{D})$. $\qquad\square$

We define a term for convergence with respect to $\beta(\mathcal{D}', \mathcal{D})$ below.

Definition 9.1.10. Convergence with respect to $\beta(\mathcal{D}', \mathcal{D})$ is said to be **distributional**.

One final comment for this section: The recent publication [5] contains results that are directly connected to concepts of distributions observed here.

Exercises

9.1.1. Prove that the expression of Equation (9.1) defines a norm and that the topology generated by the collection all such norms is not normable.

9.1.2. Prove that $\mathcal{D}$, with the metric defined in Definition 9.1.1 is not complete, as follows. Fix $\varphi \in \mathcal{D}$ such that $\mathrm{supp}(\varphi) \subset [0,1]$, $\varphi(x) > 0$ for $x \in (0,1)$, and define the sequence

$$\Phi_n(x) = \varphi(x-1) + \frac{1}{2}\varphi(x-2) + \cdots + \frac{1}{n}\varphi(x-n), \tag{9.13}$$

for each $n \in \mathbb{N}$. Prove that (Φ_n) is a Cauchy sequence with respect to the metric of Equation (9.1) but does not converge in $\mathcal{D}$.

9.1.3. Prove Theorem 9.1.1 by citing relevant results from earlier chapters.

9.1.4. Prove parts (a), (b), (c), and (f) of Theorem 9.1.2, by citing relevant results from earlier chapters. *Hint:* For part (f), use Theorem 9.1.2 and Proposition 5.3.8.

9.1.5. Prove that in fact $\mathcal{D}$ is strictly barreled (see Definition 6.3.8).

9.1.6. Complete the proof of Proposition 9.1.1.

9.1.7. Prove that for $\varepsilon > 0$, if T_ε is defined for all $\varphi \in \mathcal{D}$ by

$$\langle T_\varepsilon, \varphi \rangle = \int\limits_{|x|>\varepsilon} \frac{\varphi(x)\,dx}{x}, \tag{9.14}$$

then $T_\varepsilon \in \mathcal{D}'$.

9.1.8. Prove that for $f,g \in C(\mathbb{R})$, if $(\forall \varphi \in \mathcal{D})$, one has $\langle f,\varphi \rangle = \langle g,\varphi \rangle$, then $f = g$, where $\langle f,\varphi \rangle = \int_{\mathbb{R}} f(x)\varphi(x)dx$. *Hint:* Proceed by contradiction, assuming $f \neq g$. Use the fact that on an open interval (a,b), if

$$\psi(x) = \begin{cases} \exp[\frac{1}{a-x} + \frac{1}{x-b}], & \text{if } x \in (a,b), \\ 0 & \text{otherwise} \end{cases} \tag{9.15}$$

then $\psi \in \mathcal{D}$.

9.1.9. Suppose $\varphi \in \mathcal{D}$. Prove that if ψ is defined by the relation $x\psi(x) = \varphi(x) - \varphi(0)$, then $\psi(0) = \varphi'(0)$.

9.1.10. For each $n \in \mathbb{N}$ define f_n by

$$f_n(x) = \begin{cases} \frac{n}{2}, & \text{if } |x| < \frac{1}{n}, \\ 0 & \text{otherwise} \end{cases}. \tag{9.16}$$

Prove that (f_n) converges to δ distributionally according to Definition 9.1.10; that is, $\langle f_n, \varphi \rangle \to \varphi(0)$ as $n \to \infty$, for all $\varphi \in \mathcal{D}$.

9.1.11. For each $n \in \mathbb{N}$ let f_n be defined as follows:

$$f_n(x) = \begin{cases} n^2, & \text{if } |x| < \frac{1}{n}, \\ 0 & \text{otherwise} \end{cases}. \tag{9.17}$$

(a) Prove that (f_n) diverges distributionally according to Definition 9.1.10; that is, for each $\varphi \in \mathcal{D}$ such that $\varphi(0) \neq 0$, $\langle f_n, \varphi \rangle$ diverges in $\mathbb{R}$.

(b) (This part assumes some familiarity with convergence almost everywhere) Prove that (f_n) converges pointwise to zero almost everywhere in $\mathbb{R}$.

9.1.12. Prove that T_f as defined in Equation (9.4), is in fact, continuous.

9.1.13. Prove that δ_a of Definition 9.1.4 is continuous. *Hint:* Consider any $K \subset \Omega$ such that K is compact, and show that $|\langle \varphi, a \rangle| \le \sup\{|\varphi(x)| : x \in K\}$, for all $\varphi \in \mathcal{D}$.

9.1.14. Prove Proposition 9.1.2. *Hint:* Just check that the integration by parts correctly yields Equation (9.7) and that the integral over $\mathbb{R}$ is defined.

9.1.15. Prove that $\delta^{(n)} = (-1)^n \varphi^{(n)}(0)$.

9.1.16. Prove that $H' = \delta$ from Example 9.1.3.

9.1.17. Prove Proposition 9.1.3.

9.1.18. Let a be a nonzero constant. Prove that $\delta' \sin(ax) = -a\delta$.

9.1.19. Prove that Equation (9.11) implies Equation (9.12).

9.2 Non-Archimedean locally convex spaces

In the context of analysis, the Archimedean property is quite important. On the other hand, in general mathematical contexts one may consider structures that are more abstract than the usual fields of scalars, including those that do not satisfy the Archimedean property. In this section, we briefly explore LCS over fields that are so-called non-Archimedean. The primary reference for the LCS aspect of this section is the book by Pérez-García and Schikhof [96]. Others include: Van Rooij [125], Schneider [112], Prolla [98], and further references therein. It is also worth noting that there are many exercises throughout [91] regarding concepts from this section. Many such exercises include explanations of the concepts. The relevance of studying LCS in such an atypical context is that there are applications of non-Archimedean LCS to areas such as mathematical physics, which can be viewed in [3] as well as applications of non-Archimedean analysis to areas such as Navier–Stokes equations and wavelet theory, an example of which can be seen in [97]. To start this exploration, we now define some terminology.

Definition 9.2.1. Let K be any commutative field. A function $|\cdot| : K \to [0, \infty)$ is a **valuation** (or **absolute value function**) if for every $a, b \in K$, the following items hold:
(a) $|a| = 0$ if and only if $a = 0$.
(b) $|ab| = |a||b|$.
(c) $|a + b| \le |a| + |b|$.

For our purposes, we want any valuation to be complete, which is formalized in the next statement.

Assumption 9.2.1. Any valuation considered in this section is assumed to be complete with respect to the metric generated by that valuation.

Obviously, the fields of $\mathbb{R}$ and $\mathbb{C}$ with the standard absolute value represent commutative fields with valuations. What interests us now is when the valuation satisfies something stronger than the usual triangle inequality of item (c) of Definition 9.2.1, as enunciated next.

Definition 9.2.2. A valuation on K satisfies the **strong triangle inequality** (or **ultrametric triangle inequality**) if for every $a, b \in K$,

$$|a + b| \le \max\{|a|, |b|\}. \tag{9.18}$$

Recall that the essence of the Archimedean property in $\mathbb{R}$ is that the set of positive integers is unbounded. By replacing the traditional triangle inequality with the strong triangle inequality of Equation (9.18), observe that $|n| = |1 + 1 + \cdots + 1| = 1$, which implies that the set of positive integers *is* bounded with respect to this valuation. Such a valuation can be labeled as non-Archimedean, and this term is formalized below.

Definition 9.2.3. A valuation that does not satisfy the Archimedean property is called a **non-Archimedean valuation** and the corresponding field having a non-Archimedean valuation will be referred to as a **non-Archimedean field**.

The non-Archimedean valuation that we will use here is that given in Definition 9.2.2. To distinguish between the notation $\mathbb{K}$ that represents $\mathbb{R}$ or $\mathbb{C}$ with the standard (Archimedean) absolute value and a non-Archimedean valuation, we will use a different style of the letter K, as indicated below.

Notation 9.2.1. Items related to a non-Archimedean field will be prefixed by $\mathfrak{K}$.

Before proceeding, it behooves us to exhibit a nontrivial example of a non-Archimedean valuation, as done next.

Example 9.2.1. The p-adic valuation on $\mathbb{R}$.

Consider any positive prime number p. From number theory, any rational number q can be written in the form $q = p^k \cdot \frac{a}{b}$ where p, a, and b are integers such that p divides neither a nor b. This allows us to formally define the following.

Definition 9.2.4. The p-**adic valuation** is defined for a prime $p > 0$ by $|q|_p = p^{-k}$, where q and k are as in Example 9.2.1. The completion of the rational numbers with respect to the metric generated by $|\cdot|_p$ is referred to as the set of p-**adic numbers**, and is denoted by $(\mathbb{Q}_p, |\cdot|_p)$.

Example 9.2.1 and Definition 9.2.4 above are important in the realm of number theory. Such importance explains why functional analysis such as that of TVS over non-Archimedean fields is often referred to as p-**adic functional analysis**.

As we will see shortly, statements of definitions and results in a non-Archimedean context can seem to only differ from the Archimedean statement by the addition of the term "non-Archimedean." The next few items reveal that the differences between Archimedean and non-Archimedean are deeper than just an inclusion of an extra term. In the theorem below, refer as necessary to Definition A.3.8 regarding the concept of a zero dimensional topological space.

Theorem 9.2.1. *Let $\mathfrak{K}$ be a non-Archimedean field with a valuation denoted by $|\cdot|$. For each $a, b, c \in \mathfrak{K}$:*
(a) *If $|a| > |b|$, then $|a + b| = |a|$.*
(b) *Every point of an open (or closed) ball is in fact a center of that open (closed) ball.*
(c) *(Mercury drops property) Two balls are either disjoint or the ball of smaller radius is contained in that having the larger radius.*
(d) *As a topological space, $(\mathfrak{K}, |\cdot|)$ is zero-dimensional.*

Proof. (a) Assume $|a| > |b|$. First, this assumption implies that $|a+b| \leq \max\{|a|, |b|\} = |a|$. Second, $|a| = |a + b - b| \leq \max\{|a + b|, |b|\}$. By hypothesis, it is impossible for $|a| \leq |b|$, so it must be that $|a| \leq |a + b|$.

(b) We show this property for open balls, the case of closed balls requiring only minor adjustments. Consider an open ball $B_r(a)$ in $\Re$, and suppose $b \in B_r(a)$. For any $x \in B_r(b)$, we calculate: $|x - a| = |x - b + b - a| \leq \max\{|x - b|, |b - a|\} < r$, which tells us that $x \in B_r(a)$. Thus, $B_r(b) \subset B_r(a)$. The proof of the reverse inclusion is similar, and we conclude that $B_r(b) = B_r(a)$.

(c) For the case of open balls, consider $B_r(a)$ and $B_s(b)$ such that $0 < r < s$ and that $B_r(a) \cap B_s(b) \neq \emptyset$. Let $x \in B_r(a) \cap B_s(b)$ be arbitrary. Then by part (b), $B_r(a) = B_r(x) \subset B_s(x) = B_s(b)$.

(d) By part (b), every closed ball is open. Now consider any open ball $B_s(a)$. We will show that its complement is open. To that end, suppose $b \notin B_s(a)$. Find r such that $0 < r < s$. Then $B_r(b) \cap B_s(a) = \emptyset$. Indeed, if $B_r(b) \cap B_s(a) \neq \emptyset$, then part (c) would imply that $b \in B_r(b) \subset B_s(a)$, a contradiction. $\qquad\square$

Surely you recall from calculus that a sequence in $\mathbb{R}$, which converges to zero does not guarantee that the series of its terms converges. In the context of non-Archimedean analysis, a null sequence always generates a convergent series, as examined next.

Proposition 9.2.1. *In a non-Archimedean field, a series of terms converges if and only if the corresponding sequence converges to zero.*

Proof. Exercise 9.2.4. $\qquad\square$

The next definition is about vector spaces over non-Archimedean fields.

Definition 9.2.5. A vector space that is defined over a non-Archimedean field is called a **non-Archimedean vector space** (also $\Re$-**vector space**).

A main goal of this section is to define the necessary concepts for TVS, and in particular LCS, when defined over non-Archimedean fields. To get such a definition, we need to modify some concepts of the Archimedean context in order for such concepts to be useful in the non-Archimedean context. The first concept has to do with separability. The standard definition of a separable topological space is that the space has a countable dense subset. What is typically used in the context of TVS (such as normed spaces), however, is that showing a space is separable amounts to detecting a countable subset such that its *linear span* is dense in the space. Because $\mathbb{R}$ and $\mathbb{C}$ are both separable as scalar fields, these two concepts of separability are equivalent in the Archimedean context. In the general non-Archimedean context, we need to modify the definition, and that is the essence of the next item.

Definition 9.2.6. A normed space is of **countable type** if it contains a countable subset whose linear span is dense in the space.

For the record, there are non-Archimedean fields that are not of countable type; see [96, Remark 4.2.7, p. 176] for one such example.

Before defining a non-Archimedean LCS, we need to define the concept of convexity and take a look at some properties of such convexity. This is accomplished via the next several definitions, in which you will see some differences between the Archimedean and non-Archimedean convexity concepts.

Definition 9.2.7. For a non-Archimedean field $\mathfrak{K}$ with valuation $|\cdot|$, the **closed unit ball** is $B_{\mathfrak{K}} = \{a \in \mathfrak{K} : |a| \leq 1\}$.

Definition 9.2.8. A subset A of a non-Archimedean vector space is **absolutely convex** if for every $a, b \in B_{\mathfrak{K}}$ and for every $x, y \in A$, $ax + by \in A$.

Definition 9.2.9. A nonempty subset of a non-Archimedean vector space is **convex** if it is a translate of an absolutely convex set.

A few observations are in order. First, observe that a convex set is absolutely convex if and only if it contains 0. Second, notice that the definition of absolutely convex appears before the definition of convex. In [125, p. 87], a convex set in a non-Archimedean vector space is defined directly, using three elements; that is, a nonempty subset A of such a vector space is termed *convex* if for all $x, y, z \in A$ and all $a, b, c \in \mathfrak{K}$ for which $a + b + c = 1$, one has $ax + by + cz \in A$. It seems extraneous to have three rather than two elements involved in the definition of convexity, however, this alternate definition is equivalent to Definition 9.2.9, as is verified next. For more detailed information about this situation, see [96, Theorem 3.1.17, p. 89].

Theorem 9.2.2. *A nonempty subset A of a non-Archimedean vector space E over $\mathfrak{K}$ is convex if and only if for all $x, y, z \in A$ and all $a, b, c \in \mathfrak{K}$ that satisfy $a + b + c = 1$, the following holds:*

$$ax + by + cz \in A. \tag{9.19}$$

Proof. ($\Rightarrow$): Assume A is convex according to Definition 9.2.9. Because $A \neq \emptyset$, $A = w + D$, where $w \in E$ and $D \subset E$ is absolutely convex (by Definition 9.2.8). This allows us to write $x = w + u_1, y = w + u_2, z = w + u_3$, where $u_1, u_2, u_3 \in D$. Now calculate $ax + by + cz = a(w + u_1) + b(w + u_2) + c(w + u_3) = w(1) + au_1 + bu_2 + cu_3 \in w + D = A$.

($\Leftarrow$): Exercise 9.2.5. $\qquad\square$

The following properties hold for convex subsets of non-Archimedean vector spaces.

Proposition 9.2.2. *In a non-Archimedean vector space, convexity is preserved under the following formations:*
(a) *Linear images and inverse images*
(b) *Finite sums*

(c) *Linear subspaces and translations of such subspaces*
(d) *For the non-Archimedean field $\mathfrak{K}$, the convex sets are precisely: $\emptyset$, $\mathfrak{K}$, balls, and singletons.*

Proof. Exercise 9.2.9. □

With the concepts of convex and absolutely convex sets, we can define hulls, as done in the Archimedean case. We first need the following property.

Proposition 9.2.3. *In a vector space over a non-Archimedean field, arbitrary intersections of absolutely convex sets are absolutely convex, and arbitrary intersections of convex sets are convex.*

Proof. Exercise 9.2.10. □

Definition 9.2.10. For a subset of a non-Archimedean vector space, the **absolutely convex hull** of the subset is the intersection of all absolutely convex supersets of the given set, and the **convex hull** of the subset is the intersection of all convex supersets of the given set.

The next task is to define the concept of a LCS when the corresponding field is non-Archimedean. First is a generic definition of a TVS in such a context.

Definition 9.2.11. A **topological vector space** over a non-Archimedean field is a vector space with a topology such that vector addition and scalar multiplication are continuous functions.

Normed spaces can be defined as first examples of non-Archimedean TVS.

Definition 9.2.12. A **non-Archimedean norm** on a non-Archimedean vector space E over $\mathfrak{K}$ is a function $\|\cdot\| : E \to [0, \infty)$ such that for every $x, y \in E$ and every $a \in \mathfrak{K}$:
(a) $\|x\| = 0$ if and only if $x = 0$.
(b) $\|ax\| = |a|\|x\|$.
(c) $\|x + y\| \leq \max\{\|x\|, \|y\|\}$.

$(E, \|\cdot\|)$ is called a **non-Archimedean normed space**. A **non-Archimedean Banach space** is a complete non-Archimedean normed space.

Definition 9.2.13. A **non-Archimedean locally convex space** is a non-Archimedean topological vector space that has a base of zero neighborhoods that are convex.

In words, the definition of a non-Archimedean LCS looks like that for the Archimedean case, though we will see some differences in the paragraphs preceding the exercises of this section. It turns out that seminorms can be used to generate the topology of a non-Archimedean LCS, so we define such non-Archimedean seminorms next.

Definition 9.2.14. A **non-Archimedean seminorm** on a $\mathfrak{K}$-vector space E is a function $p : E \to [0, \infty)$ that satisfies the following properties for every $x, y \in E$ and every $a \in \mathfrak{K}$:
(a) $p(x + y) \le \max\{p(x), p(y)\}$.
(b) $p(ax) = |a|p(x)$.

With some differences (see [96, 3.1.1, p. 83; 2.1.10, p. 19]), the topology of a non-Archimedean LCS can be generated by a collection of non-Archimedean seminorms, as was done by way of Theorem 2.1.4 for the Archimedean case. For full details of the non-Archimedean version of this result, see [96, Chapter 3].

9.2.1 The Helly–Hahn–Banach theorem for non-Archimedean spaces

To give an indication of a deep non-Archimedean result, we now consider the Helly–Hahn–Banach theorem in a non-Archimedean context. A key concept related to this result involves completeness which we now examine. To start, straightforward calculations show that any metric space is complete if and only if every decreasing sequence of closed balls whose diameters shrink to zero has nonempty intersection. In the non-Archimedean context, it can happen that such an intersection is, in fact, empty, and this prompts the consideration of a stronger form of completeness. Such a stronger completeness is indicated in the next definition.

Definition 9.2.15. A non-Archimedean normed space is **spherically complete** if every decreasing sequence of balls has nonempty intersection; that is, for $B_1 \supset B_2 \supset \cdots$, where each B_n is a nonempty ball, $\bigcap_{n=1}^{\infty} B_n \ne \emptyset$.

For $\mathbb{R}$ and $\mathbb{C}$ in the Archimedean setting, the property of Definition 9.2.15 always holds. In the non-Archimedean context, this property does not always hold. In [112, pp. 3–6], you can find detailed descriptions of examples of non-Archimedean fields that are not spherically complete as well as others that are spherically complete. The importance of spherical completeness is that the Archimedean proof of the Helly–Hahn–Banach theorem indirectly assumes that such a decreasing sequence of balls is not empty. Hence, in the non-Archimedean context it can in fact happen that such an intersection is empty, as witnessed by the existence of fields that are not spherically complete. More to the point (pun intended), we will see how spherical completeness is explicitly used in the proof of Theorem 9.2.3 below. Another observation is that a spherically complete field is complete, and not surprisingly, the converse is false; see [96, Theorem 1.2.12, p. 10] for the details. With these comments in mind, we now proceed to the main result of this subsection. The result is first established for the algebraic dual and then adapted to the continuous dual.

Theorem 9.2.3 (Non-Archimedean Helly–Hahn–Banach theorem). *Suppose E is a vector space over a non-Archimedean field $\mathfrak{K}$ such that $\mathfrak{K}$ is spherically complete. Let p be a* *non-Archimedean seminorm on E. For every subspace L of E and every $f \in E^*$ such that $|f| \leq p$ on L, there exists a linear extension $\tilde{f} \in E^*$ of f such that $|\tilde{f}| \leq p$ on E.*

Before the proof, a couple of comments are relevant. First, Theorem 9.2.3 would be considered to be an analytic version. Corresponding geometric versions can be found in [96, Chapter 4]. Also, the proof here is based on [96, Theorem 4.1.1]. A different proof can be seen in [112, Proposition I.9.2, p. 51].

Proof. By way of an application of Zorn's Lemma A.1.1, we can assume that there exists $x_0 \in E \setminus L$ such that $E = L + \{ax_0 : a \in \mathfrak{K}\}$. We seek y_0 for which we define $\tilde{f}(x_0) = y_0$, and this will complete the proof. From the assumptions, such a y_0 must satisfy

$$\left|\tilde{f}(w + ax_0)\right| = \left|f(w) + ay_0\right| \leq p(w + ax_0), \tag{9.20}$$

for all $w \in L$ and all $a \in \mathfrak{K}$. One then shows that Equation (9.20) is equivalent to

$$\left|y_0 - f(w)\right| \leq p(w - x_0), \tag{9.21}$$

for all $w \in L$. This last statement is established as follows. Let $B_w = \{b \in \mathfrak{K} : |b - f(w)| \leq p(w - x_0)\}$. Given any $w_1, w_2 \in L$, for $|f(w_1 - w_2)|$, we have

$$\left|f(w_1) - f(w_2)\right| \leq p(w_1 - w_2) \leq \max\{p(w_1 - x_0), p(w_2 - x_0)\}. \tag{9.22}$$

As $w_1, w_2 \in L$ are arbitrary, it follows that for any such $w_1, w_2 \in L$, $B_{w_1} \cap B_{w_2} \neq \emptyset$. By the assumption of *spherical completeness* of $\mathfrak{K}$, we see that $\bigcap_{w \in L} B_w \neq \emptyset$, which allows us to choose $y_0 \in \bigcap_{w \in L} B_w$, and the proof is complete. $\qquad\square$

Theorem 9.2.3 above allows us to deduce some standard-looking versions of the Helly–Hahn–Banach theorem, as indicated below.

Corollary 9.2.1. *If E is a non-Archimedean normed space over a spherically complete field $\mathfrak{K}$ and L is a subspace of E, then every linear functional f that is continuous on L can be extended to $\tilde{f} \in E'$ such that $\tilde{f}|_L = f$.*

Proof. Exercise 9.2.11. $\qquad\square$

Corollary 9.2.2. *Let E be a non-Archimedean LCS over a spherically complete field $\mathfrak{K}$. Let L be a subspace of E and f a linear functional that is continuous on L. Then f can be extended to $\tilde{f} \in E'$ such that $\tilde{f}|_L = f$.*

Proof. Exercise 9.2.12. $\qquad\square$

Corollary 9.2.3. *If $\mathfrak{K}$ is spherically complete, then no infinite-dimensional normed space over $\mathfrak{K}$ is of countable type.*

Proof. Exercise 9.2.13. $\qquad\square$

If $\mathfrak{K}$ is not spherically complete then wild (hence interesting!) examples exist, a few of which are mentioned in the next paragraph. On the other hand, with a generalized concept of countable type, a non-Archimedean version of the Helly–Hahn–Banach theorem can be established for a wide class of spaces; see [96, Section 4.2, pp. 175–182] for details.

A few general comments will end this section. First, most of the types of LCS that we have seen can be defined (with relevant adjustments) in the non-Archimedean context. Constructions of new non-Archimedean LCS from old (e. g., subspaces, quotients, initial and final topologies) can be viewed in [112, Section I.5, pp. 19–26]. Moreover, types of LCS such as barreled, Montel, reflexive, as well as some other concepts such as Banach disks, are defined throughout [96]. Next, it is worth pointing out some significant differences between non-Archimedean and Archimedean LCS, as follows: Hilbert spaces are essentially nonexistent; see [96, Section 2.4, pp. 37–41] (see [96, Section 2.6, pp. 78–79] for alternative approaches to orthogonality). The vector space l_2 of square summable sequences, when equipped with a non-Archimedean norm, is not locally convex; see [96, pp. 96–97]. If $\mathfrak{K}$ is not spherically complete, then there exist Banach spaces that have trivial duals (i. e., $E' = \{0\}$); see [96, Theorem 4.1.12, p. 174]. Hence, in order to write a concise narrative regarding non-Archimedean LCS, Schneider [112] assumes the field $\mathfrak{K}$ is spherically complete.

Examples of recent research in LCS include work by Kąkol and Śliwa, in [74] regarding metrizable subspaces and quotients of spaces of continuous functions. Furthermore, recent research related to non-Archimedean LCS in the form of vector spaces over a module rather than a field can be found in three seminal works by Garetto: [51, 52], and [53], where there are definitions of concepts including those of completeness, completing webs, polars, barreled spaces, bornological spaces, and more, with applications to nonlinear distribution theory.

Exercises

9.2.1. Fill in the missing details of the proof of Theorem 9.2.1.

9.2.2. Prove that if (a_n) is a sequence in a non-Archimedean field such that $a_n \to a$ as $n \to \infty$, where $a \neq 0$, then for sufficiently large n, $|a_n| = |a|$.

9.2.3. Prove part (b) of Theorem 9.2.1 for the case of closed balls.

9.2.4. Prove Proposition 9.2.1.

9.2.5. Prove the second part of Theorem 9.2.2. *Hint:* Assume Equation (9.19) holds. Let $w \in A$ and put $D = -w + A$, with the goal of proving that D is absolutely convex. To this end, note that $0 \in D$. Next, show that for $x = -w + v_1, y = -w + v_2$, where $v_1, v_2 \in A$ and for $a, b \in B_{\mathfrak{K}}$, one has $ax + by \in -w + A = D$.

9.2.6. Give a direct proof that for a subset A of a non-Archimedean vector space, if $0 \in A$ and A is convex, then A is absolutely convex.

9.2.7. Prove that for a subset A of a non-Archimedean vector space, if A is absolutely convex then the interior of A is either A or $\emptyset$.

9.2.8. Suppose p is a non-Archimedean seminorm on a non-Archimedean vector space E. Prove the following:

    (a)   If $x, y \in E$ satisfy $p(x) \neq p(y)$, then $p(x + y) = \max\{p(x), p(y)\}$.

    (b)   The set $V = \{x \in E : p(x) < 1\}$ is absolutely convex and absorbing in E.

    (c)   If U is absolutely convex and absorbing in E, then the *Minkowski functional* of U, given by $p_U(x) = \inf\{|a| : x \in aU\}$, is a non-Archimedean seminorm on E.

9.2.9. Prove Proposition 9.2.2.

9.2.10. Prove Proposition 9.2.3.

9.2.11. Prove Corollary 9.2.1.

9.2.12. Prove Corollary 9.2.2.

9.2.13. Prove Corollary 9.2.3.

9.3 Convergence vector spaces

This section consists of replacing the topology of a TVS with something more general called a convergence structure. In this context, some results are more general and/or more efficiently obtained than in the case when topologies are present, and some results are dramatically different than in the context of a topology. For this section, the main reference where you can find further details is Beattie and Butzmann [12]. The main results of this section concern obtaining a version of the closed graph theorem in the context of convergence spaces. It will turn out that a modification of the concept of a completing web (see Definition 6.3.5) works well for obtaining said version of the closed graph theorem. As you enjoy reading this section, you may find it useful to refer, as necessary, to Section 1.2 regarding filters. Recall from Section 1.2 that a filter can be defined without the presence of a topology and that many concepts from topology can be described in terms of filters; that is, definitions in the context of filters are typically more general than comparable definitions from topology. The definition below shows how convergence structures of filters can be defined without topologies.

Definition 9.3.1 ([12, p. 2]). A function from a universal set to the power set of the collection of all filters on the set is called a **convergence structure** on X if the three items below are satisfied. Specifically, for a universal set X and $\mathcal{P}_{\mathfrak{F}}$ the power set of all filters on X, the map $\lambda : X \to \mathcal{P}_{\mathfrak{F}}$ satisfies:

(a) For every $x \in X$, the filter generated by $\{x\}$ belongs to $\lambda(x)$.
(b) For all filters $\mathfrak{F}_1, \mathfrak{F}_2 \in \lambda(x)$, the intersection $\mathfrak{F}_1 \cap \mathfrak{F}_2 \in \lambda(x)$.
(c) Given $\mathfrak{F}_1 \in \lambda(x)$, every filter on X that is finer than $\mathfrak{F}_1$ also belongs to $\lambda(x)$.

In this case, we write the convergence structure as (X, λ) and call this a **convergence space**. Moreover, rather than write $\mathfrak{F} \in \lambda(x)$, we say $\mathfrak{F}$ **converges to** x in (X, λ), using the usual notation $\mathfrak{F} \to x$.

Note that $\lambda(x)$ is a filter for each $x \in X$. In Exercise 9.3.1, you can verify that the convergence of filters in topological spaces and convergence almost everywhere are convergence structures. It is noted here that as seen in Osborne [94, Corollary A.8, p. 179] (see also [93]), convergence almost everywhere cannot be derived from a topology. The following definition of continuity represents a natural generalization of topological continuity to a convergence space definition of this concept.

Definition 9.3.2. For convergence spaces X and Y a map $f : X \to Y$ is **continuous** at $x \in X$ if $f(\mathfrak{F}) \to f(x)$ in Y whenever $\mathfrak{F} \to x$ in X. If f is continuous at every $x \in X$, we say that f is simply **continuous**.

With the previous definition in mind, we have the following important concept regarding convergence spaces (see [12, p. 3]), as described in the next two definitions.

Definition 9.3.3. For convergence spaces X and Y, $C(X, Y)$ denotes the set of all continuous functions from X to Y. When $Y = \mathbb{K}$, the notation is $C(X) = C(X, \mathbb{K})$.

Definition 9.3.4. For convergence spaces X and Y, the **evaluation map** is defined by the function

$$\omega_{X,Y} : C(X, Y) \times X \to Y, \tag{9.23}$$

given by, for all $(f, x) \in C(X, Y) \times X$, $\omega_{X,Y}(f, x) = f(x)$.

Definition 9.3.5. For convergence spaces X and Y, the **continuous convergence structure** λ_c on $C(X, Y)$ is defined by, for a filter $\mathcal{H} \in C(X, Y)$,

$$\mathcal{H} \to f \text{ in } (C(X, Y), \lambda_c) \Leftrightarrow \omega_{X,Y}(\mathcal{H} \times \mathfrak{F}) \to f(x) \tag{9.24}$$

for all $x \in X$ and all $\mathfrak{F} \to x$ in X.

Notation 9.3.1. The continuous convergence structure $(C(X, Y), \lambda_c)$ of Definition 9.3.5 is denoted by $C_c(X, Y)$. When $Y = \mathbb{K}$, the notation is $C_c(X) = C_c(X, \mathbb{K})$.

The concept of a basis at a point in a convergence space differs from the corresponding definition of a topological base at a point (see Definition A.3.7), as you will see in the next item.

Definition 9.3.6. In a convergence space, a **local basis** at a point x is a collection $\mathfrak{B}$ such that for every filter $\mathfrak{F}$ that converges to x there exists a coarser filter $\mathfrak{G}$ such that $\mathfrak{G}$ converges to x, and has a filterbase of sets from $\mathfrak{B}$.

We will also need concepts of "open" and "closed" in the context of convergence spaces, and such concepts are stated next.

Definition 9.3.7. At a point x in a convergence space, the following are defined:
(a) For filters $\mathfrak{F}$, $\mathcal{U}(x) = \bigcap\{\mathfrak{F} : \mathfrak{F} \to x\}$ is called a **neighborhood filter of** x.
(b) The sets of the collection $\mathcal{U}(x)$ are called **neighborhoods of** x.
(c) A set U is **open** if it is a neighborhood of each of points.
(d) For a subset A of a convergence space, the **adherence** of A is defined as $a(A)$, given by $a(A) = \{x : \text{there exists a filter } \mathfrak{F} \to x \text{ and } A \in \mathfrak{F}\}$.
(e) A subset A of the convergence space is **closed** if $a(A) = A$.

It is easy to check that when the convergence structure arises from a topology, then the items of Definition 9.3.7 coincide with the corresponding concepts from topology. In addition, in [12, pp. 10–11] there are explanations regarding similarities and differences between the concepts of Definition 9.3.7 in these two contexts. Exercises 9.3.3, 9.3.4, and 9.3.5 consist of verifications of some properties of such concepts.

For what we intend to do later in this section, we need the following countability concept.

Definition 9.3.8. A convergence space is **strongly first countable** if it has a countable local basis at each of its points.

For our purposes, we wish to define a convergence structure version of a TVS and in particular, a LCS. The previous definitions give us the tools to form such a definition by way of the next few items.

Definition 9.3.9. A **convergence vector space** is a vector space with a convergence structure for which vector addition and scalar multiplication are continuous. A convergence vector space is denoted by CVS.

The usual constructions of TVS can also be formed for CVS. Such constructions include subspaces, quotients, projective limits, products, inductive limits, and more, as can be viewed in [12, Chapter 3]. The development of duality of CVS is discussed in [12, Chapter 4], and discussions of types of spaces such as barreled, bornological, reflexive, and more can be seen in [12, Chapter 7]. To illustrate a couple of examples, we consider products and inductive limits. You can read about the CVS definitions of initial convergence structures, CVS products, final convergence structures, and CVS inductive limits as described in [12, Sections 1.2, 3.3, 3.4]. In the case of inductive limits, the proof of the corresponding CVS structure is more involved than in the LCS context, as seen in [12, pp. 97–98]. For brevity, we state two basic facts regarding initial and final CVS structures in the proposition below.

Proposition 9.3.1. *Consider a collection $\{E_\alpha : \alpha \in I\}$ of CVS and a CVS E. Then:*
(a) *If for each $\alpha \in I$, $f_\alpha : E \to E_\alpha$ such that each f_α is linear, then the initial CVS structure on E is the coarsest CVS structure such that each f_α is continuous. In the case of a product, one uses $f_\alpha = \pi_\alpha$, where π_α is the canonical projection of E onto E_α.*
(b) *If for each $\alpha \in I$, $g_\alpha : E_\alpha \to E$ such that each g_α is linear, then the final CVS structure on E is the finest CVS structure such that each f_α is continuous.*

Proof. Details can be found in [12, Sections 1.2, 3.3, 3.4]. $\qquad\square$

Next, we want to define the dual of a CVS.

Definition 9.3.10. Let E, F be CVS. Define

$$\mathcal{L}(E,F) = \{T : E \to F : T \text{ is linear and continuous}\}. \tag{9.25}$$

In addition, the notation $\mathcal{L}E$ represents $\mathcal{L}(E, \mathbb{K})$, where $\mathbb{K}$ is the scalar field.

It turns out that a more useful rendition of the definition of a dual space for CVS is when we consider the convergence structure inherited from the continuous convergence structure, and this rendition is stated next. As a prelude, observe that in light of Exercise 9.3.2, in each CVS the evaluation map of Definition 9.3.4 is continuous. Thus, the following definition is viable.

Definition 9.3.11. Let E, F be CVS. Then $\mathcal{L}_c(E, F)$ denotes the convergence structure on $\mathcal{L}(E, F)$ that is inherited from $C_c(E, F)$. In particular, $\mathcal{L}_c E$ represents $\mathcal{L}_c(E, \mathbb{K})$, where $\mathbb{K}$ is the scalar field. $\mathcal{L}_c E$ is referred to as the **c-dual** of the CVS E.

With Definition 9.3.11 above, the authors of [12] use $\mathcal{L}_c E$ as the main concept of the continuous dual of a CVS. It is important to note that, according to [12, Example 3.2.1(iii), p. 86], $\mathcal{L}_c E$ does not generally arise from a topology, even if E does. In [12, Chapter 4], there are more details and descriptions of c-duals of CVS. Meanwhile, our next definition concerns the concept of a LCS in the context of CVS. In particular, in [12, p. 99], one finds constructions of: $\mathcal{D}_c(\Omega)$, the CVS version of $\mathcal{D}(\Omega)$ of test functions as well as $\mathcal{L}_c \mathcal{D}_c(\Omega)$, the CVS version of $\mathcal{D}'(\Omega)$ of distributions.

Definition 9.3.12. A CVS is **locally convex** if for every filter that converges to zero, the filter generated by the convex hull of every set of the filter also converges to zero; that is, for every $\mathfrak{F} \to 0$, the filter generated by $\{\mathrm{conv}(S) : S \in \mathfrak{F}\}$ converges to zero.

As you surely recall from Chapter 8, completeness of various types play important roles in proofs of closed graph theorems for LCS. The next definition addresses the concept of completeness we will use for obtaining a closed graph theorem in the context of CVS. For other concepts of completeness of convergence spaces, see [12, p. 69]. The concept of Definition 9.3.14 below should look somewhat familiar. That is because it is similar to the concept of a completing web of Definition 6.3.5, as we will see in Subsection 9.3.1.

Prior to the definition, we need to generalize the Definition 1.2.5 to collections of sets. The setting is the following. Let $\mathfrak{G}$ be a collection of nonempty subsets of a vector space E. Denote by $\mathfrak{G}_1$ the collection of finite intersections of elements of $\mathfrak{G}$. Assume $\mathfrak{G}_1$ is nonempty. By [67, p. 76], there exists at least one filter $\mathfrak{F}$ of subsets of E that contains $\mathfrak{G}_1$. At this time, the formal definition can be stated.

Definition 9.3.13. The coarsest filter $\mathfrak{F}$ that contains $\mathfrak{G}_1$ is called the **filter generated by the sets** $\mathfrak{G}$.

The statements above basically indicate that it is possible to generate a filter from a collection of nonempty sets, as long as the conditions preceding Definition 9.3.13 are met. Unless clarification is needed, we will use the same notation for a filter generated by a collection of sets as the notation for the generation of a filter by a filterbase, namely $[\mathfrak{G}]$ denotes the filter generated by a collection $\mathfrak{G}$. With this information, the next definition makes notational sense.

Definition 9.3.14. A CVS is **ultracomplete** if it is strongly first countable and for every filter $\mathfrak{F} \to 0$ there exists a countable collection $\{W_n : n \in \mathbb{N}\}$ of sets of $\mathfrak{F}$ such that the filter generated by the sums of the sets, namely

$$\left[\left\{ \sum_{k=n}^{\infty} W_k : n \in \mathbb{N} \right\} \right], \tag{9.26}$$

converges to 0 in the CVS. In particular, given any sequence (x_n) such that $x_n \in W_n$ for each $n \in \mathbb{N}$, the series $\sum_{n=1}^{\infty} x_n$ converges in the CVS. A sequence (W_n) of this form is called a **rapid sequence**.

Proposition 9.3.2. *Every Fréchet LCS is ultracomplete as a CVS.*

Proof. Observe that the convergence structure in this case is the topological one obtained via sequences. Certainly, having a countable zero neighborhood base indicates that a Fréchet space is strongly first countable. We now choose a zero neighborhood base $\{U_n : n \in \mathbb{N}\}$ such that for each $n \in \mathbb{N}$, $U_{n+1} + U_{n+1} \subset U_n$, and the conclusion follows from the completeness of every Fréchet space. In particular, $\{U_n : n \in \mathbb{N}\}$ constitutes a rapid sequence. $\qquad\square$

Henceforth, we can say "Fréchet space" to denote such a creature as either a LCS or a CVS. The proof of the following can be seen in [12]. In this context, the definition of completeness is formed via Cauchy filters, as done in Definition 2.5.2.

Proposition 9.3.3. *Every ultracomplete CVS is complete.*

Proof. See [12, Proposition 6.1.3, p. 184]. $\qquad\square$

The ultracompleteness of a CVS is preserved under a variety of hereditary formations, similar to what was discussed in Subsection 6.3.1. Definitions of the items listed in Theorem 9.3.1 below can be seen in [12, Section 3.3]. It is worth noting that there are significant differences in defining such spaces within the realm of CVS as opposed to LCS.

Theorem 9.3.1. *Ultracompleteness is preserved under the CVS formation of closed subspaces, quotients, countable products, countable inductive limits, and countable (Hausdorff) projective limits.*

Proof. Though similar in spirit to the proof of Theorem 6.3.3, there are differences, as can be seen in the proof of [12, Theorem 6.1.6, p. 185]. Exercises 9.3.6 and 9.3.7 allow you an opportunity to work through two parts of that proof. $\qquad\square$

9.3.1 Some closed graph theorems for convergence vector spaces

We now come to the main results of this section. For the record, given a linear map $T : E \to F$, where E and F are CVS, to say that the graph G_T of T is *closed* refers to closure with respect to the convergence structures of E and F, as defined in Definition 9.3.7(e). For comparison, see Definition 8.1.2 of the traditional context of the closed graph theorem.

Theorem 9.3.2. *Suppose E is a Fréchet CVS and F an ultracomplete CVS. If $T : E \to F$ is linear with a closed graph, then T is continuous.*

Proof. See [12, Theorem 6.2.3, pp. 188–189] for full details. The steps of the proof are comparable to those of the proof of Theorem 8.1.2. □

Corollary 9.3.1. *Suppose E is a CVS that is a countable inductive limit of Fréchet spaces, and F is an ultracomplete CVS. If $T : E \to F$ is linear with a closed graph, then T is continuous.*

Proof. Exercise 9.3.8. □

A stronger closed graph theorem is next.

Theorem 9.3.3 (Beattie–Butzmann closed graph theorem for CVS). *Assume E is a CVS that is an inductive limit of Fréchet spaces and F is a CVS. Suppose there is a finer CVS structure on F under which F is ultracomplete. Then any linear map from E to F having a closed graph is continuous.*

Proof. Denote by F_0 the vector space F under the assumed finer, ultracomplete CVS structure. Let $T : E \to F$ be a linear map, the graph of which is closed in $E \times F$. Because the CVS structure of F_0 is finer than the original one on F, the graph of T is also closed in $E \times F_0$. Apply Corollary 9.3.1 to conclude that $T : E \to F_0$ is continuous. We observe that $T : E \to F$ is continuous as well. □

A discussion of the depth and applications of Theorem 9.3.3 above can be read in [12, pp. 190–191]. Included in that discussion are applications of Theorem 9.3.3 to CVS versions of test functions and distributions, denoted respectively, by $\mathcal{D}_c(\Omega)$ and its c-dual $\mathcal{L}_c\mathcal{D}'(\Omega)$, the definitions of which are found in [12, p. 99]. Meanwhile, we finish this section with a discussion of how ultracomplete CVS are connected to webs. The following is a CVS version of Definition 6.3.1.

Definition 9.3.15. Let F be a vector space and let I be the set of all finite sequences of positive integers. A structure $\mathcal{W} = \{W_\alpha : \alpha \in I\}$ is a **CVS-web** on F if the following hold:
(a) The first **layer** of $\mathcal{W}$ consists of a sequence $(W_{n_1} : n_1 \in \mathbb{N})$ such that $\bigcup \{W_{n_1} : n_1 \in \mathbb{N}\}$ is absorbing in F.
(b) Each subsequent layer is defined by sets that are determined by the preceding layer as described in Definition 6.3.1. The absorption property for the kth layer is given by

$$\bigcup_{n_k=1}^{\infty} W_{n_1 n_2 \cdots n_k} \text{ is absorbing in } W_{n_1 n_2 \cdots n_{k-1}}. \tag{9.27}$$

(c) The sets of the kth layer ($k \in \mathbb{N}$, $k \geq 2$) satisfy

$$W_{n_1 n_2 \cdots n_k} + W_{n_1 n_2 \cdots n_k} \subset W_{n_1 n_2 \cdots n_{k-1}}. \tag{9.28}$$

A sequence $S = (S_n)$ of subsets of W is a **strand** if there exists a sequence (k_n) of positive integers such that $S_n = W_{k_1 k_2 \cdots k_n}$, for each n, $k_1, \ldots, k_n \in \mathbb{N}$. The web W is a **vector web** if for any two of its strands $S^{(1)}$, $S^{(2)}$ there is a third strand S such that $[S] \subset [S^{(1)}] + [S^{(2)}]$.

As a reminder, the notation $[S]$, $[S^{(1)}]$, $[S^{(2)}]$ in the last line refers to Definition 9.3.13. Routine calculations show that the collection of all strands of a web on a vector space generate a CVS structure on that vector space. See [12, p. 192] for more details. Thus, we can give a definition to this generated CVS structure, as stated next.

Definition 9.3.16. The CVS structure generated by the strands of a CVS-web W on a vector space F is called the **web-space associated with** W, and is denoted by F_W.

The completeness aspects of webs on a CVS can also be defined, and that is the gist of the next three items.

Definition 9.3.17. If a CVS F has a CVS-web W such that the identity id : $F_W \rightarrow F$ is continuous, then F is called **webbed**.

Definition 9.3.18. A vector web W on a CVS F is **c-tight** if for any strand (S_k) of W and any sequence (x_k) in F satisfying $x_k \in S_k$ for each $k \in \mathbb{N}$, it follows that the series $\sum_{k=1}^{\infty} x_k$ converges in F, and moreover, $\sum_{n=k+1}^{\infty} x_n \in S_k$, for each $k \in \mathbb{N}$.

The completeness relation via a natural connection between F_W and c-tight webs is given next.

Proposition 9.3.4. *If W is a c-tight vector web on a vector space F, then F_W is ultracomplete.*

Proof. Because W is a local base of 0 for F_W, F_W is strongly first countable. Next, suppose $\mathfrak{F} \rightarrow 0$ in F_W. There is a strand $S = (S_k)$ of W for which $[(S_k)] \subset \mathfrak{F}$. If (x_n) is a sequence of elements of F such that $x_n \in S_n$ for each $n \in \mathbb{N}$, then $\sum_{k=n+1}^{\infty} x_k \in S_n$, thus

$$\left[\left\{ \sum_{k=n+1}^{\infty} x_k : n \in \mathbb{N} \right\} \right] \supset [S]. \tag{9.29}$$

Hence, $\sum_{k=1}^{\infty} x_k$ converges in F_W. From this, we observe that

$$\left[\left\{ \sum_{k=n}^{\infty} S_k : n \in \mathbb{N} \right\} \right] \tag{9.30}$$

is well-defined in F_W and is finer than $[S]$. Therefore, $[\{\sum_{k=n}^{\infty} S_k : n \in \mathbb{N}\}]$ converges in F_W, and we conclude that F_W is ultracomplete. $\qquad \square$

Proposition 9.3.4 now allows us to list examples of c-tight versions of some webs on LCS we saw in Section 6.3

Proposition 9.3.5. *The following CVS have c-tight vector webs as described below.*

(a) *Every Fréchet space F has a c-tight vector web $\mathcal{W}$ such that $F_{\mathcal{W}} = F$.*
(b) *Every (LF) LCS given by $F = \mathrm{ind}_n F_n$ has a c-tight vector web $\mathcal{W}$ such that $F_{\mathcal{W}}$ is the CVS inductive limit of the Fréchet spaces F_n.*
(c) *The LCS $\mathcal{D}(\Omega)$ and $\mathcal{D}'(\Omega)$, which have completing webs as LCS, have corresponding c-tight vector webs represented by $\mathcal{D}_c(\Omega)$ and $\mathcal{L}_c\mathcal{D}_c(\Omega)$, respectively.*

Proof. Parts (a) and (b) are left to you as Exercise 9.3.10. An indication of the proof of part (c) can be seen in [12, p. 193]. □

To finish this section, a CVS version of Closed Graph Theorem 8.1.2 is established.

Theorem 9.3.4 (Closed graph theorem for CVS with c-tight webs). *Assume E is a Fréchet LCS and the F is a CVS having a c-tight vector web. Then any linear map from E to F is continuous if it has a closed graph.*

Proof. Note that the vocabulary words "continuous" and "closed" are stated within the context of CVS. Anyway, let $\mathcal{W}$ be a c-tight vector web on F and let $T : E \to F$ be linear with a closed graph. By Proposition 9.3.4, $F_{\mathcal{W}}$ is ultracomplete. Apply Theorem 9.3.2 to conclude that $T : E \to F_{\mathcal{W}}$ is continuous. Thus, $T : E \to F$ is continuous as well. □

It is noteworthy to observe that the continuity of $T : E \to F_{\mathcal{W}}$ that was obtained can be a stronger conclusion than the continuity of $T : E \to F$.

Here are a few recent publications related to convergence structures and their applications: [84, 92], and [39].

Exercises

9.3.1. Consider Definition 9.3.1. Verify that the following are convergence structures:

    (a) Topological convergence of filters when the set X is endowed with a topology $\mathcal{T}$, where the definition of convergence is given in Definition 1.2.6.

    (b) (This exercise assumes some familiarity with the concept of convergence almost everywhere from measure theory). Almost everywhere convergence: Given a measure space $(\Omega, \mathcal{A}, \mu)$ consisting of a measure μ arising from a σ-algebra $\mathcal{A}$ on a set Ω, let X denote the set of all real-valued μ-measurable functions. Define λ by $\mathfrak{F} \to f$ with respect to λ if $\mathfrak{F}$ converges to f almost everywhere on Ω.

9.3.2. Prove that the evaluation map of Definition 9.3.4 is continuous.

9.3.3. Suppose X and Y are convergence spaces and that $f : X \to Y$ is a continuous function. Prove that if $A \subset X$ and $B \subset Y$, then the restriction $f : A \to B$ is continuous as a function between convergence spaces.

9.3.4. Let X be a convergence space. Prove the following properties of the adherence of a subset of X. *Hint:* The proofs consist of arguments involving convergence of filters.

    (a) $a(\emptyset) = \emptyset$.

    (b) $A \subset a(A)$ for any subset A of X.

    (c) $a(A \cup B) = a(A) \cup a(B)$ for any subsets A and B of X.

9.3.5. Let X be a convergence space. Prove the following properties of a subset U of X and a point $x \in X$. *Hint:* The proofs also consist of arguments involving convergence of filters as in Exercise 9.3.4 above.

    (a) The set U is a neighborhood of x if and only if $x \notin a(U^c)$.

    (b) The set U is open if and only if U^c is closed.

9.3.6. In reference to Theorem 9.3.1, prove that an inductive limit of a sequence of ultracomplete CVS is ultracomplete. *Hint:* Refer to Proposition 9.3.1. In [12, Proposition 3.4.2, p. 98], one sees that a filter $\mathfrak{F}$ converges to 0 in an inductive limit $E = \mathrm{ind}_n E_n$ of CVS if and only if for some $m \in \mathbb{N}$, E_m contains a filter $\mathfrak{F}_m$ such that $\mathfrak{F}_m \to 0$ in E_m and $\mathfrak{F}_m \subset \mathfrak{F}$. Take a look at [12, Theorem 6.2.3, p. 188] as necessary.

9.3.7. Regarding Theorem 9.3.1 again, prove that a countable product of ultracomplete CVS is again ultracomplete. *Hint:* Refer again to Proposition 9.3.1. Per [12, Proposition 1.2.4, p. 5], given $E = \prod_{n=1}^{\infty} E_n$, where each E_n is an ultracomplete CVS, if $\mathfrak{F} \to 0$ in E, then for each $n \in \mathbb{N}$ there is a filter $\mathfrak{F}_n \to 0$ in E_n for which $\prod_{n=1}^{\infty} \mathfrak{F}_n \subset \mathfrak{F}$. Given $k \in \mathbb{N}$, in each E_n choose a rapid sequence $(W_k^{(n)} : k \in \mathbb{N})$ from $\mathfrak{F}_k$. Put

$$W_1 = W_{1,1} \times E_2 \times E_3 \times \cdots, \tag{9.31}$$

and in general,

$$W_k = W_{k,1} \times W_{k,2} \times \cdots \times W_{k,k} \times E_{k+1} \times E_{k+2} \times \cdots. \tag{9.32}$$

Proceed in a manner similar to the proof of part (e) of Theorem 6.3.1 and Theorem 6.3.3. See [12, Theorem 6.1.6, p. 185] as necessary.

9.3.8. Prove Corollary 9.3.1.

9.3.9. Prove that a strand as described in Definition 9.3.15(c) is a filterbase.

9.3.10. Prove parts (a) and (b) of Proposition 9.3.5.

10 Pathways to research

In this chapter, we consider a few concepts and results that represent relatively recent research as it relates to concepts we have seen throughout this book. Each topic is based on a reference in which you, the reader, can delve deeper into the topic at hand. Deciding factors regarding which topics to present in this chapter include choosing topics that are relatively recent and for which some topics from previous chapters are featured in prominent, accessible ways. The topic of Section 10.1 involves a definition of spaces of distributions that have their wavefronts sets in a positive cone, as studied in [32] and [29]. Concepts such as Mackey convergence and properties of inductive limits appear. In Section 10.2, which is based on [82], we take a quick look at how differentiation and integration from calculus can be defined in a context involving LCS. The version presented here involves locally complete LCS that we saw in Chapter 7, including such spaces that are additionally bornological, from Chapter 5. The focus of Section 10.3 is on creating generalized duality pairs, the motivation coming from material covered in Chapter 4. Subseries convergence, reminiscent of property $\mathcal{K}$ is considered, among other concepts. The information in this section comes from [118]. Section 10.4 finishes this chapter with a look at Ekeland's principle, which is a type of optimization result. The main result of this section depends on properties of Banach disks. The main reference for the material of this section is [22].

10.1 Distributions having their wavefront sets in a positive cone

In this section, we take a brief look at a recent rendition of distributions. The primary reference for this section that of from Dabrowski and Brouder [32], with additional information to be found in [33] and references in these publications. The definition and basic properties of some of the concepts involved in the LCS we will consider are rather technical, such as the definition of an open or closed cone. Details can be found in [32]. Behind the many results obtained in [32] are plenty of interesting applications of LCS. We will examine a few such applications.

Definition 10.1.1. The LCS of distributions having their wavefront sets in a positive cone Γ is denoted by $\mathcal{D}'_\Gamma$.

The first observation of [32, p. 1347] is an explanation to the effect that originally $\mathcal{D}'_\Gamma$ was equipped with a convergence structure like we saw in Section 9.3, rather than a topology. On the same page [32, p. 1347], the authors define an uncountable collection of seminorms on $\mathcal{D}'_\Gamma$ that generate a topology that is equivalent to the topology that would be generated by the convergence structure. Hence, $\mathcal{D}'_\Gamma$ is a LCS under this topology. Before we begin a tour of some properties of $\mathcal{D}'_\Gamma$, a comment is in order. Several of the results here are for Fréchet spaces that are also Schwartz LCS, called **Fréchet–Schwartz** spaces. For the definition of a Schwartz LCS, refer to [67, Section

https://doi.org/10.1515/9783111392868-010

3.15, pp. 271–288]. We will not need the properties of Schwartz spaces here because the main focus will be on properties of Fréchet–Schwartz spaces as complete metrizable LCS.

Some of the properties of $\mathcal{D}'_\Gamma$ and its strong dual are summarized below. These two summaries are comparable to summaries of the LCS $\mathcal{D}$ of test functions in Theorem 9.1.1 and of its strong dual $\mathcal{D}'$ of (Schwartz–Sobolev) distributions in Theorem 9.1.2, though in this case $\mathcal{D}'_\Gamma$ is already a space of distributions. In each item of the next two theorems, information on where the result can be found in [32] is stated.

Theorem 10.1.1. *As a LCS, the following hold:*
(a) $\mathcal{D}'_\Gamma$ *is a projective limit of a sequence of strong duals of Fréchet–Schwartz LCS:* [32, *Corollary* 13, *p.* 1360].
(b) $\mathcal{D}'_\Gamma$ *satisfies the strict Mackey convergence condition, SMCC:* [32, *Lemma* 21, *p.* 1363].
(c) $\mathcal{D}'_\Gamma$ *is semi-Montel, hence, semi-reflexive:* [32, *Proposition* 28, *p.* 1368].
(d) $\mathcal{D}'_\Gamma$ *is quasicomplete with respect to all topologies consistent with the duality between* $\mathcal{D}'_\Gamma$ *and its dual.* $\mathcal{D}'_\Gamma$ *is complete with respect to any topology equivalent to or finer than the topology generated by the collection of seminorms (defined on* [32, *p.* 1347]*) but coarser than the Mackey topology:* [32, *p.* 1368].
(e) $\mathcal{D}'_\Gamma$ *is not bornological. If* Γ *is an open cone that is not closed, then* $\mathcal{D}'_\Gamma$ *is not barreled, either:* [32, *Corollary* 36, *p.* 1376].
(f) $\mathcal{D}'_\Gamma$ *is not metrizable, due to the fact that the collection of seminorms that define its topology is uncountable:* [32, *p.* 1347].

As an example of one of the results from [32], we reproduce the first part of Theorem 10.1.1(d), which is given as part of the proof of [32, Proposition 29, p. 1368]. You will see that the proof provides a splendid example of how duality properties of LCS can be applied. For the next result, recall from Notation 4.2.1 the notation A° of the polar of a set A.

Proposition 10.1.1 ([32, Proposition 29, p. 1368]). *Suppose a Hausdorff LCS E satisfies the property that $(E, \sigma(E, E'))$ is quasicomplete. Then for any topology $\mathcal{T}$ that is consistent with the duality (E, E'), $(E, \mathcal{T})$ is quasicomplete.*

Proof. Let $\mathcal{T}$ be any topology that is consistent with the duality (E, E'). Of course, $\mathcal{T}$ is finer than $\sigma(E, E')$. Let C be any subset of E that is closed and bounded with respect to $\mathcal{T}$. Then C is bounded with respect to $\sigma(E, E')$ as well. Applying the Bipolar Theorem 4.2.1, Proposition 4.2.4(e), and Theorem 4.3.2, we observe that the bipolar $C^{\circ\circ}$ of C is absolutely convex and closed with respect to both $\sigma(E, E')$ and $\mathcal{T}$. Next, $C \subset C^{\circ\circ}$ and we use this fact to apply Proposition 4.2.5, observing that C is $\sigma(E, E')$-bounded if and only if C° is absorbing in E'. Therefore, as $C^\circ = (C^{\circ\circ})^\circ$, we have that $C^{\circ\circ}$ is a closed disk in E with respect to $\mathcal{T}$. We finish by proving that C is complete with respect to $\mathcal{T}$. Indeed, on $C^{\circ\circ}$, $\mathcal{T}$ is finer than $\sigma(E, E')$ and $\mathcal{T}$ has a base of neighborhood sets that are $\sigma(E, E')$-

closed, allowing us to cite Robertson's Theorem 2.5.2 to conclude that C^{∞} is complete with respect to $\mathcal{T}$. Finally, as a closed subset of a complete set in Hausdorff LCS, C itself is $\mathcal{T}$-complete. $\qquad\square$

Some additional properties of $\mathcal{D}'_{\Gamma}$ are summarized below.

Theorem 10.1.2. *$\mathcal{D}'_{\Gamma}$ satisfies the following properties:*
(a) *$\mathcal{D}'_{\Gamma}$ is a strictly webbed LCS.*
(b) *The strict web of part* (a) *is boundedly compatible.*
(c) *$\mathcal{D}'_{\Gamma}$ satisfies Schur's property.*
(d) *The weakly bounded subsets of $\mathcal{D}'_{\Gamma}$ are strongly bounded.*

Proof. Except for part (b), details of the proof can be deduced from other properties of $\mathcal{D}'_{\Gamma}$ as a LCS, in Exercises 10.1.2 and 10.1.4. Regarding part (b), it turns out that a SMCC version of Proposition 8.2.1 can be obtained for boundedly compatible webs and the SMCC, as seen in [56]. $\qquad\square$

It is worth pointing out that part (a) of Theorem 10.1.2 shows that if Γ is an open cone that is not closed, then $\mathcal{D}'_{\Gamma}$ represents an example of a LCS that is strictly webbed but neither barreled nor bornological.

Another example of the application of results from earlier chapters is next.

Proposition 10.1.2. *The LCS $\mathcal{D}'_{\Gamma}$ has a sequential web.*

Proof. The result follows from Theorem 10.1.2(b), but we give a direct proof for the sake of illustration. The quasicompleteness of $\mathcal{D}'_{\Gamma}$ implies that it is locally Baire. As a webbed LCS that satisfies the Mackey convergence condition, we conclude that $\mathcal{D}'_{\Gamma}$ has a sequential web by Proposition 8.2.1. $\qquad\square$

The strong dual of $\mathcal{D}'_{\Gamma}$ is denoted by $\mathcal{E}'_{\Lambda}$, where Λ is the open "dual cone" of Γ: $\Lambda = (\Gamma')^{c}$ (see [32, p. 1348]). Of course, a locally convex topology needs to be described on $\mathcal{E}'_{\Lambda}$ and verified to be equivalent to the strong topology. Such a topology is constructed on [32, pp. 1353–1354]. A summary of some properties of $\mathcal{E}'_{\Lambda}$ is in the next item.

Theorem 10.1.3. *As a LCS, the following hold:*
(a) *The topology of $\mathcal{E}'_{\Lambda}$ is given by an inductive limit:* [32, pp. 1353–1354].
(b) *The inductive, Mackey, and strong topology on $\mathcal{E}'_{\Lambda}$ are equivalent:* [32, Lemma 10, p. 1356].
(c) *$\mathcal{E}'_{\Lambda}$ is in fact an inductive limit of Fréchet–Schwartz spaces:* [32, p. 1360].
(d) *$\mathcal{E}'_{\Lambda}$ is ultrabornological:* [32, Proposition 27, p. 1367].
(e) *$\mathcal{E}'_{\Lambda}$ is barreled:* [32, Proposition 28, p. 1368].
(f) *If Λ is an open cone that is not closed, then $\mathcal{E}'_{\Lambda}$ is not even locally complete:* [32, Theorem 34, pp. 1371–1376].

As with $\mathcal{D}'_{\Gamma}$, we can observe a few more properties of $\mathcal{E}'_{\Lambda}$, as listed below.

Theorem 10.1.4. *$\mathcal{E}'_\Lambda$ satisfies the following properties:*
(a) *$\mathcal{E}'_\Lambda$ is a strictly webbed, nonregular (LF) space.*
(b) *$\mathcal{E}'_\Lambda$ is not compactly regular.*
(c) *$\mathcal{E}'_\Lambda$ is not locally Baire; in particular, $\mathcal{E}'_\Lambda$ does not satisfy property $\mathcal{K}$.*

Proof. Exercise 10.1.6. $\qquad\square$

Exercises

10.1.1. Prove the first statement of item (e) of Theorem 10.1.1 that $\mathcal{D}'_\Gamma$ is not bornological. *Hint:* What can be said about the strong dual of $\mathcal{D}'_\Gamma$?

10.1.2. Prove Theorem 10.1.2(a) that $\mathcal{D}'_\Gamma$ is strictly webbed. *Hint:* $\mathcal{D}'_\Gamma$ is kind of projective limit

10.1.3. Prove part (b) of Theorem 10.1.2 that $\mathcal{D}'_\Gamma$ has a boundedly compatible web. *Hint:* Check Section 8.2.

10.1.4. Prove that $\mathcal{D}'_\Gamma$ satisfies property $\mathcal{K}$.

10.1.5. Prove part (d) of Theorem 10.1.2.

10.1.6. Prove Theorem 10.1.4, by citing relevant properties from earlier chapters.

10.2 Convenient vector spaces

This section is about using concepts and properties of LCS to define calculus concepts regarding functions that map $\mathbb{R}$ to a LCS. The primary reference for this section is the book [82] by Kriegl and Michor, mainly the first five sections of Chapter 1. Perhaps the best way to introduce this topic is to quote from the introduction of [82], namely "Our aim in this book is to treat manifolds, which are modeled on locally convex spaces..." (1). The focus in this section is to examine some specific concepts that are used in [82] that connect with concepts we have seen throughout this book. The first set of definitions shows us one way that derivatives and integrals of calculus can be defined in the context of LCS.

Definition 10.2.1. Assume E is a Hausdorff LCS. A function $c : \mathbb{R} \to E$ is called a **curve**.

Definition 10.2.2. Let E be a Hausdorff LCS. A curve $c : \mathbb{R} \to E$ is **differentiable** at $a \in \mathbb{R}$ if

$$\lim_{s \to 0} \frac{c(a + s) - c(a)}{s} \tag{10.1}$$

exists in E. In this case, we write $c'(a)$ for the limit.

Unless otherwise noted, we assume a differentiable curve is differentiable at all values $a \in \mathbb{R}$ and write c' whenever confusion is unlikely. Just like in calculus, we define notation for higher derivatives, as follows.

Definition 10.2.3. For a nonnegative integer n, a curve $c : \mathbb{R} \to E$ is n-**times differentiable** if its jth derivative exists and is continuous for $j = 0, 1, \ldots, n$. Such higher derivatives, if they exist, are written with standard notation such as c'' or $c^{(j)}$. The curve c is called a **smooth curve** if it is differentiable for all nonnegative integers.

There is corresponding notation, too, as enunciated below.

Notation 10.2.1. A curve that is n-times differentiable is also called a C^n **curve** and a smooth curve is denoted as a C^∞ **curve**. The vector space of all n-times differentiable curves is denoted by $C^n(\mathbb{R}, E)$. The vector space of all smooth curves is denoted by $C^\infty(\mathbb{R}, E)$.

An important concept in the context of curves is that of a Lipschitz condition, formalized below.

Definition 10.2.4. A curve $c : \mathbb{R} \to E$ satisfies the **Lipschitz condition** on a neighborhood $U \subset \mathbb{R}$ if the set

$$\left\{ \frac{c(a) - c(s)}{a - s} : s, a \in U, \, s \neq a \right\} \tag{10.2}$$

is bounded in E.

A more general term is given below.

Definition 10.2.5. A curve is **locally Lipschitzian** if for every $a \in \mathbb{R}$ there is a neighborhood U_a of a such that the curve satisfies the Lipschitz condition on U_a.

In Exercise 10.2.1, you can show that in fact, a locally Lipschitzian curve satisfies the Lipschitz condition on any bounded interval. Meanwhile, the following definition describes these Lipschitzian ideas in the general case.

Definition 10.2.6. A curve $c : \mathbb{R} \to E$ is called **Lip**^{k} if each of its derivatives up to order k exists and is locally Lipschitzian.

The next item indicates a connection between the concepts above and Mackey convergence.

Proposition 10.2.1. *Suppose $c : \mathbb{R} \to E$ is* Lip^1*. Then there exists a closed disk B in E such that on any bounded subset of* $\mathbb{R} \setminus \{0\}$*,*

$$\frac{1}{a}\left[\frac{c(a) - c(0)}{a} - c'(0)\right] \in B, \tag{10.3}$$

where $a \neq 0$. In particular, $\frac{c(a)-c(0)}{a} - c'(0) \in aB$*, for any $a > 0$.*

Proof. [82, Corollary 1.7, p. 13]. $\square$

Corollary 10.2.1. *The difference quotient* $\frac{c(a)-c(0)}{a}$ *of Equation (10.3) is Mackey convergent.*

Proof. For sequences, let $a_n = \frac{1}{n}$ in $\mathbb{R}$ to observe that as $a_n \to 0$, the difference quotient converges to zero in $(E_B, \| \cdot \|_B)$. In general, one can consider a net (an equivalent of a filter) of the form $(a_\alpha : \alpha \in I)$ that decreases to 0 (see [82, Lemma 1.6, p. 12] for more details). $\square$

Another concept that is part of the basic ideas in [82] is that of local completeness that we saw in Chapter 7. To put local completeness in the context of convenient vector spaces, we first define a topology in the next item.

Definition 10.2.7. The c^∞**-topology** on a LCS is the final topology with respect to all smooth curves from $\mathbb{R}$ to the LCS. The open sets in this topology are referred to as c^∞**-open.**

It is commented in [82, Definition 2.12, p. 19] that the c^∞-topology need not be compatible with the linear structure of the vector space; that is, a vector space equipped with the c^∞-topology is not necessarily a TVS. At any rate, in the context of LCS, the c^∞-topology leads to a list of conditions on a LCS that are equivalent to local completeness. But first, we need to define the Riemann integral in the context of curves, sparking fond memories of calculus. Indeed, for a curve $c : \mathbb{R} \to E$, one may define a partition of an interval $[a, b] \subset \mathbb{R}$ just like in calculus, then form the Riemann sums in E that correspond to c. The details are summarized in the next definition in which we assume, without loss of generality, that the interval $[a, b]$ is $[0, 1]$; compare with Definition 9.1.6.

Definition 10.2.8. Suppose $c : \mathbb{R} \to E$ is a curve, where we assume E is sequentially complete. If $\mathcal{P} = \{0 = t_0, t_1, t_2, \ldots, t_n = 1\}$ is a partition of $[0, 1]$ and x_j is an element of each subinterval $[t_{j-1}, t_j]$, then the **Riemann sum** with respect to c on the given partition $\mathcal{P}$ is

$$R_n = \sum_{j=1}^{n} c(x_j)(t_j - t_{j-1}). \tag{10.4}$$

If $\lim_{n \to \infty} R_n$ exists in E, then c is **Riemann integrable**, with the integral over $[0, 1]$ defined to be that limit. A curve is **locally Riemann integrable** on a nonempty open subset of $\mathbb{R}$ if the curve is Riemann integrable over every compact interval contained in the open subset.

The following result represents a list of equivalences to local completeness.

Theorem 10.2.1. *The following are equivalent for a Hausdorff LCS E:*
(a) *Any Lipschitz curve $c : \mathbb{R} \to E$ is locally Riemann integrable.*
(b) *For any $c_1 \in C^{\infty}(\mathbb{R}, E)$, there exists $c_2 \in C^{\infty}(\mathbb{R}, E)$ such that $c_2' = c_1$; that is, c_1 has an antiderivative.*
(c) *As a linear subspace of any other LCS, E is closed with respect to the c^{∞}-topology.*
(d) *If $c : \mathbb{R} \to E$ is such that for every $f \in E'$ (over $\mathbb{R}$) $f \circ c : \mathbb{R} \to \mathbb{R}$ is smooth, then c is smooth as curve from $\mathbb{R}$ to E.*
(e) *E is locally complete.*
(f) *Every continuous linear map from a normed space into E has a continuous extension to the completion of that normed space.*

Proof. [82, Theorem 2.14, pp. 20–21]. $\qquad\square$

Theorem 10.2.1 motives the next definition and indicates a justification for the title of this section.

Definition 10.2.9. A LCS that satisfies any of the equivalent conditions of Theorem 10.2.1 is called c^{∞}-**complete** or **convenient**.

10.2.1 Bornological convenient spaces

On page 20 of [82], the authors state that in many contexts a c^{∞}-complete LCS is assumed to be bornological. With this in mind, we examine some properties and examples of LCS that are locally complete and bornological.

Recall Definition 8.3.2 of a quasi-(LB) LCS and that by Theorem 8.3.1 a LCS is quasi-(LB) if and only if it has an ordered, strict web (see Definition 6.3.7 and Definition 6.3.5, respectively).

We can now prove a result that contains a necessary and a sufficient condition for a LCS to be a bornological convenient space. Surely, you will notice a connection between the result below and the Antosik–Burzyk Theorem 6.1.3.

Theorem 10.2.2. *Suppose a LCS E is a quasi-(LB) space. Then E is a bornological convenient LCS if the two conditions below are satisfied:*
(i) *Every sequentially continuous seminorm on E is continuous.*
(ii) *The LCS E satisfies property $\mathcal{K}$.*

Conversely, if E is a bornological convenient LCS and satisfies the MCC, then properties (i) *and* (ii) *both hold.*

Proof. Assume (i) and (ii) for a LCS E. A glance at Theorem 6.1.3 reveals that these assumptions imply that E is bornological. What remains is to prove that E is locally complete. Assumption (ii) inspires us to cite Theorem 7.1.6, from which we conclude that E is locally Baire. We will use this fact to prove that E is, in fact, locally complete. Let A be any bounded subset of E and find a closed disk B such that $A \subset B$ and $(E_B, \| \cdot \|_B)$ is a normed Baire space. Let $\mathcal{W}$ denote an ordered strict web on E. For a strand $\omega = (W_k)$, let $F_\omega = \bigcap\{\mathrm{span}(W_k) : k \in \mathbb{N}\}$. In [124, 7.21, p. 164], it is shown that a metrizable, complete locally convex topology $\mathcal{T}_\omega$ is generated on F_ω by the collection of sets shown in the expression of Equation (10.5) below:

$$\left\{ F_\omega \cap \frac{1}{k} W_k : k \in \mathbb{N} \right\}. \tag{10.5}$$

It also turns out $\mathcal{T}_\omega$ is finer than the relative topology induced on F_ω by E, which you can verify in Exercise 10.2.2. Because $(E_B, \| \cdot \|_B)$ is a Baire space and id $: E_B \to E$ is continuous, we may apply the localization theorem that corresponds to Valdivia's Closed Graph Theorem 8.3.2, (see [124, Theorem 7.6, p. 164]), which indicates that there is a particular strand ω of $\mathcal{W}$ such that id $: (E_B, \| \cdot \|_B) \to (F_\omega, \mathcal{T}_\omega)$ is continuous, implying that B is a Banach disk in $(F_\omega, \mathcal{T}_\omega)$ by the completeness of $(F_\omega, \mathcal{T}_\omega)$. Moreover, id $: (F_\omega, \mathcal{T}_\omega) \to E$ is continuous, and the conclusion that E is locally complete follows.

Now assume E is a bornological convenient LCS and satisfies the MCC. Property (i) holds for E by Proposition 6.2.4. Next, Proposition 8.2.1 tells us that because E satisfies the MCC, is locally complete, and has a strict web, the web is sequential. Hence, given any null sequence (x_n), there exists a strand (W_k) of $\mathcal{W}$ such that for each $k \in \mathbb{N}$, there is $x_{n_k} \in W_k$. Of course, $\mathcal{W}$ is completing so the series $\sum_{k=1}^{\infty} x_{n_k}$ converges in E, and we have proven that E satisfies property $\mathcal{K}$. $\qquad\square$

Some examples distinguish between the properties assumed in Theorem 10.2.2, which we consider next. First, we symbolize the property that E is a quasi-(LB) LCS, by denoting this property as (iii). This leads to the following.

Proposition 10.2.2. *Consider the conditions:*
(i) *Every sequentially continuous seminorm on E is continuous.*
(ii) *The LCS E satisfies property $\mathcal{K}$.*
(iii) *The LCS E is a quasi-(LB) space.*

Then no two of these properties are sufficient to conclude that E is a bornological convenient LCS.

Proof. In Exercise 10.2.3, you can prove that for any combination of the two conditions (i), (ii), and (iii) there is a LCS that satisfies two of these conditions but is not a bornological convenient LCS. $\square$

If a LCS is not bornological, then we can consider the ultrabornological topology $\mathcal{T}_{\mathrm{ub}}$ associated with the given LCS; see Definition 5.2.6. In this case, the following can be proved. See [57] for more details.

Proposition 10.2.3. *Suppose $(E, \mathcal{T})$ has an ordered strict web that is sequential. Then $(E, \mathcal{T}_{\mathrm{ub}})$ is a bornological convenient LCS that satisfies property $\mathcal{K}$.*

Proof. Exercise 10.2.4. $\square$

We finish this section with the comment that throughout the book [82], Mackey convergence appears several times. Moreover some other concepts we have seen appear. As a few instances, webbed LCS are assumed in [82, Theorem 5.24, p. 66], and the property of a LCS being Mackey first countable (see Definition 2.2.8) is assumed in [82, Proposition 15.3, p. 159] and [82, Theorem 22.17, p. 236].

Exercises

10.2.1. Prove that a locally Lipschitzian curve $c : \mathbb{R} \to E$ satisfies the Lipschitz condition on any bounded interval of $\mathbb{R}$. *Hint:* Show that for an increasing sequence (t_n) in $\mathbb{R}$, $\frac{1}{t_n - t_1}[c(t_n) - c(t_1)]$ is contained in the convex balanced hull of a finite union of bounded subsets in E.

10.2.2. Prove that the topology $\mathcal{T}_\omega$ described by Equation (10.5) is finer than the relative topology by E induced on F_ω.

10.2.3. Verify the following, regarding Proposition 10.2.2:

    (a)   An incomplete normed space satisfying property $\mathcal{K}$ satisfies conditions (i) and (ii) but is not a bornological convenient LCS. The existence of such a space was indicated in Chapter 6.

    (b) $(l_1, \sigma(l_1, l_\infty))$ satisfies conditions (ii) and (iii) but is not a bornological convenient LCS.

    (c)   A nonregular (LB) space satisfies conditions (i) and (iii) but is not a bornological convenient LCS. *Hint:* Construct an ordered, strict web on such a space and then take a look at Theorem 8.2.2.

10.2.4. Prove Proposition 10.2.3. *Hint:* For the first part, it is easy to see that $(E, \mathcal{T})$ satisfies property $\mathcal{K}$, and hence, Theorem 10.2.2 implies that $(E, \mathcal{T})$ is locally complete. Then observe that $(E, \mathcal{T})$ and $(E, \mathcal{T}_{\mathrm{ub}})$ have the same bounded subsets. To show that $(E, \mathcal{T}_{\mathrm{ub}})$ satisfies property $\mathcal{K}$, show that $(E, \mathcal{T}_{\mathrm{ub}})$ satisfies the MCC and then apply Theorem 10.2.2.

10.2.5. While it seems that a strictly webbed ultrabornological LCS must be at least locally complete, describe an example where this conclusion fails.

10.2.6. Describe examples of the following types of LCS related to those seen in this section:

    (a)   A bornological LCS that satisfies property $\mathcal{K}$ and the MCC but is neither locally complete nor a quasi-(LB) space.

    (b)   A locally complete LCS that is not a quasi-(LB) space.

10.3 Abstract duality

Let us fondly recall fabulous Chapter 4 on duality. One of the main concepts in that chapter was that of the definition of pairings (see Definition 4.2.1), in which two vector spaces F and G are paired by way of a binary form b, denoted by the expression $(F, G; b)$. In that context, $b : F \times G \to \mathbb{K}$. A typical habit in mathematics is to ask the proverbial question of "what if," and this question arises as follows: What if we assume sets of the product are more general than vector spaces and something more general than the field $\mathbb{K}$ in the range? One such generalization of duality pairings was done by Swartz, Li, and Cho. In this section, we summarize some basic ideas and results of this work, the main reference being [118]. By the way, in [118, pp. 1–3], you will find a summary of other efforts to generalize the concept of pairings and duality. The primary definition of this section is next.

Definition 10.3.1. Let E and F be nonempty sets and let G be a Hausdorff abelian topological group. Let $b : E \times F \to G$ be a function. The triple E, F, and G is called an **abstract duality pair** (or **abstract triple**).

See Definition A.3.2 of a topological group, which is a generalization of a TVS. More information about topological groups can be seen in [69]. Meanwhile, below is the notation for an abstract duality pair.

Notation 10.3.1. An abstract duality pair of Definition 10.3.1 is denoted by $(E, F; G)$.

Observe that, just like in the case of duality, if $(E, F; G)$ is an abstract duality pair with respect to $b : E \times F \to G$, then $(F, E; G)$ is an abstract duality pair with respect to $\tilde{b} : F \times E \to G$, where $\tilde{b}(y, x) = b(x, y)$, for all $x \in E, y \in F$. In the sequel, there are three examples. Many more examples, including of contexts such as measurable functions, can be seen in [118, Chapter 1].

> **Example 10.3.1.** The duality of TVS from Chapter 4 as an abstract duality pair.

Naturally, the concept of duality from Chapter 4 is a motivating example.

> **Example 10.3.2.** Spaces of continuous functions in an abstract duality pair.

Let E be a topological space and G a Hausdorff Abelian topological group. We use $F = C(E, G)$, the vector space of continuous functions from E to G. Define $b : E \times C(E, G) \to G$ to be the evaluation map, namely for all $x \in E$ and all $f \in C(E, G)$, $b(x, f) = f(x)$. Then $(E, C(E, G); G)$ is an abstract duality pair. Notice that E need not be a vector space, so this example comes from outside the realm of the duality of Chapter 4.

> **Example 10.3.3.** Abelian topological groups as abstract duality pairs.

Let E and F be Abelian topological groups and G a Hausdorff Abelian topological group such that there exists a function $b : E \times F \to G$ that is additive in both variables. Then $(E, F; G)$ is an abstract duality pair.

Much of the material in [118] involves subseries convergence, which we considered in Chapter 6 (specifically, Definition 6.1.2). Two comparable definitions, adapted for the context of abstract duality pairs, are next. The definitions are stated in the context of an abstract duality pair $(E, F; G)$ even though only the topology of G is needed. This for consistency with Definition 10.3.5 that follows afterward.

Definition 10.3.2. Let $(E, F; G)$ be an abstract duality pair. A **formal series** $\sum_j x_j$ is a sum of elements of G. Such a sum is **subseries convergent** if for every subsequence (n_j) of $\mathbb{N}$, the series $\sum_{j=1}^{\infty} x_{n_j}$ converges in G.

It is possible to consider more general subsets of $\mathbb{N}$, as clarified in the next definition.

Definition 10.3.3. For an abstract duality pair $(E, F; G)$ and an infinite subset σ of $\mathbb{N}$, one writes $\sum_{j \in \sigma} x_j = \sum_{j=1}^{\infty} x_{n_j}$, where the elements of σ are arranged in a subsequence $\{n_j : j \in \mathbb{N}\}$ of $\mathbb{N}$. If σ is finite, there is no ambiguity in the calculation of the sum over σ.

Of the subseries convergence in [118], of particular importance is the convergence with respect to a generalized version of the weak topology (see Definition 4.2.3 for duality of LCS). The definition below illustrates such a concept of weak topology in the context of abstract duality pairs.

Definition 10.3.4. For an abstract duality pair $(E, F; G)$, $w(E, F)$ represents the coarsest topology on E such that the collection of maps

$$\{b_y(\cdot, y) : E \to G : y \in F\}, \tag{10.6}$$

consists of maps that are continuous.

Definitions 10.3.3 and 10.3.4 above lead to the definition below.

Definition 10.3.5. Let $(E, F; G)$ be an abstract duality pair. A sequence (x_j) or formal series $\sum_j x_j$ of elements of E is $w(E, F)$ **subseries convergent** if for every $\sigma \subset \mathbb{N}$ there exists $x_\sigma \in E$ such that $\sum_{j \in \sigma} b(x_j, y) = b(x_\sigma, y)$, for every $y \in F$.

In typical fashion of reducing notation, we write $w(E, F)$ subseries convergence using the notation below.

Notation 10.3.2. For a subseries convergent sequence as stated in Definition 10.3.5, we write

$$\sum_{j \in \sigma} x_j = x_\sigma. \tag{10.7}$$

Observe that there need not be a definition of addition in E, because the calculation is in G via b; that is, we calculate $b(x_j, y)$ in G. Furthermore, note that in the context of a LCS E and its dual $F = E'$, with $G = \mathbb{K}$, then $w(E, F)$ convergence is nothing more than weak convergence, in other words, with respect to $\sigma(E, E')$.

For the next part of this section, we consider a specific kind of result, mainly the property that whenever a sequence is subseries convergent for a weak topology the sequence is subseries convergent for a finer (such as the original) topology. We have already seen a comparable result in Theorem 6.1.2 in which every weakly $\mathcal{K}$-convergent sequence in a normed space converges as a sequence with respect to the normed topology. The difference here is that we check for *subseries* convergence in a finer topology. The next several definitions set up the terminology of convergence that we will use in the sequel. These definitions are from [118, pp. 25–27].

Definition 10.3.6. In an abstract duality pair, a result such that any subseries convergent sequence in a weak topology is subseries convergent in a finer topology is referred to informally as an **Orlicz–Pettis theorem.**

We will also need the following topological definitions (see [118, p. 25]).

Definition 10.3.7. In a topological space a subset is:
(a) **Sequentially conditionally compact** if every sequence in the set has a subsequence that is Cauchy.
(b) **Sequentially relatively compact** if every sequence in the set has a subsequence that converges to an element of the space.

The definition below expresses the abstract duality pairs concept of convergence upon rearranging terms of a series.

Definition 10.3.8. Let $(E, F; G)$ be an abstract duality pair. Then
(a) A formal series $\sum_j x_j$ is **unordered convergent** if

$$\lim_D \sum_{j \in \sigma} x_j \tag{10.8}$$

converges, where $D = \{\sigma \subset \mathbb{N} : \sigma \text{ is finite}\}$ is directed by set inclusion.
(b) For $\sigma \subset \mathbb{N}$ and $A \subset F$, a series $\sum_{j \in \sigma} b(x_j, y)$ **converges uniformly** on A if for every closed neighborhood U of the group identity of G, there exists $N \in \mathbb{N}$ such that

$$\sum_{j \in \sigma} b(x_j, y) \in U, \tag{10.9}$$

whenever $y \in A$ and $\min\{j : j \in \sigma\} > N$. If the set A is sequentially conditionally $w(F, E)$ compact, we will refer to this convergence as **strong unordered convergence.** Note that "strong" in this context implies "finer" in the topological sense.

(c)   A collection $\{\sum_j x_{\alpha,j} : \alpha \in I\}$ of series is **uniformly unordered convergent** if

$$\lim_D \sum_{j \in \sigma} x_{\alpha,j} \tag{10.10}$$

converges uniformly for each $\alpha \in I$.

Note that in general the set D in Definition 10.3.8 is uncountable, so the limit is in terms of a net, an uncountable version of convergence that is distinct yet equivalent to convergence of filters; see [129, Section 3.4, pp. 39–42] for more details. With the vocabulary of the definitions above, we now state a version of the Orlicz–Pettis theorem that we will use. It is one of several that are obtained in [118]. For a discussion of the Orlicz–Pettis theorem in the context of normed spaces, see [27, pp. 149–152].

Theorem 10.3.1 (Orlicz–Pettis theorem for abstract duality pairs). *In an abstract duality pair $(E, F; G)$, if $\sum_j x_j$ is $w(E, F)$ subseries convergent, then on any sequentially conditional $w(F, E)$ compact subset of F, $\sum_{j \in \sigma} x_j$ is strongly unordered convergent.*

Proof. See [118, Theorem 2.2, pp. 26–27]. $\square$

A comment on the proof of Theorem 10.3.1 is that a key step of that proof is the application of the Basic Matrix Theorem 6.1.1.

The preliminary item below is about defining bilinear maps and the concept of separate continuity of such maps.

Definition 10.3.9. A **bilinear map** is a function $b : E \times F \to G$ that is linear in both variables, where E, F, and G are vector spaces.

Notice that Definition 10.3.9 is a generalization of a bilinear form; see Definition 4.2.1 for comparison. Naturally, if E, F, and G are TVS, then a definition of continuity would be welcomed, and that definition is next.

Definition 10.3.10. A bilinear map from two TVS to a third TVS is **continuous** (also **jointly continuous**) if it satisfies the definition of continuity as a function from a product of two TVS into a third TVS.

The result below will not be surprising; it states that continuity of bilinear maps on TVS need only be checked at the origin.

Proposition 10.3.1. *For TVS E, F, and G, a bilinear map $b : E \times F \to G$ is continuous if and only if it is continuous at $(0, 0)$.*

Proof. Exercise 10.3.1. $\square$

On the other hand, it can occur that a bilinear map is continuous with respect to each variable separately, which we define next.

Definition 10.3.11. For TVS E, F, G, a bilinear map $b : E \times F \to G$ is **separately continuous** if it is continuous in each variable.

Another way of expressing Definition 10.3.11 is to say that for each fixed $y \in F$, $b_y : E \to G$ is continuous and for each fixed $x \in E$, $b_x : F \to G$ is continuous. Clearly, a continuous bilinear map is separately continuous, but the converse is false, as you can verify in Exercise 10.3.2(b).

We can now obtain a few significant results in the context of abstract duality pairs.

Theorem 10.3.2. *Suppose E is a LCS, F is a TVS and G is a LCS, as an abstract duality pair. Assume E is sequentially complete. If $b : E \times F \to G$ is a separately continuous bilinear map, then b is bounded; that is, for every $A \subset E$ and $B \subset F$ such that A and B are bounded, the image $b(A, B) = \{b(x,y) : x \in A, y \in B\}$ is bounded in G.*

Proof. Suppose the conclusion fails. Then there exist bounded sets $A \subset E$ and $B \subset F$ as well as a continuous seminorm p on G such that

$$\sup\{p(x,y) : x \in A, y \in B\} = \infty. \tag{10.11}$$

In particular, we may find for each $k \in \mathbb{N}$, $x_k \in A$ and $y_k \in B$ for which

$$p(x_k, y_k) > 2^{2k}. \tag{10.12}$$

By the boundedness of A, the series $\sum_{k=1}^{\infty} x_k 2^{-k}$ converges in E. Furthermore, the sequential completeness of E implies that $\sum_{k=1}^{\infty} x_k 2^{-k}$ is subseries convergent. The separate continuity with respect to x implies that $\sum_{k=1}^{\infty} x_k 2^{-k}$ is also $w(E, F)$ subseries convergent. Next, $B \subset F$ is also bounded, which means $y_k 2^{-k} \to 0$ in F, as $k \to \infty$. Applying the assumption of separate continuity of b, this time with respect to y, tells us that $y_k 2^{-k} \to 0$ with respect to $w(F, E)$. It follows that $\{y_k 2^{-k} : k \in \mathbb{N}\}$ is $w(F, E)$ sequentially relatively compact. We now apply the Orlicz–Pettis Theorem 10.3.1 to conclude that

$$\sum_{j=1}^{\infty} b(x_j 2^{-j}, y_k 2^{-k}) \tag{10.13}$$

is strongly unordered convergent; that is, the series in Equation (10.13) converges uniformly for each $k \in \mathbb{N}$. In particular, $\lim_{k\to\infty} b(x_k 2^{-k}, y_k 2^{-k}) = 0$ in G. The continuity of p reveals that $p(x_k 2^{-k}, y_k 2^{-k}) \to 0$ as $k \to \infty$. The previous statement is pleasantly problematic because it contradicts Equation (10.12), and completes the proof. $\square$

For the next result, we need a concept of continuity of bilinear maps that is "intermediate" between the concepts of continuous and separately continuous, due to the fact that not every separately continuous bilinear map is continuous. Such a concept is the gist of the next definition. You will notice a similarity to the $\mathcal{S}$-topologies of Chapter 4.

Definition 10.3.12. Suppose E, F, and G are TVS. Let $\mathcal{S}$ be a collection of bounded subsets of F. A bilinear map $b : E \times F \to G$ is **(left) hypocontinuous** (with respect to $\mathcal{S}$) if for every $W \in \mathcal{N}_{0,G}$ and for every $B \in \mathcal{S}$ there exists $U \in \mathcal{N}_{0,E}$ such that $b(x,y) \in W$ whenever $x \in U$ and $y \in B$. An analogous statement can be made for defining **(right) hypocontinuous**. The adjectives "left" and "right" are suppressed when the context is clear.

For a concept comparable to sequential continuity, we have the following.

Definition 10.3.13. For TVS E, F, and G, a bilinear map $b : E \times F \to G$ is **sequentially (left) hypocontinuous** if whenever (x_k) is a null sequence in E and $B \subset F$ is bounded, then $b(x_k, y) \to 0$ uniformly in G as $k \to \infty$, for every $y \in B$.

Of course, one may specify left or right sequential hypocontinuity as necessary for clarity. The concept of sequential hypocontinuity is used in the next result. The concept of Mackey convergence of Definition 6.2.1 appears as well.

Theorem 10.3.3. *Assume E is a LCS, F is a TVS and G is a LCS, as an abstract duality pair $(E, F; G)$. Suppose E is sequentially complete and satisfies the MCC. If $b : E \times F \to G$ is a bilinear separately continuous map, then b is (left) sequentially hypocontinuous.*

Proof. We are to prove that if $x_k \to 0$ in E and $B \subset F$ is bounded, then $b(x_k, y) \to 0$ uniformly in G for every $y \in B$. As we seek to accomplish this sequentially, it suffices to show that $b(x_k, y_k) \to 0$ uniformly in G, where $y_k \in B$. Given $x_k \to 0$ in E, we apply the assumption that E satisfies the MCC to find a sequence (a_k) of positive numbers such that $a_k \to \infty$ and $a_k x_k \to 0$ in E. In particular, $\{b(a_k x_k, y_k) : k \in \mathbb{N}\}$ is bounded in G. Hence,

$$\frac{1}{a_k} b(a_k x_k, y_k) = b(x_k, y_k) \to 0 \tag{10.14}$$

in G. The sequential completeness of E allows us to cite Theorem 10.3.2, concluding that $b(x_k, y_k) \to 0$ uniformly in G. The proof is completed. $\qquad\square$

Theorem 10.3.3 implies a result of Mazur and Orlicz [87] for general TVS, as indicated below. Note that in the corollary below, the MCC in a general TVS is expressed via sequences rather than with disks; that is, for any $x_k \to 0$ in E, there exists (a_k) in $(0, \infty)$ such that $a_k \to \infty$ and $a_k x_k \to 0$ in E. Recall from comments preceding Proposition 6.2.2, that the MCC in a TVS is sometimes referred to as the TVS being braked.

Corollary 10.3.1. *Assume E and F are TVS and G is a LCS, as an abstract duality pair $(E, F; G)$. Suppose E is sequentially complete and satisfies the MCC. Then b is in fact a continuous bilinear map.*

Proof. Exercise 10.3.4. $\qquad\square$

In particular, Corollary 10.3.1 applies to any complete, metrizable TVS E.

Theorem 10.3.4 (Principle of uniform boundedness for abstract duality pairs). *Suppose E, F, and G are LCS as an abstract duality pair. Assume E and F are sequentially complete and that E satisfies the MCC. Let H be a collection of separately continuous bilinear maps from $E \times F$ to G that is pointwise bounded. Then H is uniformly bounded; that is, for every $A \subset E$ and $B \subset F$ such that A and B are bounded, $H(A, B)$ is bounded in G.*

Proof. [118, Theorem 2.73, pp. 45–46]. □

Exercises

10.3.1. Prove Proposition 10.3.1. *Hint:* Suppose a bilinear map $b : E \times F \to G$ is continuous at a point $(x_0, y_0) \neq (0,0)$. Given any zero neighborhood W in G, one obtains $U \in \mathcal{N}_{0,E}, V \in \mathcal{N}_{0,F}$ such that for any $(x,y) \in U \times V, b(x_0 + x, y_0 + y) - b(x_0, y_0) \in W$. Then let $W_1 \in \mathcal{N}_{0,G}$ such that $W_1 + W_1 + W_1 \subset W$, and find balanced zero neighborhoods U_1 in E and V_1 in F for which $x \in U_1$ and $y \in V_1$ imply $b(x,y) \in W_1$. Next, use the fact that U_1 and V_1 are absorbing in E and F, respectively, to find the desired zero neighborhoods U and V.

10.3.2. Consider the bilinear map $b : c_{00} \times c_{00} \to \mathbb{R}$ (see Definition A.5.1 regarding c_{00}) given by: for all $((x_k), (y_k)) \in c_{00} \times c_{00}, b((x_k), (y_k)) = \sum_{k=1}^{\infty} x_k y_k$.

    (a)   Explain why b is defined for all pairs $((x_k), (y_k)) \in c_{00} \times c_{00}$.

    (b)   Prove that b is separately continuous but not continuous. *Hint:* Show that if $e^{(k)}$ represents the unit vector (a sequence with 1 in the kth entry and zeros for all other entries), then the sequence $\frac{1}{\sqrt{k}} \sum_{k=1}^{\infty} x_k \to 0$ in c_{00} but for the pair $((x_k), (x_k)) \in c_{00} \times c_{00}$ one has $b(x_k, x_k) = 1$ for every $k \in \mathbb{N}$.

10.3.3. Prove that every continuous bilinear map is hypocontinuous.

10.3.4. Prove Corollary 10.3.1.

10.4 Ekeland's variational principle

Ekeland's variational principle is an optimization result in which there exist solutions to some nonlinear problems that are in some sense nearly optimal. The original version by Ekeland [42] involves a function $f : X \to \mathbb{R} \cup \{\infty\}$ that satisfies certain conditions (defined below), where X is a complete metric space.

Theorem 10.4.1 (Ekeland's variational principle for complete metric spaces). *Assume (X, d) is a complete metric space. Let $f : X \to \mathbb{R} \cup \{\infty\}$ be lower semicontinuous and bounded below. Let $x_0 \in X$. Given any $\varepsilon > 0$, there exists $v \in X$ such that:*
(a) $f(v) \le f(x_0) - \varepsilon d(x_0, v)$.
(b) *For every $x \in X$, such that $x \ne x_0, f(v) < f(x) + \varepsilon d(x, v)$.*

Our goal here is to prove a version of this result in the case where X is replaced by a locally complete LCS (see Definition 7.1.1). General references for Ekeland's variational principle include [34, 132], and the aforementioned [42]. The main reference in the context of LCS for this section is that of Bosch, García, and García [22]. The definitions below have been modified by assuming a Hausdorff LCS in place of a metric space. Part (c) below is an adaptation of Definition A.3.26 to the context of LCS.

Definition 10.4.1. Assume E is a Hausdorff LCS, and $f : E \to \mathbb{R} \cup \{+\infty\}$.
(a) The **effective domain** of f is $\{x \in E : f(x) \ne +\infty\}$. The function f is **proper** if its effective domain is nonempty.
(b) f is **bounded below** if $\inf\{f(x) : x \in E\}$ is finite.
(c) f is **lower semicontinuous at** $x_0 \in E$ if for every y such that $y < f(x_0)$ there is a neighborhood U of x_0 such that $y < f(w)$ for all $w \in U. f$ is **lower semicontinuous** if the previous statement holds for every $x_0 \in E$.

We will need a concept of perturbations, which is stated next.

Definition 10.4.2. A **perturbation** is a function $\varphi : [0, \infty) \to [0, \infty)$ that is continuous, subadditive (i. e., $\varphi(x + y) \le \varphi(x) + \varphi(y)$), nondecreasing, and satisfies $\varphi(0) = 0$.

The function $\varphi(t) = t$ is a trivial example of a perturbation, and you can indicate a few others in Exercise 10.4.1.

We are now ready to state and prove the main result of this section.

Theorem 10.4.2 (Ekeland's variational principle for locally complete spaces). *Assume E is a Hausdorff locally complete LCS. Let $f : E \to \mathbb{R} \cup \{\infty\}$ be proper, lower semicontinuous, and bounded below. Denote by $\mathcal{P}$, the set of perturbations of Definition 10.4.2. Let $\varphi \in \mathcal{P}$ and $x_0 \in E$. Given any Banach disk B containing x_0, there exists $x^* \in E$ such that:*
(a) $f(x^*) + \varphi(\|x^* - x_0\|_B) \le f(x_0)$.
(b) *For every $x \in E$, such that $x \ne x_0, f(x^*) \le f(x) + \varphi(\|x^* - x\|_B)$.*

Proof. Let B be a Banach disk as assumed, and x_0 such that $x_0 \in B$. Put

$$C = \{x \in E_B : f(x) + \varphi(\|x - x_0\|_B) \le f(x_0)\}. \tag{10.15}$$

It is easy to check that $C \ne \emptyset$. In Exercise 10.4.4, you can verify that C is closed in $(E_B, \|\cdot\|_B)$. Now define a function $g : E_B \to \mathbb{R} \cup \{\infty\}$ by

$$g(x) = \begin{cases} f(x), & \text{if } x \in C \\ \infty, & \text{if } x \in E_B \setminus C \end{cases}. \tag{10.16}$$

Observe that g is bounded below. That g is also lower semicontinuous is left as Exercise 10.4.3. Choose any $y_1 \in C$ and construct by induction a sequence (y_k) in C such that the following inequalities hold:

$$\begin{aligned} &\inf\{g(x) + \varphi(\|y_k - x\|_B) : x \in E_B\} \\ &\le g(y_{k+1}) + \varphi(\|y_k - y_{k+1}\|_B) \\ &\le \inf\{g(x) + \varphi(\|y_k - x\|_B) : x \in E_B\} + 2^{-k}. \end{aligned} \tag{10.17}$$

We proceed to use these inequalities to prove that (y_k) is Cauchy in $(E_B, \|\cdot\|_B)$. To this end, using $x = y_k$ in the inequalities shown in (10.17) reveals that (y_k) satisfies

$$g(y_{k+1}) \le g(y_{k+1}) + \varphi(\|y_k - y_{k+1}\|_B) \le g(y_k) + 2^{-k}. \tag{10.18}$$

Hence, $g(y_{k+1}) \le g(y_k) + 2^{-k}$; that is, $(g(y_k))$ is bounded below and "almost" decreasing. Next, by passing to a subsequence if necessary, we may assume that $(g(y_k))$ converges. Thus, $\varphi(\|y_k - y_{k+1}\|_B) \le g(y_k) - g(y_{k+1}) + 2^{-k}$. In Exercise 10.4.5, you can verify that for $n > m$,

$$\varphi(\|y_n - y_m\|_B) \le g(y_n) - g(y_m) + \sum_{j=m+1}^{n} \frac{1}{2^j}, \tag{10.19}$$

thus,

$$\|y_n - y_m\|_B \le \varphi^{-1}\left(g(y_n) - g(y_m) + \frac{1}{2^m} - \frac{1}{2^n}\right). \tag{10.20}$$

Because $g(y_n) - g(y_m) + \frac{1}{2^m} - \frac{1}{2^n} \to 0$ as $n, m \to \infty$, we have shown that (y_k) is Cauchy in $(E_B, \|\cdot\|_B)$. As B is a Banach disk, (y_k) converges in $(E_B, \|\cdot\|_B)$ to some element, call it x^*. Of course, $x^* \in C$. The definition of the set C implies that item (a) holds. So as to prove (b), observe that by the inequalities shown in (10.17), if $x \in E_B$ (with $x \ne x^*$), we obtain

$$g(y_{k+1}) + \varphi(\|y_k - y_{k+1}\|_B) \le g(x) + \varphi(\|y_k - x\|_B) + 2^{-k}. \tag{10.21}$$

In the inequality displayed in (10.21), let $k \to \infty$ to obtain $g(x^*) \le g(x) + \varphi(\|x - x^*\|_B)$. To establish (b), there are three cases.

<u>Case 1</u>: $x \in C$. In this case, $g(x) = f(x)$, so $f(x^*) = g(x^*) \le f(x) + \varphi(\|x - x^*\|_B)$, as you can check in Exercise 10.4.6.

<u>Case 2</u>: $x \in E_B \setminus C$. By the definition of C, $f(x) + \varphi(\|x - x_0\|_B) > f(x_0)$. Now, $x^* \in C$, which implies that $f(x^*) + \varphi(\|x^* - x_0\|_B) \le f(x_0)$. The last sentence indicates that with $x \ne x_0$,

$$
\begin{aligned}
f(x^*) + \varphi(\|x^* - x_0\|_B) \le f(x_0) &< f(x) + \varphi(\|x - x_0\|_B) \\
&\le f(x) + \varphi(\|x - x^*\|_B) + \|x^* - x_0\|_B \\
&\le f(x) + \varphi(\|x^* - x\|_B) + \varphi(\|x^* - x_0\|_B).
\end{aligned}
\tag{10.22}
$$

Therefore, $f(x^*) < f(x) + \varphi(\|x^* - x\|_B)$.

<u>Case 3</u>: $x \in E \setminus E_B$. Exercise 10.4.7. $\qquad\square$

Other works connected with this result include [101, 64, 25], and [23].

Exercises

10.4.1. Give two or three examples of perturbations (other than $\varphi(t) = t$), including at least one that is bounded and one that is unbounded.

10.4.2. Prove that the set C in Equation (10.15) is nonempty.

10.4.3. Prove that the function g of Equation (10.16) is lower semicontinuous.

10.4.4. Prove that the set C in Equation (10.15) is closed in $(E_B, \|\cdot\|_B)$.

10.4.5. Verify that for $n > m$ the inequality shown in (10.19) holds. *Hint:* Use the triangle inequality and the subadditivity of φ.

10.4.6. Verify the claim made in Case 1 of the proof of Theorem 10.4.2.

10.4.7. Verify the claim made in Case 3 of the proof of Theorem 10.4.2.

11 Other resources

This "chapter" consists of a summary of other resources you may consider. The first category of introductory resources consists of those that are at approximately at the level of this book. They contain distinct presentations of most of the topics covered here as well as topics not included in this book. The second category of advanced material consists of those that go to a higher level of discussion, both of topics covered in this book and of topics not found here. The categories of being introductory versus advanced are not mutually exclusive. Resources labeled as introductory do contain some advanced discussions and resources labeled as advanced contain discussions that are understandable at an introductory level. In my experience, I have found that reading about the same concept in a variety of distinct sources and levels allows me to form my own understanding of that concept. Moreover, you may find some hints for some exercises in these references. The third category of additional background material consists of items that can be used mainly as resources regarding concepts that are assumed in this book, such as topology, linear algebra, and normed/Banach spaces. It is worth noting that references focused on normed and Banach spaces often have some material on the topic of locally convex spaces, which you may find useful. The fact that there are many types of locally convex spaces leads to many kinds of relationships between those types. Apart from how some locally convex spaces are related to others is useful, it is equally important to know counterexamples that distinguish between types of spaces. Some relationships and counterexamples appear in this book, however, the information is not complete; that is, there are many more relationships and counterexamples that put the general area of locally convex spaces into sharper focus. The fourth category lists a few resources in which you can find such information. The fifth category is on more general references such as historical works related to locally convex spaces. Finally, this collection of references is based on works with which I have some familiarity, and is by no means exhaustive.

- **Other introductory resources:** (In order of my familiarity) Horváth [67], Narici and Beckenstein [91], Robertson and Robertson [103], Osborne [94], Rudin [107], Swartz [116], Treves [119], Voigt [127], Bonet et al. [20].
- **Advanced resources:** (In alphabetical order) Bogachev and Smolyanov [18], Edwards [41], Jarchow [70], Kelly et al. [76], Köthe [79] and [80], Pérez Carreras and Bonet [95], Schaefer [111], Valdivia [123], and Wilansky [130].
- **Other background material resources:** (normed and Banach spaces, topology, etc. in no particular order) [4, 14, 31, 37, 47, 65, 89, 90, 115, 129].
- **Relationships and counterexamples:** (In alphabetical order) [67, Tables, pp. 441–442], [77, 88], [95, Tables, p. 507], [129, Tables of Theorems and Counterexamples, pp. 329–265], [130, Tables, pp. 267–282].
- **A few historical and general resources:** (Bio of Banach, etc., in alphabetical order) [21, 8, 38, 85, 86, 108, 113, 114, 120, 131].

https://doi.org/10.1515/9783111392868-011

A Background Information

A.1 Sets and functions

Notation A.1.1. The set $\{1, 2, \ldots, n\}$ is denoted by $\underline{n}$.

Definition A.1.1. On a set S, a relation $\sim$ is an **equivalence relation** if all of the following hold: ($\forall a, b, c \in S$), $a \sim a$ (reflexive property), $a \sim b \Rightarrow b \sim a$ (symmetric property), and $a \sim b$ and $b \sim c$ imply $a \sim c$ (transitive property). An equivalence relation partitions the set S into **equivalence classes**, $\overset{\bullet}{a}$ defined by, for $a \in S$, $\overset{\bullet}{a} = \{x \in S : x \sim a\}$, that is, $x \in \overset{\bullet}{a} \Leftrightarrow x \sim a$.

Theorem A.1.1. *DeMorgan's properties: For any collection $\{A_\alpha : \alpha \in I\}$,*

$$\left(\bigcup_{\alpha \in I} A_\alpha\right)^c = \bigcap_{\alpha \in I} A_\alpha^c, \quad \left(\bigcap_{\alpha \in I} A_\alpha\right)^c = \bigcup_{\alpha \in I} A_\alpha^c. \tag{A.1}$$

Definition A.1.2. In a Cartesian product $X = \prod\{X_\alpha : \alpha \in I\}$, a subset A of X is of the form $A = \prod\{A_\alpha : \alpha \in I\}$, where $A_\alpha \subset X_\alpha$, for each $\alpha \in I$.

Theorem A.1.2. *Let X and Y be sets and $f : X \to Y$ a function. For collections of sets $\{A_\alpha : \alpha \in I\}$ in X and $\{B_\alpha : \alpha \in I\}$ in Y, the following relations hold:*
(a) $f(\bigcup_{\alpha \in I} A_\alpha) = \bigcup_{\alpha \in I} f(A_\alpha)$.
(b) $f(\bigcap_{\alpha \in I} A_\alpha) \subset \bigcap_{\alpha \in I} f(A_\alpha)$. *This inclusion can be strict.*
(c) $f^{-1}(\bigcup_{\alpha \in I} B_\alpha) = \bigcup_{\alpha \in I} f^{-1}(B_\alpha)$.
(d) $f^{-1}(\bigcap_{\alpha \in I} B_\alpha) = \bigcap_{\alpha \in I} f^{-1}(B_\alpha)$.

Definition A.1.3. A function $f : X \to Y$ is:
(a) **Injective** (or is an **injection**, or is **one-to-one**) if for each $a, b \in X$, whenever $f(a) = f(b)$, it follows that $a = b$.
(b) **Surjective** (or is a **surjection**, or is **onto**) if for each $y \in Y$, there exists at least one $x \in X$ such that $f(x) = y$.
(c) **Bijective** (or is a **bijection**) if f is both injective and surjective.

Proposition A.1.1. *Let $f : X \to Y$ be a function from a set X to a set Y. Let $A \subset X$ and $B \subset Y$. The following relations hold:*
(a) *$A \subset f^{-1}(f(A))$ and equality holds if f is injective.*
(b) *$f(f^{-1}(B)) \subset B$ and equality holds if f is surjective.*

Lemma A.1.1 (Zorn's lemma; see [81, Section 4.1, pp. 210–213]). *Suppose a nonempty partially ordered set satisfies the property that every totally ordered subset (every chain) has an upper bound. Then the set contains a maximal element.*

https://doi.org/10.1515/9783111392868-012

Note. Results that make use of Zorn's lemma are considered *deep* because Zorn's lemma is equivalent to the Axiom of Choice. Thus, applying Zorn's lemma amounts to reaching down to the foundation of axiomatic mathematics.

A.2 Linear algebra

Any unspecified concepts or results that are not listed here can be found, for example, in Friedberg, Insel, and Spence [47], or your favorite linear algebra book.

Notation A.2.1. The span of a set S of elements of a vector space is denoted by $\mathrm{span}(S)$.

Definition A.2.1. A vector space V is the **linear direct sum** of subspaces L and M if $L + M = V$ and $L \cap M = \{0\}$. The notation is $V = L \oplus M$.

Definition A.2.2. A **Hamel basis** of a vector space is a collection of vectors that is linearly independent and spans the vector space.

Theorem A.2.1 (A deep result equivalent to Zorn's lemma; see [89, Theorem 1.1.1, p. 6]). *Every vector space has a Hamel basis.*

Definition A.2.3. Let E and F be vector spaces over a field $\mathbb{F}$ and $T : E \to F$ a function. Then:
(a) T is a **linear map** if for every $x, y \in E$, $a \in \mathbb{F}$, $T(x + y) = T(x) + T(y)$, and $T(a \cdot x) = a \cdot T(x)$. Other terms include **linear transformation** and **linear operator**. One can easily prove that a function $T : E \to F$ is linear if and only if $(\forall x, y \in E)(\forall a \in \mathbb{F})$, $T(a \cdot x + y) = a \cdot T(x) + T(y)$.
(b) The **nullspace** (also: **kernel**) of T is defined as $T^{-1}(\{0\}) = \{x \in E : T(x) = 0\}$, where 0 represents the zero vector of F.
(c) If T is a bijection (see Definition A.1.3), then T is called a **linear isomorphism**, and in this case, E and F are referred to as being **linearly isomorphic**.

Theorem A.2.2 (A couple of standard results).
(a) *For linear $T : E \to F$, $T^{-1}(\{0\})$ is a vector subspace of E. T is injective if and only if $T^{-1}(\{0\}) = \{0\}$.*
(b) *Any $m \times n$ matrix A of real numbers is a linear map from $\mathbb{R}^n$ to $\mathbb{R}^m$. In fact, a function $A : \mathbb{R}^n \to \mathbb{R}^m$ is a linear map if and only if A can be represented as an $m \times n$ matrix of real numbers. This is also true for complex numbers with adjusted details.*

A.3 General topology

This section contains information about topology that is, from time to time, referenced in the main text. My primary resource for this appendix is Richmond [102], and there are others you may wish to consult such as Munkres [90], Starbird and Su [115], Kelley, [75],

Hocking and Young [65], and others. A concise, well-written classic is that of Bushaw [31], which includes some discussion of filters in uniform spaces (which are more general than topological vector spaces).

Definition A.3.1. A **topology** on a universal set X is a collection $\mathcal{T}$ of subsets of X such that all three of the following hold:

(a) The empty set and X belong to $\mathcal{T}$.

(b) The collection $\mathcal{T}$ is closed under arbitrary unions; that is, if $\{U_\alpha : \alpha \in I\}$ is any collection of elements of $\mathcal{T}$, then $\bigcup_{\alpha \in I} U_\alpha \in \mathcal{T}$.

(c) The collection $\mathcal{T}$ is closed under finite intersections; that is, if $U_1, U_2, \ldots, U_k \in \mathcal{T}$, then $\bigcap_{j=1}^{k} U_j \in \mathcal{T}$.

Sets belonging to $\mathcal{T}$ are called **open**. A set is **closed** if its complement is open. A set X with a topology $\mathcal{T}$ is called a **topological space**. Topological spaces are denoted by $(X, \mathcal{T})$.

Definition A.3.2. A **topological group** is a group that carries a topology for which the group operations (multiplication and inversion) are continuous.

Definition A.3.3. Given a topological space $(X, \mathcal{T})$ and a subset A of X, the **relative topology** (also **subspace topology induced topology**) is defined by $\{U \cap A : A \in \mathcal{T}\}$. We say that $\mathcal{T}$ **induces** this topology on A. The notation for the relative topology is $\mathcal{T}\,|_A$.

Definition A.3.4. A **neighborhood** of x in $(X, \mathcal{T})$ is a set U such that there exists $V \in \mathcal{T}$ for which $x \in V \subset U$.

Definition A.3.5. Let $(X, \mathcal{T})$ be a topological space. A topological **base** or (**basis**) of $\mathcal{T}$ is a collection $\mathcal{B}$ of subsets of X such that every set in $\mathcal{T}$ is a union of elements of $\mathcal{B}$.

Theorem A.3.1. *A collection $\mathcal{B}$ of subsets of a set X is a base for some topology on X if and only if the following two items hold:*

(a) *The union of all sets of $\mathcal{B}$ is X; that is, $\bigcup\{B : B \in \mathcal{B}\} = X$.*

(b) *Given $B_1, B_2 \in \mathcal{B}$ and any $x \in X$ such that $x \in B_1 \cap B_2$, $(\exists B_3 \in \mathcal{B})$ such that $x \in B_3 \subset B_1 \cap B_2$.*

Definition A.3.6. A collection $\mathcal{S}$ of subsets of a set X is a **subbase** or (**subbasis**) of a topology $\mathcal{T}$ on X if the collection of finite intersections of the sets of $\mathcal{S}$ is a base of $\mathcal{T}$. In this case, arbitrary unions of finite intersections of sets of $\mathcal{S}$ generate $\mathcal{T}$.

Definition A.3.7. A **neighborhood base at a point** $x \in X$, denoted by $\mathcal{N}_x$, is a collection of neighborhoods of the point x such that every neighborhood of x (Definition A.3.4) from $\mathcal{T}$ contains a set from $\mathcal{N}_x$.

Definition A.3.8. A subset of a topological space is **clopen** if it is both open and closed. A topological space that has a base of clopen sets is called **zero-dimensional**.

Definition A.3.9. The **discrete topology** is the collection of all subsets of the universal set. The **trivial topology** consists of the empty set and the universal set.

Definition A.3.10. A sequence (x_n) in a topological space **converges** to a point x, if for every neighborhood U of x, there exists $N = N_U \in \mathbb{N}$ such that for all $n \geq N$, $x_n \in U$.

Definition A.3.11. Suppose $(X, \mathcal{T})$ is a topological space. For $A \subset X$, the **interior** $\mathring{A}$ of A is the union of all open subsets of A (also denoted by $\mathrm{int}(A)$). A point x is an **interior point** of A if there exists an open set U such that $x \in U \subset A$. The **closure,** $\overline{A}$ **of** A is the intersection of all closed sets that contain A (also denoted by $\mathrm{cl}(A)$). A set is **sequentially closed** if whenever a sequence of points from the set converges, the limit of the sequence belongs to the set. A point x is a **closure point** of A if every neighborhood of x intersects A. A point x is a **boundary point** of A if every neighborhood of x intersects both A and its complement A^c.

Notation A.3.1. The closure of a set A with respect to a particular topology $\mathcal{T}$ is denoted by $\overline{A}^{\mathcal{T}}$.

Definition A.3.12. In a topological space X, a subset A of a set B is **dense in** B if $\overline{A} = B$. If $B = X$, and we simply say A is **dense**. A topological space X is **separable** if it contains a countable, dense subset.

Definition A.3.13. Given two topologies $\mathcal{T}_1$ and $\mathcal{T}_2$ on a set X, the topology $\mathcal{T}_2$ is **finer than** $\mathcal{T}_1$ (or $\mathcal{T}_1$ is **coarser than** $\mathcal{T}_2$) if $\mathcal{T}_2$ contains all of the sets of $\mathcal{T}_1$. That is, $\mathcal{T}_1 \subset \mathcal{T}_2$. Topologies $\mathcal{T}_1$ and $\mathcal{T}_2$ are **equivalent** if $\mathcal{T}_2$ is finer than $\mathcal{T}_1$ and $\mathcal{T}_1$ is finer than $\mathcal{T}_2$. In this case, we write $\mathcal{T}_1 \cong \mathcal{T}_2$. When two topologies are equivalent on a set, we say that the topologies **agree**.

Definition A.3.14. A topological space $(X, \mathcal{T})$ is **Hausdorff** if given any two unequal elements x, y in X, there exist open sets U containing x, and V containing y, such that $U \cap V = \emptyset$.

Definition A.3.15. A function is **open** (or is an **open map**) if it satisfies the property that the image of every open set is open.

Definition A.3.16 (See [67, p. 73]). Consider a set X and a collection $\{(X_\alpha, \mathcal{T}_\alpha) : \alpha \in I\}$ of topological spaces. For each $\alpha \in I$, let $f_\alpha : X \to X_\alpha$ be a function. The coarsest topology on X such that every function f_α is continuous is called the **initial topology** with respect to the collection $\{f_\alpha : \alpha \in I\}$.

Definition A.3.17. Consider a product $\Pi_{\alpha \in I}(X_\alpha, \mathcal{T}_\alpha)$ of topological spaces $(X_\alpha, \mathcal{T}_\alpha)$, with corresponding projections $\pi : X \to X_\alpha$. The **product topology** is the coarsest topology that makes all projections continuous. The product topology is generated by a base consisting of sets U_α such that $U_\alpha \in \mathcal{T}_\alpha$ for each $\alpha \in I$, and $U_\alpha = X_\alpha$ for all but finitely many α. Whenever $(\forall \alpha \in I)\, X_\alpha = X$, the notation is X^I.

Theorem A.3.2. *For $X = \Pi_{\alpha \in I} X_\alpha$ with the product topology, the following properties hold:*

(a) *For any topological space Y, suppose a function $f : Y \to \Pi_{\alpha \in I} X_\alpha$ is given by $f(w) = (f_\alpha(w))$ $(\alpha \in I)$, where $f_\alpha : Y \to E_\alpha$. Then f is continuous if and only if each f_α is continuous.*

(b) *The product $X = \Pi_{\alpha \in I} X_\alpha$ is Hausdorff if and only if each space X_α is Hausdorff.*

(c) *For any $A \subset \Pi_{\alpha \in I} X_\alpha$, $\overline{A} = \Pi_{\alpha \in I} \overline{A}_\alpha$; in particular, $A = \Pi_{\alpha \in I} A_\alpha$ is closed in X if and only if A_α is closed in X_α for each $\alpha \in I$.*

Definition A.3.18. A topological space is **connected** if it cannot be expressed as a disjoint union of two nonempty open sets. That is, $(X, \mathcal{T})$ is connected if it is not possible to write $X = U \cup V$, where U and V are disjoint nonempty open sets. The definition of a subset being connected is similar.

Definition A.3.19. Let $(X, \mathcal{T})$ be a topological space. Let A be a subset of X. An **open cover** of A is a collection $\mathcal{O} = \{U_\alpha : \alpha \in I\}$ of open sets such that $A \subset \bigcup\{U_\alpha : \alpha \in I\}$. A is **compact** if every open cover of A can be reduced to a finite collection that covers the set; that is, a finite **subcover** of A. In other words, there exists a finite subcollection $U_{\alpha_1}, U_{\alpha_2}, \ldots, U_{\alpha_k}$ from $\mathcal{O}$, such that $A \subset \bigcup_{j=1}^{k} U_{\alpha_j}$.

Theorem A.3.3. *Two comparable compact Hausdorff topologies are equivalent. That is, suppose $\mathcal{T}_1$ and $\mathcal{T}_2$ are Hausdorff topologies on a set X such that $\mathcal{T}_2$ is finer than $\mathcal{T}_1$. If a subset A is compact with respect to $\mathcal{T}_2$, then $\mathcal{T}_1$ and $\mathcal{T}_2$ are equivalent on the set A.*

Theorem A.3.4. *Compact subsets of Hausdorff topological spaces are closed.*

Theorem A.3.5 (Tychonov's theorem). *Arbitrary products of compact topological spaces are compact with respect to the product topology.*

Definition A.3.20. A subset of a topological space is **relatively compact** if its closure is compact.

Definition A.3.21. A topological space is **locally compact** if each point of the space is contained in a relatively compact neighborhood.

Definition A.3.22. Let $(X, \mathcal{T}_X)$ and $(Y, \mathcal{T}_Y)$ be topological spaces. A function $f : X \to Y$ is **continuous at** x if for every neighborhood V of $f(x)$ there is a neighborhood U of x such that $f(U) \subset V$. The function f is **continuous** on X if f is continuous at each point $x \in X$. The function f is **sequentially continuous at** x if whenever a sequence (x_n) converges to x in X, the sequence $(f(x_n))$ converges to $f(x)$ in Y. If f is sequentially continuous at every $x \in X$, we say f is simply **sequentially continuous**.

Theorem A.3.6 (Some equivalences of continuity). *Consider topological spaces $(X, \mathcal{T}_X)$ and $(Y, \mathcal{T}_Y)$ and a function $f : X \to Y$. The following are equivalent:*

(a) *f is continuous.*

(b) *For any set $V \in \mathcal{T}_Y, f^{-1}(V) \in \mathcal{T}_X$ ("inverse images of open sets are open").*

(c) *For any closed subset $C \subset Y$, $f^{-1}(C)$ is closed in X ("inverse images of closed sets are closed").*
(d) *For any $A \subset X$, $f(\overline{A}) \subset \overline{f(A)}$ in Y.*

Definition A.3.23. A **homeomorphism** between two topological spaces is a bijection (see Definition A.1.3) such that the function and its inverse are both continuous. In this case, the two topological spaces are referred to as **homeomorphic**.

Theorem A.3.7. *For a function between topological spaces the following are equivalent:*
(a) *The function is a homeomorphism.*
(b) *The function maps closed sets to closed sets.*
(c) *The function maps open sets to open sets.*

Theorem A.3.8. *The continuous image of a compact set is compact. The continuous image of a connected set is connected.*

Definition A.3.24. A subset of a topological space is **nowhere dense** (also **rare**) if the interior of the closure of the set is empty ($(\overset{\circ}{\overline{A}}) = \varnothing$). A subset of the topological space is **meager** (also **first category**) if it is a countable union of rare sets. A set that is not meager is called (unsurprisingly) **nonmeager** (also **second category**).

Definition A.3.25. A topological space that is nonmeager is called a **Baire space**.

Theorem A.3.9 (Baire category theorem). *A complete metric space (see Definition A.4.1 below) is a Baire space.*

Theorem A.3.10. *The following condition on a topological space $(X, \mathcal{T})$ are equivalent:*
(a) *$(X, \mathcal{T})$ is a Baire space.*
(b) *If X is the countable union of closed sets, then at least one of those sets has nonempty interior.*
(c) *A countable intersection of open dense subsets of X is dense.*

Definition A.3.26. Let X be a topological space. A function $f : X \to \mathbb{R}$ is **lower semi-continuous at** $x_0 \in X$ if for every $\varepsilon > 0$ there is a neighborhood U of x_0 such that $f(x) > f(x_0) - \varepsilon$ for all $x \in U$. f is **lower semicontinuous** if the previous statement holds for every $x_0 \in X$.

A.4 Metric spaces

Definition A.4.1. Let X be a nonempty set. A **metric** on X is a function $d : X \times X \to \mathbb{R}$ that satisfies the following properties: For every $x, y, z \in X$,
(a) $d(x, y) \geq 0$.
(b) $d(x, y) = 0$ if and only if $x = y$.

(c) $d(y, x) = d(x, y)$.

(d) (*Triangle inequality*) $d(x, y) \leq d(x, z) + d(z, y)$.

A set X with a metric d is denoted by (X, d), and one refers to it as a **metric space**.

Definition A.4.2. A topological space is **metrizable** if its topology can be generated by a metric. In a metric space (X, d), given $a \in X$ and $r > 0$, the **open ball centered at** a **of radius** r is $B_r(a) = \{x : d(x, a) < r\}$.

Theorem A.4.1. *The collection of open balls $B_r(a)$, for $a \in X, r > 0$ forms a base for a topology on X, called the **metric topology**.*

Definition A.4.3. Let (X, d) be a metric space.

(a) A sequence (x_n) is a **Cauchy sequence** if, given any $\varepsilon > 0$, there is an $N \in \mathbb{N}$ such that for all $m, n \geq N$, $d(x_n, x_m) < \varepsilon$. A **complete** metric space is one in which every Cauchy sequence converges to an element of the space.

(b) A sequence (x_n) is **bounded** if it can be contained in a ball of some fixed radius. That is, the sequence (x_n) is bounded if and only if for some $a \in X$ there exists $M > 0$ such that for all $n \in \mathbb{N}$ $x_n \in B_M(a)$.

Proposition A.4.1. *In a metric space (X, d), $A \subset X$ is closed if and only if, for any sequence (x_n) from A, if $x_n \to x$, then $x \in A$.*

Theorem A.4.2. *For metric spaces (X, d_X) and (Y, d_Y), a function $f : X \to Y$ is continuous at x if and only if f is sequentially continuous at x (see Definition A.3.22).*

A.5 Normed spaces and related results

Background references for this section include [89, 4], or check your favorite book or resource on normed and Banach spaces.

Definition A.5.1. Let X be a vector space over a field $\mathbb{F}$. A **norm** on X is a function $\| \cdot \| : X \to [0, \infty)$ that satisfies all of the following properties:

(a) $(\forall x \in X)$, $\|x\| \geq 0$, and $\|x\| = 0 \iff x = 0$ (the zero vector in X).

(b) $(\forall x \in X)(\forall a \in \mathbb{F})$ $\|a \cdot x\| = |a| \cdot \|x\|$.

(c) (*Triangle inequality*) $(\forall x, y \in X)$ $\|x + y\| \leq \|x\| + \|y\|$.

A vector space X that has a norm is denoted by $(X, \| \cdot \|)$ (or $(X, \| \cdot \|_X)$ when necessary to clarify), and is referred to as a **normed (vector) space**, or **normed linear space**.

Proposition A.5.1. *A norm generates a metric given by: For every pair of elements, x, y, $d(x, y) = \|x - y\|$.*

In light of Proposition A.5.1, topological properties of metric spaces from Section A.4 apply to normed spaces.

Definition A.5.2. Definition A.4.3 applied to normed spaces leads to: A complete normed space is called a **Banach space**.

Definition A.5.3. In a normed space $(X, \| \cdot \|)$, the **open ball** of radius r, centered at a, is $B_r(a) = \{x : \|x - a\| < r\}$. The **open unit ball** is defined as $\{x : \|x\| < 1\}$. A **closed ball** of radius r, centered at a, is defined as $\overline{B}_r(a) = \{x : \|x - a\| \leq r\}$.

Theorem A.5.1. *Suppose* $T : (E, \| \cdot \|_E) \longrightarrow (F, \| \cdot \|_F)$ *is a linear map. Then the following are equivalent:*
(a) *T is continuous at every point of E.*
(b) *T is continuous at 0.*
(c) *For every sequence $x_n \to 0$ in E, $T(x_n) \to 0$ in F.*
(d) *T is bounded on the closed (or open) unit ball B; that is, $(\exists M > 0)(\forall x \in B) \|T(x)\|_F \leq M$.*
(e) *T is bounded on all of E; that is, $(\exists C > 0)$ such that $(\forall x \in E) \|T(x)\|_F \leq C\|x\|_E$.*

Definition A.5.4. Suppose $(E, \|\cdot\|)$ is any normed space. Another norm, say $\|\cdot\|^\sim$, is **equivalent to** $\| \cdot \|$ if there exist positive numbers a, b such that $(\forall x \in E) \, a\|x\| \leq \|x\|^\sim \leq b\|x\|$.

Theorem A.5.2 ([89, Theorem 1.4.15, p. 31]). *Suppose $(E, \| \cdot \|)$ over $\mathbb{K}$ (where $\mathbb{K}$ represents either $\mathbb{R}$ or $\mathbb{C}$) is finite-dimensional, of dimension n. The following hold:*
(a) *The topologies of all norms on E are equivalent to each other, and generate a topology equivalent to that of the Euclidean norm on $\mathbb{K}^n$.*
(b) *$(E, \| \cdot \|)$ is a Banach space.*
(c) *$(E, \| \cdot \|)$ satisfies the **Heine–Borel property**: A subset of E is compact if and only if it is closed and bounded.*

Theorem A.5.3 ([89, Theorem 1.4.23, p. 31]). *Suppose $(E, \|\cdot\|)$ is any normed space. If every closed, bounded set in E is compact then E must be finite-dimensional.*

Definition A.5.5. For normed spaces E, F, the collection $\mathcal{B}(E, F)$ is the set of all $T : (E, \| \cdot \|_E) \longrightarrow (F, \| \cdot \|_F)$ such that T is a continuous (bounded) linear map. On $\mathcal{B}(E, F)$, we have a norm defined by: $\|T\|_\mathcal{B} = \sup\{\|T(x)\|_F : \|x\|_E \leq 1\}$. When $F = \mathbb{K} (= \mathbb{R}$ or $\mathbb{C})$, we have the **dual space of** E, denoted by E', and use the norm $\|T\|' = \sup\{|T(x)| : \|x\|_E \leq 1\}$. Note that $\| \cdot \|'$ is called the **operator norm** on E'.

Theorem A.5.4 ([89, Theorem 1.6.9, p. 45] Banach–Steinhaus theorem for Banach spaces). *Let H be a nonempty collection of bounded linear functions from a Banach space E to a normed space F. If $\sup\{\|T(x)\| : T \in H\}$ is finite for each $x \in E$, then $\sup\{\|T\|' : T \in H\}$ is finite, where $\| \cdot \|'$ is the operator norm of Definition A.5.5.*

A.5.1 Some specific normed and Banach spaces

Example A.5.1.

(a) The vector space c_{00} of all **finitely nonzero sequences**. That is, c_{00} is the set of all sequences (x_n) for which there exists $N \in \mathbb{N}$ depending on (x_n), such that $x_n = 0$, for all $n \geq N$. The space c_{00} is a normed space under the norm $\| \cdot \|_\infty$ defined by $(\forall x = (x_n) \in c_{00})\ \|x\|_\infty = \max\{|x_j| : j \in \mathbb{N}\}$.

(b) The vector space c_0 of all sequences of scalars that converge to zero. The space c_0 is a Banach space under the norm defined by $(\forall x = (x_n) \in c_0)\ \|x\|_\infty = \sup\{|x_j| : j \in \mathbb{N}\}$.

(c) For p such that $1 \leq p < \infty$, the space l_p is defined as the set of all $x = (x_n)$ with $x_n \in \mathbb{K}$ for each n such that $\sum_{n=1}^{\infty} |x_n|^p$ converges. Each space l_p is a Banach space under the norm defined by

$$\|x\|_p = \left[\sum_{n=1}^{\infty} |x_n|^p \right]^{\frac{1}{p}}. \tag{A.2}$$

For $p = \infty$, l_∞ is the vector space of all bounded sequences and is a Banach space under the norm: $\|x\|_\infty = \sup\{|x_n| : n \in \mathbb{N}\}$.

(d) Let μ be a positive measure on a σ-algebra $\mathcal{A}$ of subsets of a set Ω. For $1 \leq p < \infty$, $L_p(\Omega, \mathcal{A}, \mu)$ represents the vector space of all (equivalence classes of) μ-measurable scalar-valued functions f for which $\int_\Omega |f|^p\, d\mu < \infty$. Then $L_p(\Omega, \mathcal{A}, \mu)$ is a Banach space under the norm $\|f\|_p$, given by

$$\|f\|_p = \left[\int_\Omega |f|^p\, d\mu \right]^{\frac{1}{p}}. \tag{A.3}$$

In the case of $p = \infty$, $L_\infty(\Omega, \mathcal{A}, \mu)$ is defined as the set of all μ-measurable functions f on Ω such that there exists $M \geq 0$ depending on f, for which $|f(x)| \leq M$ μ-almost everywhere. In this case, $L_\infty(\Omega, \mathcal{A}, \mu)$ is a Banach space under the essential sup norm defined by

$$\|f\|_\infty = \inf\big\{M : |f(x)| \leq M \text{ for } \mu\text{-almost all } x \in \Omega\big\}. \tag{A.4}$$

Bibliography

[1] A. A. Albanese. Bounded linear maps between (LF)-spaces. *Rev. R. Acad. Cienc. Exactas Fís. Nat., Ser. A Mat.*, 97(1):13–31, 2003.

[2] A. A. Albanese, J. Bonet, and W. J. Ricker. Linear operators on the (LB)-sequence spaces $ces(p-)$, $1 < p \leq \infty$. In *Descriptive topology and functional analysis. II*, volume 286 of *Springer Proc. Math. Stat.*, pages 43–67. Springer, Cham, 2019.

[3] S. Albeverio, A. Khrennikov, and V. M. Shelkovich. *Theory of p-adic distributions: Linear and nonlinear models*, volume 370 of *London Mathematical Society Lecture Note Series*. Cambridge University Press, Cambridge, 2010.

[4] C. D. Aliprantis and O. Burkinshaw. *Principles of real analysis*. Academic Press, Inc., San Diego, CA, 3rd edition, 1998.

[5] J. A. Álvarez López, Y. A. Kordyukov, and E. Leichtnam. Topology of the space of conormal distributions. *J. Pseudo-Differ. Oper. Appl.*, 15(3):Paper No. 47, 68, 2024.

[6] P. Antosik and J. Burzyk. Sequential conditions for barrelledness and bornology. *Bull. Polish Acad. Sci. Math.*, 35:457–459, 1987.

[7] P. Antosik and C. Swartz. *Matrix methods in analysis*, volume 1113 of *Lecture Notes in Mathematics*. Springer-Verlag, Berlin, 1985.

[8] S. Banach. *Théorie des opérations linéaires. Monografje Matematyczne, T. 1*. Z subwencji Funduszu kultury narodowej, Warszawa, 1932.

[9] S. Banach. *Théorie des opérations linéaires*. Chelsea Publishing Co., New York, 1955.

[10] M. J. Barany, A.-S. Paumier, and J. Lützen. From Nancy to Copenhagen to the World: The internationalization of Laurent Schwartz and his theory of distributions. *Hist. Math.*, 44(4):367–394, 2017.

[11] W. R. Bauer and R. H. Benner. The non-existence of a Banach space of countably infinite hamel dimension. *Am. Math. Mon.*, 78(8):895–896, 1971.

[12] R. Beattie and H.-P. Butzmann. *Convergence structures and applications to functional analysis*. Kluwer Academic Publishers, 2002.

[13] G. Beer and M. I. Garrido. Bornologies and locally Lipschitz functions. *Bull. Aust. Math. Soc.*, 90(2):257–263, 2014.

[14] S. K. Berberian. *Lectures in functional analysis and operator theory*, volume No. 15 of *Graduate Texts in Mathematics*. Springer-Verlag, New York-Heidelberg, 1974.

[15] K. D. Bierstedt. An introduction to locally convex inductive limits. In *Functional analysis and its applications (Nice, 1986), ICPAM Lecture Notes*, pages 35–133. World Sci. Publishing, Singapore, 1988.

[16] K. D. Bierstedt and J. Bonet. Dual density conditions in (DF)-spaces. I. *Result. Math.*, 14(3–4):242–274, 1988.

[17] K. D. Bierstedt and J. Bonet. Stefan Heinrich's density condition for Fréchet spaces and the characterization of the distinguished Köthe echelon spaces. *Math. Nachr.*, 135:149–180, 1988.

[18] V. I. Bogachev and O. G. Smolyanov. *Topological vector spaces and their applications. Springer Monographs in Mathematics*. Springer, Cham, 2017.

[19] J. Bonet. A question of Valdivia on quasinormable Fréchet spaces. *Can. Math. Bull.*, 34(3):301–304, 1991.

[20] J. Bonet, D. Jornet, and P. Sevilla-Peris. *Function spaces and operators between them*. Springer, 2023.

[21] C. Bosch and E. Fernández Bermejo. *Análisis funcional*. Universidad Autónoma de México, 1989.

[22] C. Bosch, A. García, and C. L. García. An extension of Ekeland's variational principle to locally complete spaces. *J. Math. Anal. Appl.*, 328(1):106–108, 2007.

[23] C. Bosch, C. L. García, T. E. Gilsdorf, C. Gómez-Wulschner, and R. Vera. Fixed points of set-valued maps in locally complete spaces. *Fixed Point Theory Appl.*, Paper No. 13, 2017.

https://doi.org/10.1515/9783111392868-013

[24] C. Bosch, C. L. García, T. E. Gilsdorf, C. Gómez-Wulschner, and R. Vera. On docile spaces, Mackey first countable spaces, and sequentially Mackey first countable spaces. *Int. J. Math. Anal.*, 11(8):377–388, 2017.

[25] C. Bosch and A. García-Martínez. Local completeness and Pareto optimization. *Nonlinear Anal., Theory Methods Appl.*, 73(4):1098–1100, 2010.

[26] C. Bosch, T. E. Gilsdorf, and C. Gómez-Wulschner. Mackey first countability and docile locally convex spaces. *Acta Math. Sin. Engl. Ser.*, 27:737–740, 2011.

[27] C. Bosch and C. Swartz. *Functional calculi*. World Scientific Publishing Co. Pte. Ltd., Hackensack, NJ, 2013.

[28] N. Bourbaki. *Espaces vectoriels topologiques: Chapitres 1à 5*, volume 4. Springer Science & Business Media, 2007.

[29] C. Brouder, N. V. Dang, and F. Hélein. Continuity of the fundamental operations on distributions having a specified wave front set (with a counterexample by Semyon Alesker). *Stud. Math.*, 232(3):201–226, 2016.

[30] J. Burzyk, C. Klís, and Z. Lipecki. On metrizable Abelian groups with a completeness-type property. *Colloq. Math.*, 49(1):33–39, 1984.

[31] D. Bushaw. *Elements of general topology*. John Wiley & Sons, Inc., New York-London, 1963.

[32] Y. Dabrowski and C. Brouder. Functional properties of Hörmander's space of distributions having a specified wavefront set. *Commun. Math. Phys.*, 332(3):1345–1380, 2014.

[33] Y. Dabrowski and M. Kerjean. Models of linear logic based on the Schwartz ε-product. *Theory Appl. Categ.*, 34:1440–1525, 2019.

[34] D. G. de Figueiredo. *Lectures on the Ekeland variational principle with applications and detours*, volume 81 of *Tata Institute of Fundamental Research Lectures on Mathematics and Physics*. Tata Institute of Fundamental Research, Bombay; by Springer-Verlag, Berlin, 1989.

[35] M. De Wilde. Réseaux dans les espaces linéaires à semi-normes. *Mem. Soc. R. Sci. Liege, Collect. in-8, (5)*, 18(2):144, 1969.

[36] M. De Wilde. *Closed graph theorems and webbed spaces*, volume 19 of *Research Notes in Mathematics*. Pitman, Boston, Mass.-London, 1978.

[37] J. Diestel. *Sequences and series in Banach spaces*, volume 92 of *Graduate Texts in Mathematics*. Springer-Verlag, New York, 1984.

[38] J. Dieudonné. *History of functional analysis. Notas de Matemática. [Mathematical Notes]*. North-Holland Publishing Co., Amsterdam-New York, 1981. North-Holland Mathematics Studies, 49.

[39] S. Dolecki. *A royal road to topology—convergence of filters*. World Scientific Publishing Co. Pte. Ltd., Hackensack, NJ, 2024.

[40] L. Drewnowski. Resolutions of topological linear spaces and continuity of linear maps. *J. Math. Anal. Appl.*, 335(2):1177–1194, 2007.

[41] R. E. Edwards. *Functional analysis*. Dover Publications, Inc., New York, 1995. Theory and applications, Corrected reprint of the 1965 original.

[42] I. Ekeland. On the variational principle. *J. Math. Anal. Appl.*, 47(2):324–353, 1974.

[43] J. C. Ferrando. Descriptive topology for analysts. *Rev. R. Acad. Cienc. Exactas Fís. Nat., Ser. A Mat.*, 114(2):Paper No. 107:34, 2020.

[44] J. C. Ferrando, S. Gabriyelyan, and J. Kąkol. Bounded sets structure of $C_p(X)$ and quasi-(DF)-spaces. *Math. Nachr.*, 292(12):2602–2618, 2019.

[45] K. Floret. Bases in sequentially retractive limit spaces. *Stud. Math.*, 38:221–225, 1970.

[46] K. Floret. Folgenretraktive sequenzen lokalkonvexer Räume. *J. Reine Angew. Math.*, 259:65–85, 1973.

[47] S. Friedberg, A. Insel, and L. Spence. *Linear algebra*. Pearson, 5th edition, 2018.

[48] S. Gabriyelyan. Locally convex spaces and Schur type properties. *Ann. Acad. Sci. Fenn., Math.*, 44(1):363–378, 2019.

[49] A. García-Martínez. Webbed spaces, double sequences, and the Mackey convergence condition. *Int. J. Math. Math. Sci.*, 22(3):521–524, 1999.

[50] A. García-Martínez. Strictly webbed spaces and regularity properties of inductive limits. *Int. J. Math. Math. Sci.*, 27(11):695–699, 2001.

[51] C. Garetto. Topological structures in Colombeau algebras: Investigation of the duals of $\tilde{\mathcal{G}}_c(\Omega)$, $\tilde{\mathcal{G}}(\Omega)$ and $\tilde{\mathcal{G}}_S(\mathbb{R}^n)$. *Monatshefte Math.*, 146(3):203–226, 2005.

[52] C. Garetto. Topological structures in Colombeau algebras: Topological $\tilde{\mathbb{C}}$-modules and duality theory. *Acta Appl. Math.*, 88(1):81–123, 2005.

[53] C. Garetto. Closed graph and open mapping theorems for topological $\tilde{\mathbb{C}}$-modules and applications. *Math. Nachr.*, 282(8):1159–1188, 2009.

[54] T. E. Gilsdorf. Mackey convergence and quasi-sequentially webbed spaces. *Int. J. Math. Math. Sci.*, 14(1):17–26, 1991.

[55] T. E. Gilsdorf. Regular inductive limits of $\mathcal{K}$-spaces. *Collect. Math.*, 42(1):45–49, 1991.

[56] T. E. Gilsdorf. Boundedly compatible webs and strict Mackey convergence. *Math. Nachr.*, 159:139–147, 1992.

[57] T. E. Gilsdorf. Strictly webbed convenient locally convex spaces. *Int. J. Math. Anal.*, 1(13–16):775–782, 2007.

[58] T. E. Gilsdorf. Inductive limits of quasi-locally Baire spaces. *Asian-Eur. J. Math.*, 15(4):Paper No. 2250074, 5, 2022.

[59] T. E. Gilsdorf. Valdivia's lifting theorem for non-metrizable spaces. *Topol. Appl.*, 317:Paper No. 108160, 13, 2022.

[60] T. E. Gilsdorf and J. Kąkol. On some non-Archimedean closed graph theorems. In *p-adic functional analysis (Nijmegen, 1996)*, volume 192 of *Lecture Notes in Pure and Appl. Math.*, pages 153–158. Dekker, New York, 1997.

[61] C. Gomez-Wulschner and J. Kučera. Sequentially complete inductive limits and regularity. *Czechoslov. Math. J.*, 54(3)(3):697–699, 2004.

[62] A. Grothendieck. *Espaces vectoriels topologiques*. Universidade de São Paulo, Instituto de Matemática Pura e Aplicada. São Paulo, 1954.

[63] A. Grothendieck. Produits tensoriels topologiques et espaces nucléaires. In *Séminaire Bourbaki, vol. 2*, Exp. No. 69, pages 193–200. Soc. Math. France, Paris, 1995.

[64] A. H. Hamel and C. Zălinescu. Minimal element theorems revisited. *J. Math. Anal. Appl.*, 486(2):123935, 28, 2020.

[65] J. Gi. Hocking and G. S. Young. *Topology*. Dover Publications, Inc., New York, 1988. Reprint of the 1961 original.

[66] H. Hogbe-Nlend. *Bornologies and functional analysis*, volume 26 of *North-Holland Mathematics Studies*. North-Holland Publishing Co., Amsterdam-New York-Oxford, 1977. Translated from the French by V. B. Moscatelli, Notas de Matemática, No. 62. [Mathematical Notes].

[67] J. Horváth. *Topological vector spaces and distributions: Volume 1*. Dover Publications, Inc., New York, 2012. Unabridged corrected republication of the 1966 original.

[68] T. Husain. *The open mapping and closed graph theorems in topological vector spaces*. Clarendon Press, Oxford, 1965.

[69] T. Husain. *Introduction to topological groups*. Dover Publications, Inc., New York, 2018. Reprint of the W. B. Saunders, Philadelphia, 1966 edition.

[70] H. Jarchow. *Locally convex spaces. Mathematische Leitfäden. [Mathematical Textbooks]*. Springer, 2014.

[71] J. Kąkol and S. A. Saxon. Montel (DF)-spaces, sequential (LM)-spaces and the strongest locally convex topology. *J. Lond. Math. Soc.*, 66(2):388–406, 2002.

[72] J. Kąkol, S. A. Saxon, and A. R. Todd. Docile locally convex spaces. In *Topological algebras and their applications*, volume 341 of *Contemp. Math.*, pages 73–78. Amer. Math. Soc., Providence, RI, 2004.

[73] N. J. Kalton, N. T. Peck, and J. W. Roberts. *An F-space sampler*, volume 89 of *Lecture Note Series-London Mathematical Society*. Cambridge University Press, 1984.

[74] J. Kąkol and W. Śliwa. On metrizable subspaces and quotients of non-Archimedean spaces $C_p(X, \mathbb{K})$. *Rev. R. Acad. Cienc. Exactas Fís. Nat., Ser. A Mat.*, 114(3):Paper No. 125, 23, 2020.

[75] J. L. Kelley. *General topology*. Dover Publications Inc., New York, 2017. Reprint of the original 1955 edition.

[76] J. L. Kelley, I. Namioka, W. F. Donoghue Jr., K. R. Lucas, B. J. Pettis, E. T. Poulsen, G. B. Price, W. Robertson, W. R. Scott, and K. T. Smith. *Linear topological spaces*. Springer, 1963.

[77] S. M. Khaleelulla. *Counterexamples in topological vector spaces*, volume 936 of *Lecture Notes in Mathematics*. Springer-Verlag, Berlin-New York, 1982.

[78] C. Kliś. An example of noncomplete normed (K)-space. *Bull. Acad. Pol. Sci., Sér. Sci. Math. Astronom. Phys.*, 26(5):415–420, 1978.

[79] G. Köthe. *Topological vector spaces I*, volume 159 of *Die Grundlehren der Mathematischen Wissenschaften*. Springer-Verlag New York, Inc., New York, 1969. Translated from the German by D. J. H. Gårling.

[80] G. Köthe. *Topological vector spaces. II*, volume 237 of *Grundlehren der Mathematischen Wissenschaften*. Springer-Verlag, New York-Berlin, 1979.

[81] E. Kreyszig. *Introductory functional analysis with applications. Wiley Classics Library*. John Wiley & Sons, Inc., New York, 1989.

[82] A. Kriegl and P. W. Michor. *The convenient setting of global analysis*, volume 53 of *Mathematical Surveys and Monographs*. American Mathematical Society, Providence, RI, 1997.

[83] S. R. Lay. *Convex sets and their applications*. Robert E. Krieger Publishing Co., Inc., Malabar, FL, 1992. Revised reprint of the 1982 original.

[84] D. Lorenz and N. Worliczek. Necessary conditions for variational regularization schemes. *Inverse Probl.*, 29(7):075016, 19, 2013.

[85] G. W. Mackey. On infinite-dimensional linear spaces. *Trans. Am. Math. Soc.*, 57:155–207, 1945.

[86] G. W. Mackey. On convex topological linear spaces. *Trans. Am. Math. Soc.*, 60:519–537, 1946.

[87] O. Mazur and W. Orlicz. Über folgen linearen Operationen. *Stud. Math.*, 4:152–157, 1933.

[88] K. McKennon and J. M. Robertson. *Locally convex spaces*, volume 15 of *Lecture Notes in Pure and Applied Mathematics*. Marcel Dekker, Inc., New York-Basel, 1976.

[89] R. E. Megginson. *An introduction to Banach space theory*, volume 183 of *Graduate Texts in Mathematics*. Springer, New York, 1998.

[90] J. R. Munkres. *Topology*. Prentice Hall, Inc., Upper Saddle River, NJ, 2nd edition, 2023.

[91] L. Narici and E. Beckenstein. *Topological vector spaces*, volume 296 of *Pure and Applied Mathematics (Boca Raton)*. CRC Press, Boca Raton, FL, 2nd edition, 2011.

[92] L. Nel. *Continuity theory*. Springer, Cham, 2016.

[93] E. T. Ordman. Convergence almost everywhere is not topological. *Am. Math. Mon.*, 73(2):182–183, 1966.

[94] M. S Osborne. *Locally convex spaces*, volume 269 of *Graduate Texts in Mathematics*. Springer, 2014.

[95] P. Pérez Carreras and J. Bonet. *Barrelled locally convex spaces*. Elsevier, 1987.

[96] C. Pérez-García and W. H. Schikhof. *Locally convex spaces over non-Archimedean valued fields*, volume 119 of *Cambridge Studies in Advanced Mathematics*. Cambridge University Press, Cambridge, 2010.

[97] E. Pourhadi, A. Khrennikov, R. Saadati, K. Oleschko, and M. de Jesús Correa Lopez. Solvability of the p-adic analogue of Navier-Stokes equation via the wavelet theory. *Entropy*, 21(11):Paper No. 1129, 20, 2019.

[98] J. Prolla. *Topics in functional analysis over valued division rings. Notas de Matemática. [Mathematical Notes]*. North-Holland Publishing Co., Amsterdam-New York, 1982. North-Holland Mathematics Studies, 77.

[99] V. Pták. Completeness and the open mapping theorem. *Bull. Soc. Math. Fr.*, 86:41–74, 1958.

[100] J.-H. Qiu. Quasi-fast completeness and inductive limits of webbed spaces. *J. Math. Res. Expo.*, 18(1):55–59, 1998.

[101] J.-H. Qiu and F. He. Ekeland variational principles for set-valued functions with set perturbations. *Optimization*, 69(5):925–960, 2020.

[102] T. Richmond. *General topology—an introduction. De Gruyter Textbook*. De Gruyter, Berlin, 2020.

[103] A. P. Robertson and W. Robertson. *Topological vector spaces*. Cambridge University Press 2nd edition, 1973.

[104] W. Robertson. Completions of topological vector spaces. *Proc. Lond. Math. Soc. (3)*, 8:242–257, 1958.

[105] W. Robertson. On the closed graph theorem and spaces with webs. *Proc. Lond. Math. Soc.*, 3(4):692–738, 1972.

[106] S. Rolewicz. *Metric linear spaces*. PWN Warsaw, 2nd edition, 1984.

[107] W. Rudin. *Functional analysis. International Series in Pure and Applied Mathematics*. McGraw-Hill, Inc., New York, 2nd edition, 1991.

[108] A. I. Saichev and W. Woyczynski. *Distributions in the physical and engineering sciences. Vol. 1. Applied and Numerical Harmonic Analysis*. Birkhäuser/Springer, 2018. Distributional and fractal calculus, integral transforms and wavelets, Reprint of the 1997 original.

[109] L. M. Sánchez Ruiz. Topological vector spaces without local convexity conditions. In *Functional analysis with current applications in science, technology and industry (Aligarh, 1996)*, volume 377 of *Pitman Res. Notes Math. Ser.*, pages 37–48. Longman, Harlow, 1998.

[110] S. A. Saxon and L. M. Sánchez Ruiz. Reinventing weak barrelledness. *J. Convex Anal.*, 24(3):707–762, 2017.

[111] H. H. Schaefer and M. P. Wolff. *Topological vector spaces*, volume 3 of *Graduate Texts in Mathematics*. Springer-Verlag, 2nd edition, 1999.

[112] P. Schneider. *Nonarchimedean functional analysis. Springer Monographs in Mathematics*. Springer-Verlag, Berlin, 2002.

[113] L. Schwartz. *Théorie des distributions. (1951)*. Hermann, 1951.

[114] L. Schwartz. *Un mathématicien aux prises avec le siècle*. Odile Jacob, 1997.

[115] M. Starbird and F. Su. *Topology through inquiry*, volume 58 of *AMS/MAA Textbooks*. MAA Press, Providence, RI, 2019.

[116] C. Swartz. *An introduction to functional analysis*, volume 157 of *Monographs and Textbooks in Pure and Applied Mathematics*. Marcel Dekker, Inc., New York, 1992.

[117] C. Swartz. *Infinite matrices and the gliding hump*. World Scientific Publishing Co., Inc., River Edge, NJ, 1996.

[118] C. Swartz. *Abstract duality pairs in analysis*. World Scientific Publishing Co. Pte. Ltd., Hackensack, NJ, 2018.

[119] F. Trèves. *Topological vector spaces, distributions and kernels*. Dover Publications, Inc., Mineola, NY, 2006. Unabridged republication of the 1967 original.

[120] F. Tréves, G. Pisier, and M. Yor. Laurent Schwartz (1915–2002). *Not. Am. Math. Soc.*, 50(9):1072–1084, 2003.

[121] B. Tsirulnikov. Barrelledness and dual strong sequences in locally convex spaces. *Bull. Soc. R. Sci. Liège*, 73(1):7–19, 2004.

[122] M. Valdivia. On bounded sets which generate Banach spaces. *Arch. Math. (Basel)*, 23:640–642, 1972.

[123] M. Valdivia. *Topics in locally convex spaces*, volume 67 of *Notas de Matemática. [Mathematical Notes]*. North-Holland Publishing Co., Amsterdam-New York, 1982. North-Holland Mathematics Studies.

[124] M. Valdivia. Quasi-LB-spaces. *J. Lond. Math. Soc. (2)*, 35(1):149–168, 1987.

[125] A. C. M. van Rooij. *Non-Archimedean functional analysis*, volume 51 of *Monographs and Textbooks in Pure and Applied Mathematics*. Marcel Dekker, Inc., New York, 1978.

[126] C. T. M. Vinagre. Webbed locally **K**-convex spaces. *Math. Jpn.*, 44(2):331–341, 1996.

[127] J. Voigt. *A course on topological vector spaces. Compact Textbooks in Mathematics*. Birkhäuser/Springer, Cham, 2020.

[128] L. Waelbroeck. *Topological vector spaces and algebras*, volume 230 of *Lecture Notes in Mathematics*. Springer-Verlag, Berlin-New York, 1971.

[129] A. Wilansky. *Topology for analysis*. Dover Publications, Inc., New York, 2008. Unabridged republication of the 1970 original.

[130] A. Wilansky. *Modern methods in topological vector spaces*. Dover Publications, Inc., New York, 2013. Unabridged republication of the 1978 original.

[131] K. Yosida. *Functional analysis. Classics in Mathematics*. Springer-Verlag, Berlin, 1995. Reprint of the sixth (1980) edition.

[132] C. Zălinescu. *Convex analysis in general vector spaces*. World Scientific Publishing Co., Inc., River Edge, NJ, 2002.

Index

https://doi.org/10.1515/9783111392868-014

9 783111 390796